24th Offshore Symposium 2019

Houston, Texas, USA
20 February 2019

ISBN: 978-1-5108-8381-9

Printed from e-media with permission by:

Curran Associates, Inc.
57 Morehouse Lane
Red Hook, NY 12571

Some format issues inherent in the e-media version may also appear in this print version.

Printed by Curran Associates, Inc. (2019)

Society of Naval Architects and Marine Engineers (SNAME) Texas Section
99 Canal Center Plaza, Suite 310
Alexandria, VA 22314
USA

Phone: +1-703-997-6701

www.sname.org/texassection

Additional copies of this publication are available from:

Curran Associates, Inc.
57 Morehouse Lane
Red Hook, NY 12571 USA
Phone: 845-758-0400
Fax: 845-758-2633
Email: curran@proceedings.com
Web: www.proceedings.com

The 24th Offshore Symposium

Feb 20, 2019 | Houston, Texas

www.snamesymposium.com

SNAME
INTERNATIONAL COMMUNITY FOR MARITIME AND OCEAN PROFESSIONALS

Time	Room - Pecan	Room - Elm	Room - Red Oak Ballroom A
7:00am - 8:30am	**Breakfast and Registration - Live Oak**		
7:30am - 8:30am	**Ethics Course,** presented by TBPE		
8:30am - 10:00am Room: Red Oak Ballroom A-B	**Panel Discussion:** **"Adapting Mooring Design and Standards for the Next Generation of Offshore Platforms"**		
10:00am - 10:20am	**Coffee Break: Sponsored by GustoMSC**		
10:20am - 11:50am	*Morning Session I: Soil Structure Interaction*	*Morning Session II: Analytical Mooring*	*Morning Session III: Deepwater Optimization*
	Jack-up Spud Can Foundation Fixity for Various Clay Strength Profiles *J. S. Templeton, III* ***SAGE USA, Inc*** 1	**Numerical Study on OPB/IPB Interlink Angle with Underwater Chain Stopper System** *Woo Chul Chung, Heon Yong Kang, Moo Hyun Kim* ***Texas A&M University*** 39	**Fueling Strategy for Offshore Facilities - a Study** *Anup Srivastava* ***Chevron*** 78
	Challenges of Using Residual Curvature Method in Deep Water Pipelines *Tanner J. Powell, Qiang Bai, Mark Brunner, Shawn Searle* ***Genesis Oil and Gas, TechnipFMC*** 13	**High Strength and High Stiffness Rope for Permanent Deepwater Mooring: Technology Development and Challenges** *Xiaohong Chen, Sue Wang, Mark Ausborn, Zhen Li* ***ABS, Exxonmobil*** 52	**Investigating Opportunities and Challenges for Unmanned Floating Production Facilities** *Anders Martin Moe, Gaurav Singhal, Astrid Barros, Paul Erb* ***Wood, Woodside Energy LTD*** 94
	Design and Installation of Dynamically Installed Piles for Deepwater Mooring *X.Y. Zhang, F. Yong, Y.P. Li, J.T. Yi, F.H. Lee, X. Chen, S.Q. Wang* 28 ***ABS, National University of Singapore, Hohai University, Chongqing University***	**Dissection, Measurement and Analysis of a Recovered Damaged Flexible Riser** *Colin C. Duffett* ***Wood PLC*** 66	**Roadmap for Selecting Cost Efficient Deepwater Investments** *Shiladitya Basu, Tirtharaj Bhaumik* ***KBR/Granherne, KBR/GVA*** 106
11:50am - 1:20pm Room : Magnolia	**Luncheon Keynote:** **"Blue Oceans: Offshore and Ship Research for Future Maritime Challenges"** *Dr. Bas Buchner, President, MARIN (Maritime Research Institute Netherlands)* **SNAME Membership Awards**		
1:20pm - 1:30pm	**Coffee Break**		
1:30pm - 3:00pm	*Afternoon Session I: Student Program*	*Afternoon Session II: Mooring Design*	*Afternoon Session III: SNAME T&R Panels*
	Student Poster Session	**Review of Offshore Mooring System Life Extension - Challenges and Recommendations** *Aifeng Yao, Qinghe Zhang* ***Chevron, ABS*** 123	**Overview:** **OC-10 Ocean Renewables** **OC-8 Wind Technologies**
		Assessment of Nearshore FSRU/FLNG Mooring *Amitava Guha, Sue Wang* ***ABS*** 139	**Panel Session:** **OC 9 Robustness Checks for Floating Systems** 171
		The Evolution of Deepwater Disconnectable Moorings *Jason D. Pasternak, Daniel J. Alonso, John T. Shelton III, Ginny R. Whisenhunt* ***Delmar Systems, Inc.*** 154	
3:00pm - 3:20pm	**Coffee Break: Sponsored by Halton Marine Americas**		
3:20pm - 4:50pm	*Closing Session I: Improving Operation Standards*	*Closing Session II: Hydrodynamics*	*Closing Session III: Offshore Structure Reliability*
	Explosion Proof Technology Application on Tier 3 and Tier 4 Final Engines (Electronic engines application) *Joseph Tovlin* ***MIRETTI*** N/A	**A Distributed Vorticity Model for Flow Past a 3-D Hydrofoil** *Chunlin Wu, Spyros A. Kinnas* ***University of Texas at Austin*** 224	**Demonstrating Target Safety Level for Life Extension of Offshore Structures** *Piotr Szalewski, Grzegorz Malinowski, Jenny Yan Lu* ***DNV GL Houston*** 271
	Hybrid Induction - Oxygen Cutting Process for Faster, More Efficient, and Less Costly Scrapping of Off-Shore Petroleum Platforms *Jerald E. Jones, Valerie L. Rhoades, James R. Dydo* ***EnergynTech, Inc., Gatekey Engineering*** 193	**Experimental Investigation of a Dropped Model Rocket in the Towing Tank of the University of New Orleans** *Hallie Collopy, Xiaochuan Yu, Yi Li, Ken Hagimoto, Haozhan Meng, Scott Arbeiter* ***University of New Orleans*** 243	**A Comprehensive Framework to Develop Resilient Offshore Structures** *Alberto Morandi* ***GustoMSC US, Inc.*** 287
	Virtual Multimodal Automated Object Detection with Deep Neural Networks *Nikolaos Mitsakos, Sanat Upadhyay, Manos Papadakis* ***University of Houston, LOLAARK LLC*** 202	**Applying 3D Viscous/Inviscid Interaction Method to Predict Flows Around Wings and Propellers** *Weikang Du, Spyros A. Kinnas* ***University of Texas at Austin*** 254	**Investigating Impact of Revised DNVGL-OTG-13 Guideline on Air Gap Predictions for Semisubmersibles** *Gaurav Singhal, Jun Cheng, Joao Pessoa, Anders Martin Moe* ***Wood*** 302
4:50pm - 6:30pm	**Reception: Sponsored by SOFEC**		

Formatting Issue - Best Available Paper/Graphic

SNAME
THE INTERNATIONAL COMMUNITY
FOR MARITIME AND OCEAN PROFESSIONALS

JACK-UP SPUD CAN FOUNDATION FIXITY FOR VARIOUS CLAY STRENGTH PROFILES

J. S. TEMPLETON, III
SAGE USA, INC.

Proceedings of the 24th Offshore Symposium, February 20th 2019, Houston, Texas
Texas Section of the Society of Naval Architects and Marine Engineers

ABSTRACT

For site specific assessment of jack-up rigs, existing standards and practices require establishment of certain spud can fixity parameters. Among the required parameters are the maximum moment capacity and the maximum horizontal capacity. These parameters are established by use of site-specific soils data, spud can geometry and preload values. Logically the parameter formulations should provide for improvement in horizontal and moment capacity with increased penetration. Prior to the issue of ISO 19905-1 there was no such provision. However, work sponsored by the International Association of Drilling Contractors' Jack-Up Committee (IADC) provided technical basis for inclusion of penetration-dependent foundation performance parameters to rectify this situation, particularly in the case of clay soils. The result was substantial increase in foundation parameter values for deep penetrations in clay. Subsequent hindcast studies using the new penetration-dependent parameters have reported very good agreement with actual field performance of instrumented rigs, and the IADC-recommended parameter formulations were subsequently adopted in ISO 19905-1 and in SNAME Bulletin 5-5.

The IADC recommendations were based, in part, of results of finite element analysis of spud-can/soil interaction. Those analyses modeled part of the spud can installation as well as the subsequent moment and horizontal loading. Both parts were done largely with both small strain and finite strain assumptions. More recent finite element work by on Zhang, Biennen and Cassidy (2013, 2014) used newer modeling techniques to substantiate more conservative recommendations.

The current revision draft of ISO 19905-1 prescribes the original IADC-based recommendations only for the case of uniform soil strength, and it prescribes recommendations based on Zhang, Biennen and Cassidy (2013, 2014) for the case of a strength profile increasing strictly linearly from negligible strength at the mudline. This situation leaves a gap in recommendations for more commonly encountered strength profiles intermediate to these two extremes.

The present paper reports results of new IADC-sponsored finite element analyses done with full Eulerian modeling for both the spud can installation and subsequent moment and horizontal loading. The new results support recommendations needed for the more commonly encountered strength profiles intermediate to the extreme cases now covered by the current revision draft of ISO 19905-1.

Keywords: SNAME, Offshore, Symposium, Jack-up, spud can, fixity, clay

Proceedings of the 24th Offshore Symposium, February 20th 2019, Houston, Texas
Texas Section of the Society of Naval Architects and Marine Engineers

INTRODUCTION

The objective of the subject study is to provide new recommendations for important parameters used in the calculation of spud can foundation performance for deep penetrations in commonly encountered clay soil conditions. Without new technical work in this area the next edition of ISO 19905-1 would provide, for clay soils, separate fixity recommendations only for the extreme cases of (strictly) normally consolidated or uniform soil strength conditions. The clay strength profiles most commonly encountered offshore are intermediate to these two extreme cases. Recommendations specific to the intermediate cases should enable more accurate site assessments.

For site specific assessment of jack-up rigs, ISO 19905-1 (2012) requires establishment of certain spud can fixity parameters. Among the required parameters are the maximum moment capacity and the maximum horizontal capacity. These parameters are established by use of site-specific soils data, spud can geometry and preload values. Logically the parameter formulations should provide for improvement in horizontal and moment capacity with increased penetration. Prior to the issue of ISO 19905-1 there was no such provision. However, work sponsored by the International Association of Drilling Contractors' Jack-Up Committee (IADC) provided technical basis for inclusion of penetration-dependent foundation performance parameters to rectify this situation, particularly in the case of clay soils. The result was substantial increase in foundation parameter values for deep penetrations in clay. Subsequent hindcast studies by Templeton, Lewis and Brekke (2009) and Templeton and Lewis (2011) using the new deep penetration parameters showed very good agreement with actual field performance of instrumented rigs, and the IADC-recommended parameter formulations were subsequently adopted in ISO 19905-1 (see, Wong 2012) and in corrigenda to SNAME Bulletin 5-5 (2008).

The IADC recommendations had been based, in part, of results of finite element analysis of spud-can/soil interaction. Those analyses modeled part of the spud can installation as well as the subsequent moment and horizontal loading. Both parts were done with both small strain and finite strain assumptions. More recent finite element work by the Zhang, Bienen and Cassidy (2013, 2014) used newer modeling techniques to substantiate more conservative recommendations specific to normally consolidated clay.

The current revision draft of ISO 19905-1 prescribes the original IADC-based recommendations only for the case of uniform soil strength, and it prescribes the recommendations based on Zhang, Bienen and Cassidy (2013, 2014) for the case of a normally consolidated strength profile (increasing strictly linearly from negligible strength at the mudline). This situation leaves a gap in recommendations for more commonly encountered strength profiles intermediate to these two extremes.

The clay conditions assumed in the Zhang, Bienen and Cassidy (2013, 2014) work are not representative of those commonly encountered offshore. The strength profile assumed involved negligible strengths near the mudline while offshore clays have substantial near-mudline strength. This is important because, in the Zhang, Bienen and Cassidy (2013, 2014) analyses weak soil from near the mudline is stuck to the spud can and carried down to the penetration zone. This effect is apparently realistic, and the weakness of the soil carried downward limits both the horizontal and moment capacities.

The recommendations presently in ISO 19905 for deep penetrations in clay were based on work done by SAGE for the IADC jack-up Committee beginning in 2002. That technical work was done with conventional finite element analysis methods of the time and is somewhat outdated by today's standards – although large strain methods were included to simulate part of the installation process as well as the subsequent in-place performance. The Zhang, Bienen and Cassidy (2013, 2014) work used newer methods which allowed analysis of the entire spud can penetration process to produce soil stress and stain field data for use in a separate analysis of the in-place performance.

The present study makes use of the Combined Eulerian-Lagrangian (CEL) capabilities currently available in the ABAQUS program which allow the entire process of penetration and subsequent in-place performance to be handled in a single analysis employing large strain theory throughout.

WORK PERFORMED

Analyses with the Abaqus program were used to provide numerical solutions for the deep penetration and subsequent in-place loading of spud cans in clay soils with strength profiles involving various strengths at depth. . Abaqus (Ref 4) is a large, nonlinear, general purpose, commercial finite element analysis program with substantial soil mechanics capabilities. It has a long history of successful use in offshore geotechnical analysis (Ref 5). The analyses reported here were performed using the Combined Eulerian-Lagrangian capabilities available in Abaqus.

Traditionally, finite element analysis has been accomplished using what is now called a Lagrangian approach. In this approach, nodes and element boundaries are tied to material, and a fixed body of material is contained in each element. The mesh deforms with the material, and the amount of material deformation that can be achieved is ultimately limited by errors resulting from element deformation. With an Eulerian approach the nodes and element boundaries are fixed in space and material flows across element boundaries as material deformations occur without any element deformation. In this way arbitrarily large material deformations can be achieved without element performance problems. In Abaqus Combined Eulerian-Lagrangian (CEL) analysis, Lagrangian meshing for bodies with limited deformation can be combined with Eulerian meshing for bodies with large deformations, and contacting surfaces of the two kinds of bodies can interact. In addition, large deformation theory (including large strain formulations for both stress and strain as well as proper treatment of large rotations and the effects of essential geometry change) is used throughout all steps of the CEL analyses in both Lagrangian and Eulerian bodies.

In the analyses reported here, the spud can was modeled as a Lagrangian body, and the soil was modeled as an Eulerian body. The soil and spud can were connected with a contact surface. The general arrangement for such modeling is shown in Figure 1 (after Vazquez et al., 2017). Note that this figure shows a half-space model taking advantage of the symmetry of the spud can and the fact that the motion studied is confined to a plane (the x-z plane in Figure 1). This symmetry restriction (typical of the models used in the present study) is proper for vertical (z-direction) and horizontal (x-direction) motions as well as for rocking rotations (in the x-z plane). The following are additional details of the analyses reported here:

- The spud can geometry was based on a Letourneau 116-C with the forged tip removed and the corners rounded. These simplifications improved the contact surface performance.
- A no-slip specification was used on the contact surface.
- The spud can was made rigid.
- The (clay) soil was modeled as an elastic-plastic solid with a Tresca yield condition and shear strength generally variable as a function of depth. Typically the soil shear strength was made constant over a limited depth near the mudline and given a normally-consolidated rate of increase with depth below that and continuing well below the embedment depth.
- The soil was given submerged unit weight consistent with interpreted submerged unit weights corresponding to the interpreted soil strength profiles reported here and sufficient to cause nearly complete backfill above the spud can.

Care was taken to ensure that the modeling of the soil in the Eulerian body was done in such a manner that near-mudline soil could be adhered to the surface of the spud can and drug downward below the can during penetration. Figure 2 displays this behavior. Figure 2a shows a model plot of soil and spud can prior to penetration with the soil initially within approximately ¼ spud can diameter colored red and the initially deeper soil colored blue. Figure 2b shows a plot of the same model after penetration with the same soil colorations. These plots show clearly that a significant body of the soil originally near the mudline has been transported downward beneath the penetrated spud can. The moment or horizontal resistance should logically be affected by the strength of the soil transported downward.

Figure 3 shows contour plots of plastic strain from CEL analysis of a penetrating spud can. Figure 3a shows the condition after a slight penetration. Figure 3b shows the condition after substantial penetration in the same analysis. For each of the production analyses of the study, the spud can was first penetrated to a particular penetration, typically in the range of 1.0 to 1.5 spud can diameters. Next, the vertical load was reduced to one half of the amount needed for that penetration, and finally either a horizontal or rotation loading was applied while the vertical load was maintained constant at the reduced level.

A parametric study was made of the effect of mesh fineness by way of an automatic mesh refinement technique. Starting with a base mesh similar to that shown in Figure 3, three levels of refinement were investigated. The density (number of elements per unit volume) of Eulerian soil elements near the spud can was increased, sequentially, by factors of 8 (Level 1 refinement), 64 (Level 2 refinement) and 512 (Level 3 refinement). The results at refinement Levels 2 and 3 were virtually the same. With results at these levels taken as a reference, the results at refinement Level 1 were generally off by amounts on the order of 1%, and the results for the base (unrefined) model were generally off by amounts on the order of 10%. Since the Level 2 and Level 3 refinement analyses required extremely long computational times (up to many days in the case of Level 3), production analyses for this study were performed at the base level and at refinement Level 1.

The work of the present study included not only the performance of CEL analyses of penetration and subsequent in-place loading but also identification of a number of soil strength profiles commonly encountered in jack-up rig installations. The identified strength profiles were generally intermediate to normally consolidated and uniform strength conditions.

RESULTS

Figure 4. shows a number of soil strength profiles for clay. Sites 1 through 3 are from SAGE files. Sites 4 through 12 were reported by Menzies and Roper (2012). The additional cases are idealized. Figure 5 shows the same strength profiles with the strengths normalized to their values at a penetration depth of 50 ft. The leftmost blue line is a normally consolidated profile with zero strength at the mudline, similar to that used in the Zhang, Bienen and Cassidy (2013, 2014). The rightmost line is a uniform strength case. The remainder of the lines are realistic cases, most of them are for actual sites from a previous IADC study. The model developed in the present study is capable of modeling any of these profiles.

Figure 6 shows results for the moment capacity factor as a function of the relative strength of the near-mudline soil. The specific measure of relative strength used is the ratio of the soil shear strength near the mudline (specifically the average strength within a depth of ¼ spud can diameter down from the mudline) to the strength similarly near the spud can embedment. On this figure, the point for the uniform strength case was selected with complete backfill. Nearly complete backfill was observed in most of the analysis case results. Figure 7 shows similar results for the horizontal capacity factor. Also shown on both figures are capacity factor values from the current draft revision of ISO 19905-1 for normally consolidated and uniform strength clay. The current draft revision prescriptions for uniform strength clay are the same as the presently issued standard prescriptions for all clay soils. These are described by Wong et al. (2012), and are based on Templeton, Brekke and Lewis (2005), and Templeton (2006, 2007, 2009). The current draft revision prescriptions for normally consolidated clay are based on Zhang, Bienen and Cassidy (2013, 2014). Also shown on both figures are straight lines representing linear interpolations from the normally consolidated to uniform strength values. In legend entries on both of these figures the qualification, S=1, indicated sensitivity of 1.0. In the normally consolidated case, factors are provided in ISO 19905-1 for other values of sensitivity (although the dependence is weak). In the uniform strength case sensitivity should be considered in the establishment of the factors per ISO19905-1 formulae. The analysis cases were performed for S=1.

In the case of horizontal factors (Figure 7) two normally consolidated points are shown. One (marked by an open circle) gives a value from the draft 19905-1 revision. The second (marked by a filled circle) gives a corresponding value adjusted to account for the influence of effect of spud can area aspect ratio (specifically the ratio of the horizontal projected area, A_s, to the vertically projected area, A). The adjustment is needed because the draft 19905-1 revision prescription for the normally consolidated case specifically account for the aspect ratio influence, but those for the normally consolidated case do not. The area aspect ratio for the CEL FEA analyses and for the uniform strength point in Figure 7 was 0.42. The unadjusted horizontal capacity factor for the normally consolidated case was taken to be

0.2. The form of the adjustment made was:

$$C_{h\text{-}nc\text{-}adj} = C_{h\text{-}nc}\,(0.11 + 0.39\,A_{s2}/A)\,/\,(0.11 + 0.39\,A_{s1}/A) \qquad \text{(equation 1)}$$

where:
- $C_{h\text{-}nc}$ is the unadjusted horizontal capacity factor for the normally consolidated case
- $C_{h\text{-}nc\text{-}adj}$ is the adjusted horizontal capacity factor for the normally consolidated case
- A_{s1}/A is the horizontal-to-vertical projected spud can area ratio corresponding to the unadjusted factor
- A_{s2}/A is the horizontal-to-vertical projected spud can area ratio corresponding to the adjusted factor.

This form for the aspect ratio influence is the same form (based on Templeton, 2009) that is embedded in the horizontal capacity prescription of the current issued ISO19905-1 for all clays as well as those of the present draft prescription for the uniform strength case.

The results in Figures 6 and 7 indicate that both the moment and horizontal capacity factors increase with increasing strength of the near-mudline soil – a very useful and also intuitive result. On the low end (i.e. at low near-mudline strengths) the trend of the results tend to confirm the present draft ISO19905-1 revision values. On the high end, the data appear to trend toward the uniform strength case value. I the case of both the moment and the horizontal, the CEL FEA results could be reasonably well characterized by the linear interpolations shown.

The interpolation for the moment factor is given by:

$$C_m = C_{m\text{-}nc} + (C_{m\text{-}u} - C_{m\text{-}nc})(S_{u\text{-}nml}/S_{u\text{-}ned} - S_{u\text{-}nmlnc}/S_{u\text{-}nednc})/(1 - S_{u\text{-}nmlnc}/S_{u\text{-}nednc}) \qquad \text{(equation 2)}$$

where:
- C_m is the moment capacity factor
- $C_{m\text{-}nc}$ is the moment capacity factor for the normally consolidated case
- $C_{m\text{-}u}$ is the moment capacity factor for the uniform strength case
- $S_{u\text{-}ned}$ is the average (undrained shear) strength within ¼ (spud can) diameter below the embedment depth
- $S_{u\text{-}nml}$ is the average (undrained shear) strength within ¼ (spud can) diameter below the mudline
- $S_{u\text{-}nednc}$ is the average strength within ¼ diameter below the embedment depth, normally consolidated
- $S_{u\text{-}nmlnc}$ is the average strength within ¼ diameter below the mudline, normally consolidated.

The interpolation for the horizontal factor is given by:

$$C_h = C_{h\text{-}nc\text{-}adj} + (C_{h\text{-}u} - C_{h\text{-}nc\text{-}adj})(S_{u\text{-}nml}/S_{u\text{-}ned} - S_{u\text{-}nmlnc}/S_{u\text{-}nednc})/(1 - S_{u\text{-}nmlnc}/S_{u\text{-}nednc}) \qquad \text{(equation 3)}$$

where:
- C_h is the horizontal capacity factor
- $C_{h\text{-}nc\text{-}adj}$ is the adjusted horizontal capacity factor for the normally consolidated case (equation 2)
- $C_{h\text{-}u}$ is the horizontal capacity factor for the uniform strength case
- $S_{u\text{-}ned}$ is the average (undrained shear) strength within ¼ (spud can) diameter below the embedment depth
- $S_{u\text{-}nml}$ is the average (undrained shear) strength within ¼ (spud can) diameter below the mudline
- $S_{u\text{-}nednc}$ is the average strength within ¼ diameter below the embedment depth, normally consolidated
- $S_{u\text{-}nmlnc}$ is the average strength within ¼ diameter below the mudline, normally consolidated.

In connection with both equation 2 and equation 3 it is useful to note that in any strictly normally consolidated case (with strength strictly proportional to depth) the ratio, $S_{u\text{-}nmlnc}/S_{u\text{-}nednc}$, is dependent only upon the ratio of embedment depth to spud can spud can diameter, specifically:

$$S_{u\text{-}nmlnc}/S_{u\text{-}nednc} = 1/(8 + D/B) \qquad \text{(equation 4)}$$

where:

D is the embedment depth

B is the spud can diameter.

The specific number, 8, appears in the denominator as a consequence of strength averaging over depth ranges of B/4 below the mudline and below the embedment depth.

In the application of both equation 2 and equation 3 the influence of soil sensitivity should be recognized via proper selection and calculation of values of the factors for the normally consolidated and uniform strength cases.

Conclusions and Recommendations

From the work of this study it is observed that the CEL FEA analysis technique is quite capable of modeling the penetration and subsequent in-place performance of a spud can in clay with various strength profiles. The analyses performed in this study clearly include the phenomenon of relatively weak near-mudline soil being transported downward beneath the spud can, and the results show that the subsequent moment and horizontal capacities are directly related to the strength of the soil brought down.

We recommend that that for cases of clay sites with strength profiles intermediate to normally consolidated and uniform, the moment capacity and horizontal capacity factors, C_m and C_h should be interpolated between factors for the two extreme cases, as via equations 2 and 3 of this paper.

Although the analysis results presented here seem to suggest that in some cases C_m and C_h might achieve values greater than presently prescribed by ISO19905-1, proper substantiation of that possibility would require further work.

Acknowledgements

The work presented in this paper was sponsored by the International Association of Drilling Contractors. Jack-up Operators Committee. The IADC is gratefully acknowledged for its sponsorship of the work and for permission to publish this paper. On behalf of IADC, Mr. David R. Lewis, P.E. of Lewis Engineering Group and Mr. Mason Melkowits, P.E. of Ensco provided considerable technical guidance during the course of the work.

The CEL modeling techniques used in this study drew significantly upon techniques used in prior studies with significant contributions to CEL modelling and analysis techniques by SAGE engineers F. B. Biegler and M. J. Barrett. Ms. Biegler developed the CEL models used in this study.

Mr. David Menzies of MatthewsDaniel, a Bureau Veritas Group Company, provided important input regarding the selection of realistic soil profiles.

Mr. Bryan Lane, P.E., of JBL Technologies provided valuable consulting regarding the use of multi-core parallel processing with automatic mesh refinement in Abaqus CEL analyses.

References

ISO 19905-1 (2012 & folg.). Site-specific assessment of mobile offshore units – Part 1: Jack-ups. International Organization for Standardization, issued standard, second edition and draft revision.

Menzies, D. & Roper, R. (2008). Comparison of jackup rig spudcan penetration methods in clay. OTC 19545, Proceedings, Offshore Technology Conf., Houston.

SIMULIA (2016). Abaqus User's Guide, Dassault Systemes, Providence, RI.

SNAME 5-5A (1994 and subsequent editions). Guidelines for site-specific assessment of mobile jack-up units. Technical & Research Bulletin 5-5A, Society of Naval Architects and Marine Engineers

Templeton, J. S., III, Brekke, J. N. & Lewis, D.R. (2005). Spud can fixity in clay, final findings of a study for IADC. Proceedings, 10th Int. Conf., The Jack-up Platform, City University, London.

Templeton, J. S., III (2006). Jackup foundation performance in clay. OTC 18367, Proceedings, Offshore Technology Conf., Houston.

Templeton, J. S., III (2007) Spud Can Fixity in Clay, Findings from Additional Work in a Study for IADC, Proceedings, Eleventh International Conference, The Jackup Platform, City University, London.

Templeton, J. S., III (2009). Spud can fixity in clay, further results from a study for IADC. Proceedings, 12th Int. Conf., The Jack-up Platform, City University, London.

Templeton, J. S., III and Lewis, D. R. (2011) Hysteretic Damping in Jack-up Dynamics. Proceedings, Thirteenth International Conference, The Jack-up Platform, City University, London

Templeton, J. S. III, Lewis D. R. and Brekke, J. N. (2009), Assessment of Jack-Up Survival in Severe Storms," Paper No. OTC 20255, Proceedings, Forty-First Offshore Technology Conference, Houston, Texas

Vazquez, J. H., Grasso, B. D., Gamino, M.A. and Templeton, J. S., III (2017) Using CEL to Account for Seabed Deformation Effects for Jack-ups Going of Location. In: Proceedings of the 22nd Offshore Symposium. Houston, Texas..

Wong, P. C., Templeton, J. S. III, Purwana O. A., Hofstede, H., Cassidy. M. J., Hossain, M.S. and Martin, C.M. (2012) Foundation Modeling and Assessment in the New ISO Standard 19905-1, OTC 23521, Proceedings, Offshore Technology Conf., Houston.,

Zhang, Y., Bienen, B. and Cassidy, M. J. (2013) On the VHM capacity of a buried spudcan in soft clay as it relates to the new ISO 19905-1 guidelines, Proc. 14th Int. Conf. Jack-Up Platform – Design, Construction and Operation, London, UK,

Zhang, Y., Bienen, B. and Cassidy, M. J. (2014) Jack-up push-over analyses featuring a new force resultant model for spudcans in soft clay, Ocean Engineering, Vol. 81, pp. 139-149,

Formatting Issue - Best Available Paper/Graphic

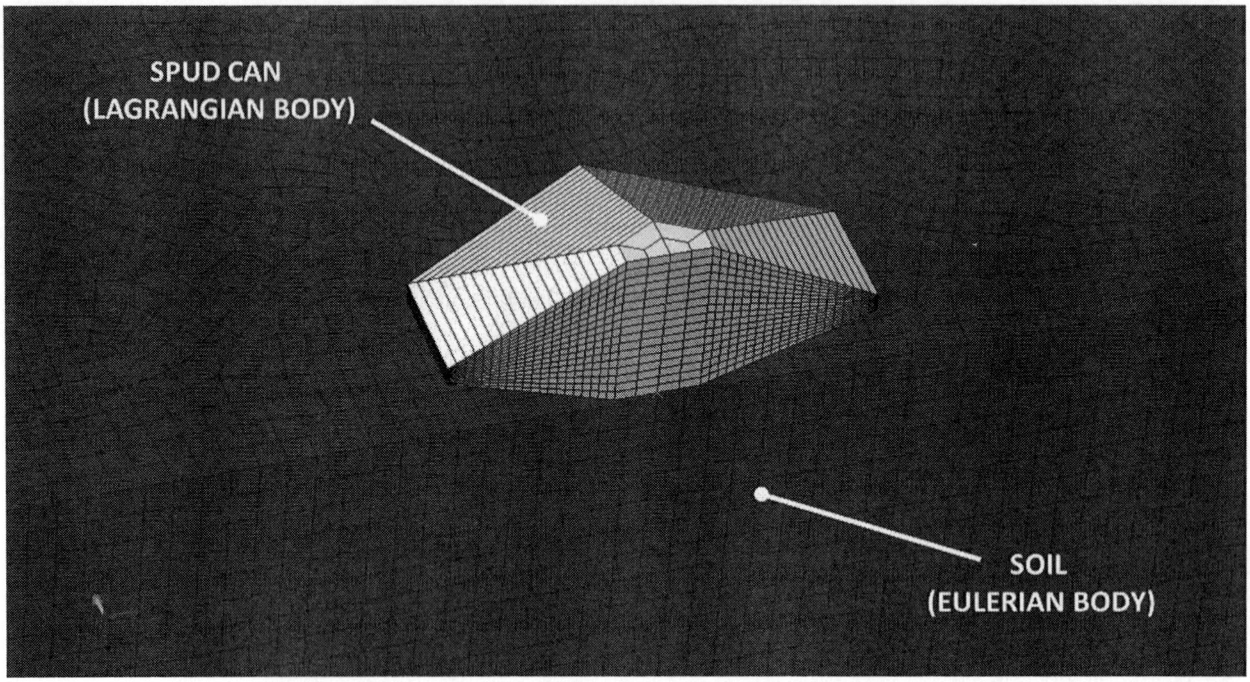

Figure 1 Lagrangian Spud Can and Eulerian Soil

Formatting Issue - Best Available Paper/Graphic

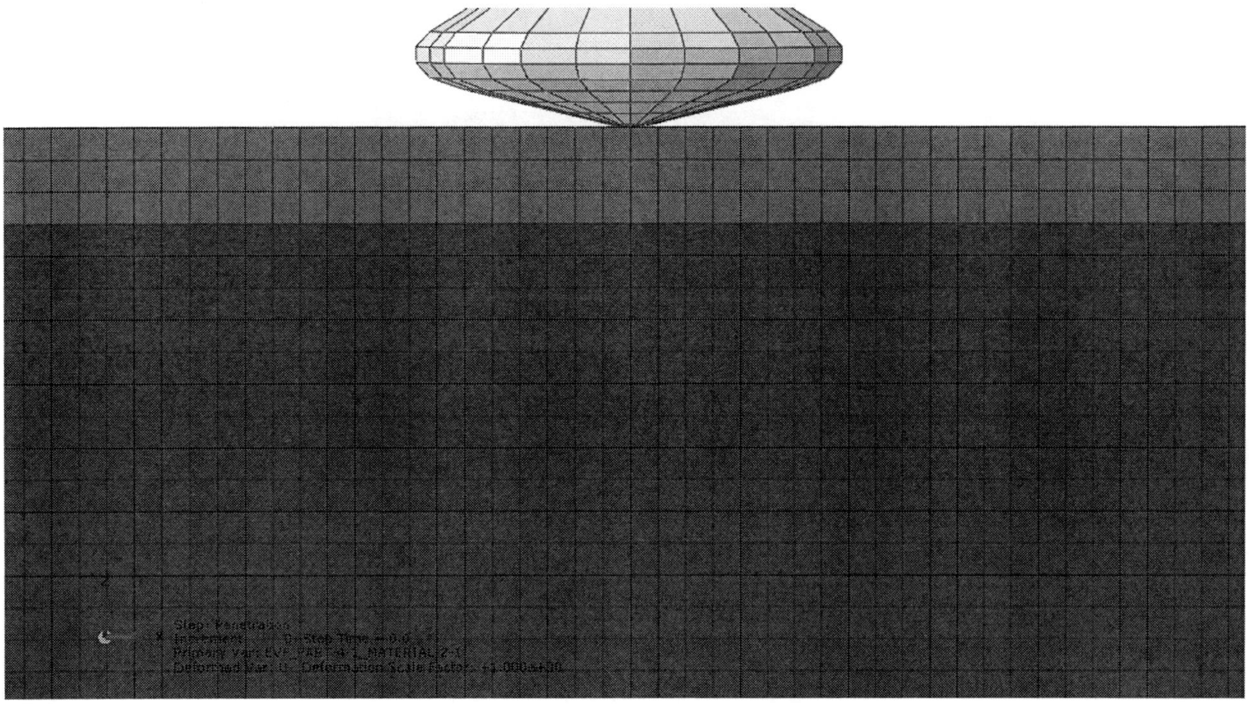

Figure 2a Soil configuration prior to penetration

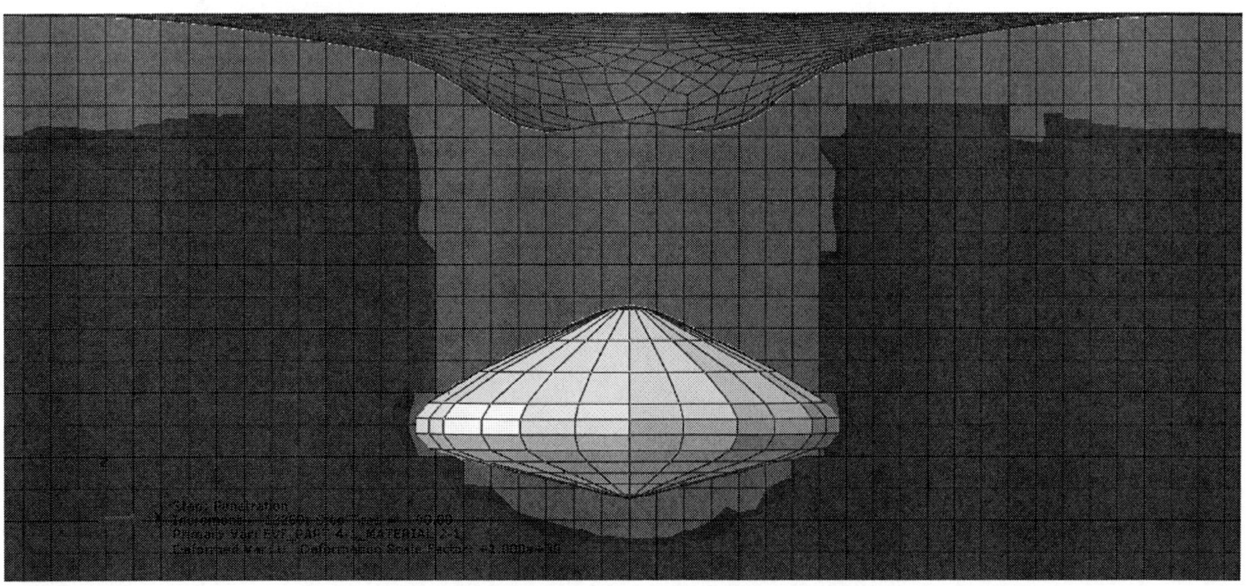

Figure 2b Soil configuration with penetration

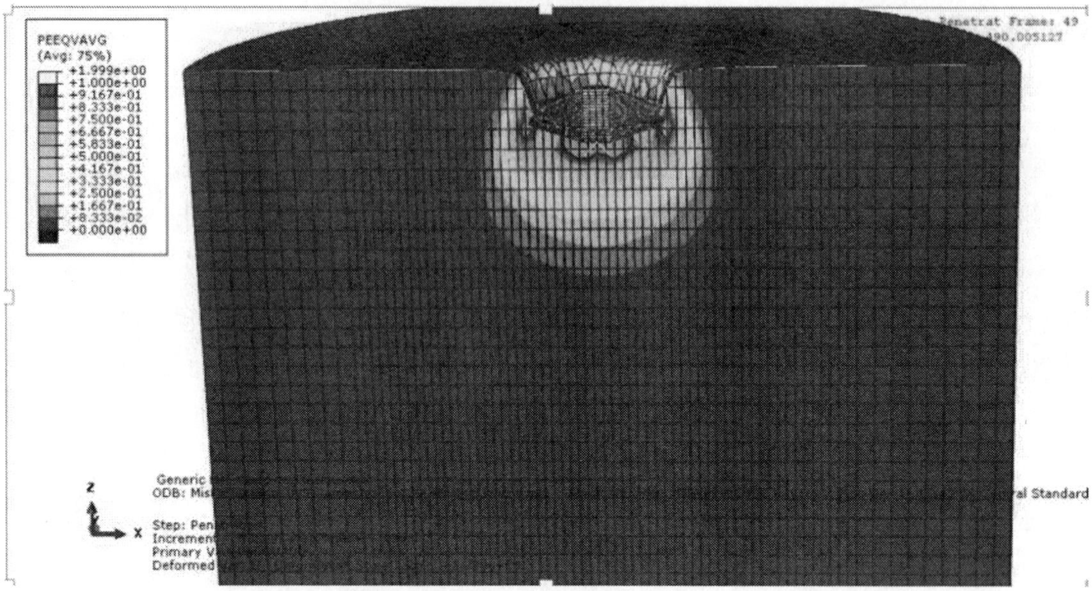

Figure 3a Plastic strain contours with moderate penetration

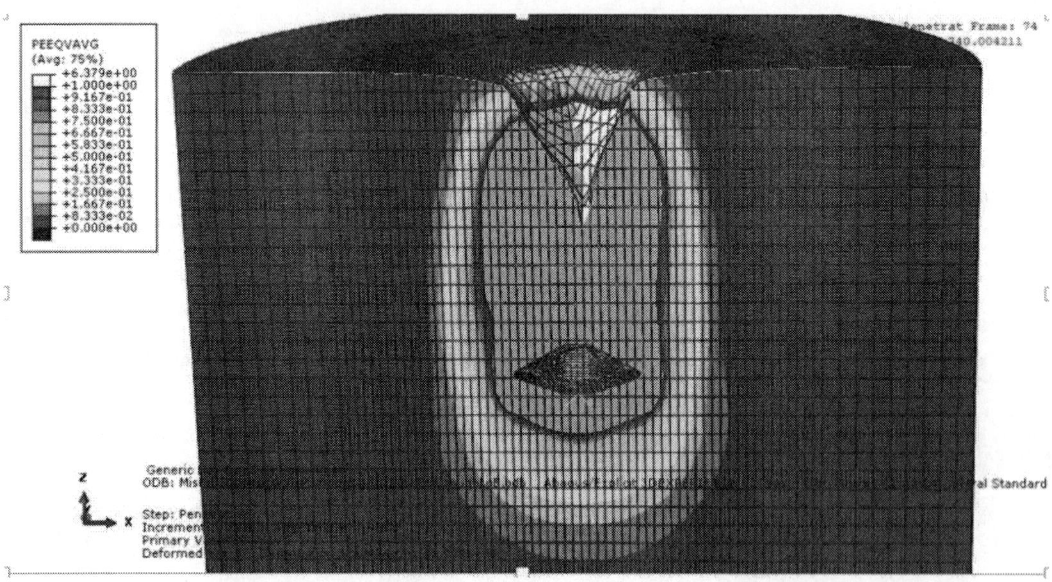

Figure 3b Plastic strain contours with deep penetration

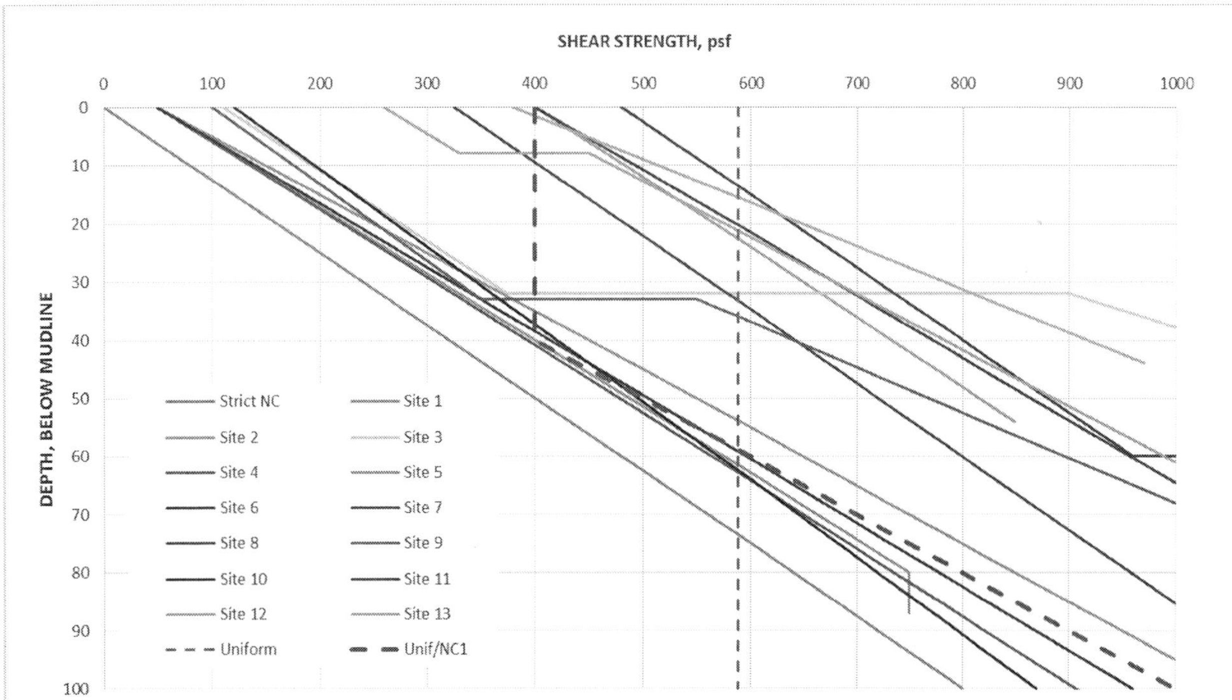

Figure4 Soil strength profiles

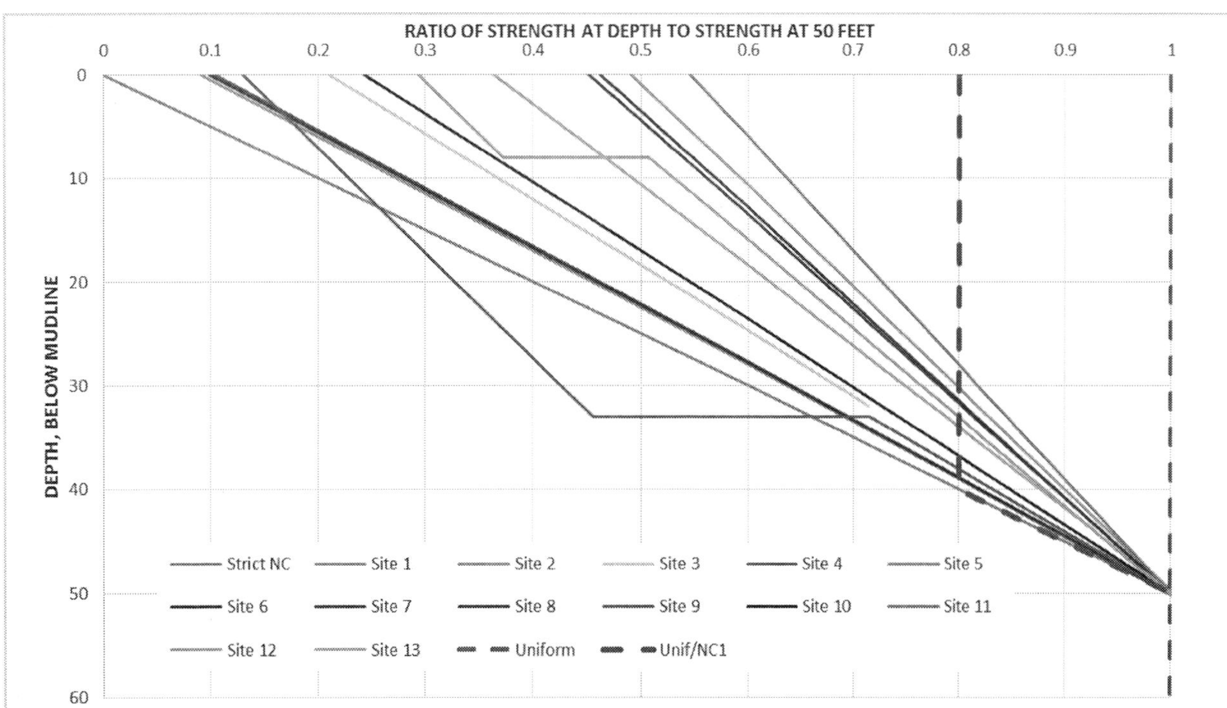

Figure 5 Normalized soil strength profiles

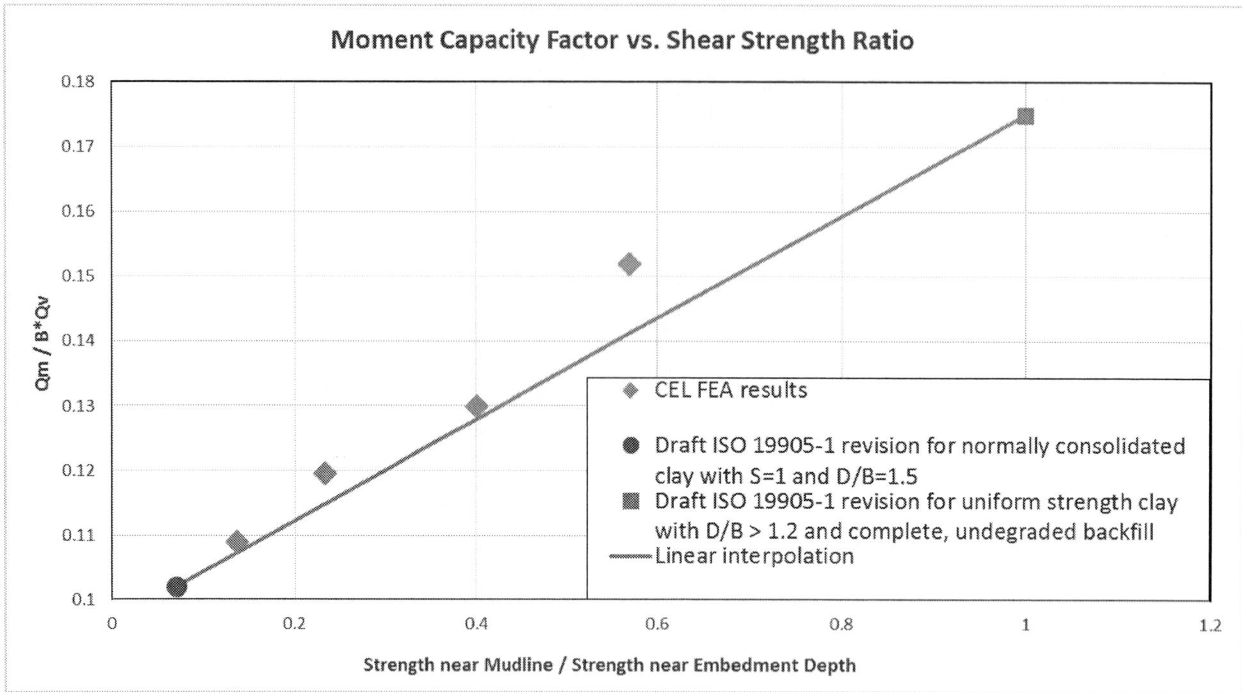

Figure 6 Normalized moment capacity vs. soil shear strength ratio

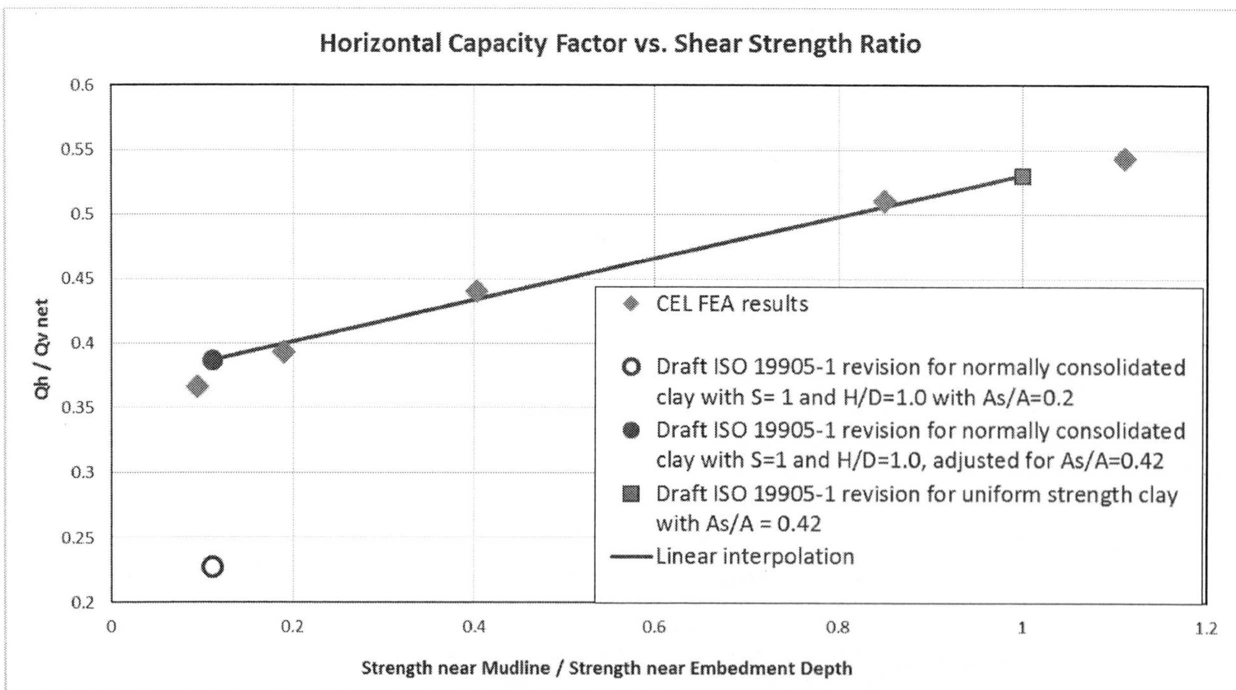

Figure 7 Normalized horizontal capacity vs. soil shear strength ratio

SNAME
THE INTERNATIONAL COMMUNITY
FOR MARITIME AND OCEAN PROFESSIONALS

CHALLENGES OF USING RESIDUAL CURVATURE METHOD IN DEEP WATER PIPELINES

TANNER J. POWELL
SENIOR PIPELINE ENGINEER, GENESIS

QIANG BAI
SENIOR PRINCIPLE ENGINEER, GENESIS

MARK BRUNNER
MANAGER PROJECT ENGINEERING, GENESIS

SHAWN SEARLE
DIRECTOR, FRONT END / SYSTEM ENGINEERING MANAGEMENT USA, TECHNIPFMC

Proceedings of the 24th Offshore Symposium, February 20th, 2019, Houston, Texas
Texas Section of the Society of Naval Architects and Marine Engineers

ABSTRACT

Uncontrolled Lateral Buckling (LB) of subsea High-Pressure, High-Temperature (HPHT) pipelines can cause high plastic deformation, which could lead to premature failure during operation if it is not mitigated. Conventional LB mitigation methodologies can be very expensive to implement and, in some cases, can have unsatisfactory "trigger" reliability for subsea pipelines. The Residual Curvature Method (RCM) is a relatively new and efficient method for LB mitigation and has, to date, been utilized for shallow water pipelines installed by the Reel-lay method. When compared to other lateral buckling mitigation methods, RCM offers advantages, such as:

- *reduced pipeline installation cost;*
- *low buckle-initiation (trigger) load, and therefore high buckle reliability, which help prevent "rogue" (unplanned) buckles;*
- *lower post-buckle strains and fatigue damage compared to other mitigation methods, which may eliminate the need for expensive, fatigue-driven welding criteria at the planned buckle locations;*

However, in the application of RCM for deep water pipelines, there are some design aspects, such as high top-tension, residual bottom-tension, soft soil, high soil lateral resistance, and orientation of in-place RC sections that present technical challenges. This paper presents the benefits as well as the feasibility of using RCM for lateral-buckle mitigation for deep-water HPHT pipelines.

To justify the implementation of the RCM for deep-water pipelines, advanced non-linear finite-element (FE) analysis is used to assess the installation of the pipeline with RC sections and to determine the achievable in-situ RC parameters for a pipeline on the seabed. This paper addresses the technical challenges of performing detailed analyses, including a sensitivity analysis of RC section performance related to lateral soil resistance and in-lace orientation of the RC pipe section, in order to confirm the suitability of RCM for lateral-buckle mitigation in deep-water pipelines in soft Gulf of Mexico clays.

Keywords: Lateral Buckling, HPHT, Reel-lay, Residual Curvature Method (RCM), Deepwater, Orientation of in-place RC section

INTRODUCTION

One of the most challenging aspects of deep-water pipeline design is thermo-mechanical design, or the quantification of a pipeline's response to internal temperature and pressure affects. The associated response of the pipeline against the seabed, appurtenances, and terminations will also impact the behavior of the pipeline. Proper thermo-mechanical design helps ensure the continued function and longevity of the pipeline system through identifying and mitigating peak loads and stresses to stay within allowable design limits. The number and range of cyclic thermal loads and the effects on cumulative fatigue damage and mitigation of excess fatigue damage is also key. For pipelines with high contents pressure and temperature, expansion along the seabed, opposed by axial and lateral soil friction, first develops the critical buckling force (triggering force) and is followed by the formation of a lateral buckle. Left unmitigated, "rogue" buckles (as shown in **Figure 1** and **Figure 2**) form in the pipeline system resulting in buckle formations of uncontrolled shape, strain, and over-the-life fatigue damage.

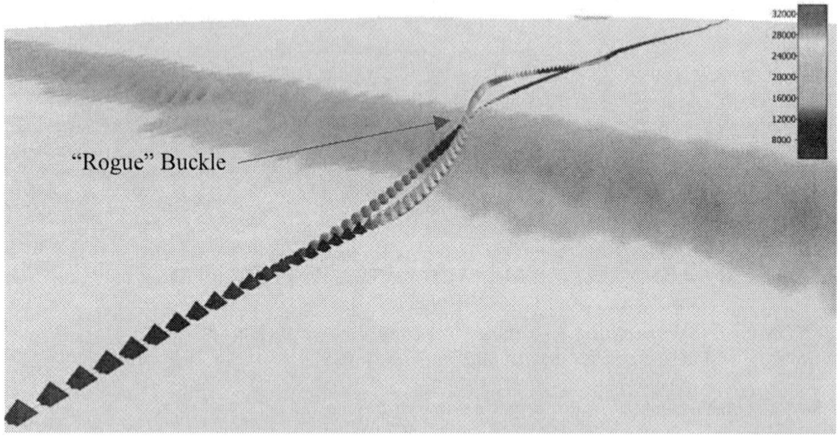

Figure 1 "Rogue" Buckle Forming at a Free-Span Shoulder

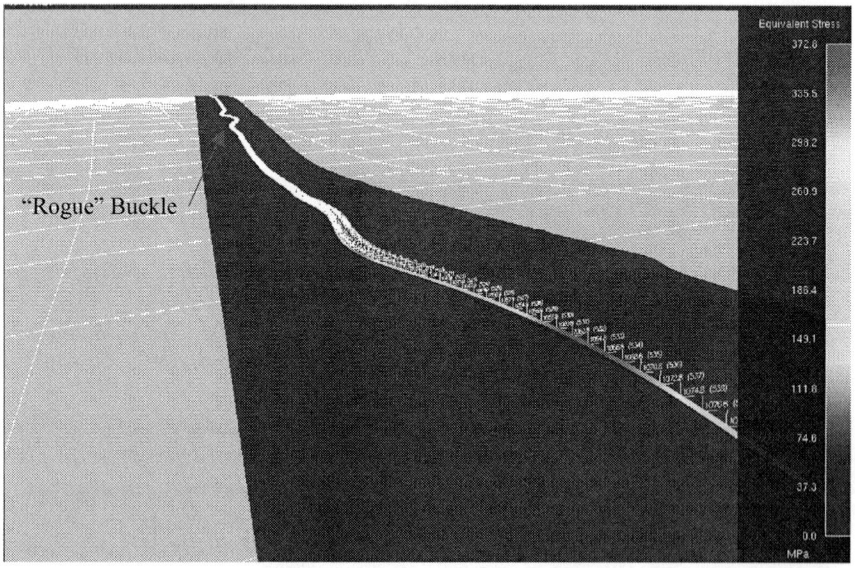

Figure 2 "Rogue" Buckle Forming in a Gas Flowline due to Seabed Profile

To limit the peak bending load, reduce the stress ranges for various operation cycles, and ensure consistent buckle behavior over several stress cycles, buckle initiators are planned at specific locations. These locations are selected to ensure the distance between buckle initiators result in reliable buckling and that the length of pipe contributing to the feed-in at the buckle is limited so that the stress range and peak strain in the planned buckle region is acceptable.

Several buckle initiation devices have been developed and successfully employed around the world to date. The more common of these include sleepers, distributed-buoyancy segments, and snake-lay. The Residual Curvature Method (RCM), originally patented by Statoil [1], is a less common mitigation method, but a novel one that can be utilized most practically in pipelines installed by Reel-lay in shallow water. Of these buckle initiation methods, RCM is the only method which requires negligible additional equipment or pipe length and a minimal increase of offshore time for installation. An RC section is installed by making an adjustment to straightener settings on the reel-lay vessel, which prevents the residual curvature from the reeling process from being straightened, so curvature remains in the pipe and is ultimately installed on the seabed, as shown in **Figure *3***. The simplicity and efficiency of RCM for reeled pipeline buckle mitigation underscores the importance of reviewing the suitability of RCM for use on any project where it may offer benefits. The straightener system controls the pipeline straightening and residual curvature as it exits the vessel. The evaluation of the suitability of RCM may be divided into three phases: Phase 1, assessment of vessel capabilities; Phase 2, Installation of the pipeline through catenary and sagbend, and Phase 3 In-place pipeline on seabed.

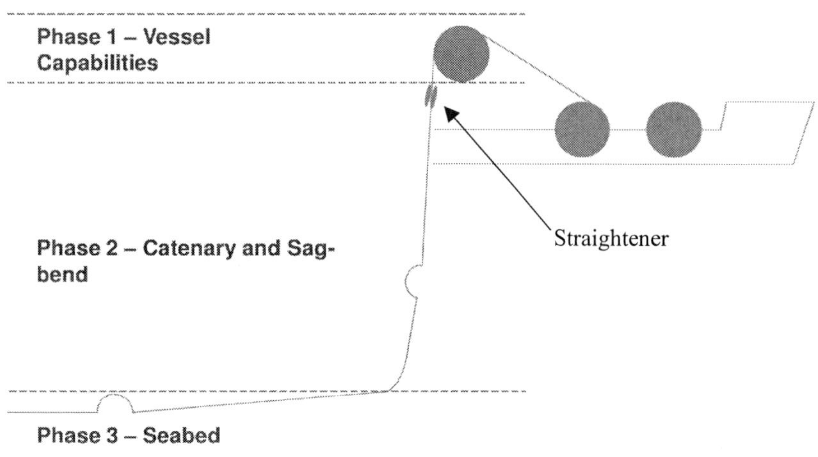

Figure 3 Installation of RCM by Reel-lay for Deepwater Pipeline [2]

Reference 3 presents the settings required for implementation of the RCM for pipe-in-pipe systems with the Deep Energy, in which the vessel capabilities for developing RC sections are detailed.

Previous research and work [4, 5, 6] have shown that during pipeline installation, the pipeline segment in the sag-bend section is subject to rotation about the pipe longitudinal centerline. Pipeline rotation is, in part, caused by a residual torsional moment in the sagbend that occurs between the laybarge and the seabed during pipe lay that is associated with the plastic strain of the reeled and un-straightened pipe. The torsional load in the pipe may also result from lateral currents acting unevenly along the suspended pipe span. Gravity introduces another torsional moment as it releases the potential energy stored by the residual curvature section and the pipe twists into a new equilibrium position. The equilibrium position, due to the combined torsional load, is held by friction at the seabed that is supported by the pipe sagbend and the vessel tensioner. Some residual torsion is "locked in" after completion of pipelay. Even though the pipeline rotation due to torsion can be predicted by engineering analysis, there are uncertainties associated with the parameters relating to the phenomenon, especially pipe-soil interactions at the touchdown region.

Rotation of RC Section during Installation

In a separate study, the lowering of the RC section from the tensioner onto the seabed is investigated using Flexcom software by modelling the pipe-lay operation from the Deep Energy [5]. The analysis focuses on the three stages of the installation process that the RC section progresses through: 1) RC section at vessel departure (top of the

catenary), 2) RC section in the sagbend and 3) RC section at the touch down (TD) on the seabed.
When the RC section is leaving the vessel, it undergoes the highest axial tension and imposes the highest reverse bending (straightening by gravity, not by the tensioner) in the RC section. The axial tension at the sagbend is significantly lower compared to the top tension. However, due to the bending moment at the sagbend location, the RC section will behave differently than a normal pipe without an RC section. The different behavior is due to the misalignment between the RC section bending plane and the sagbend bending plane associated with a normal near-catenary shape. The RC section may rotate downwards, minimizing the system's elastic strain energy, as shown in **Figure 4**, or it may it may rotate to some other plane.

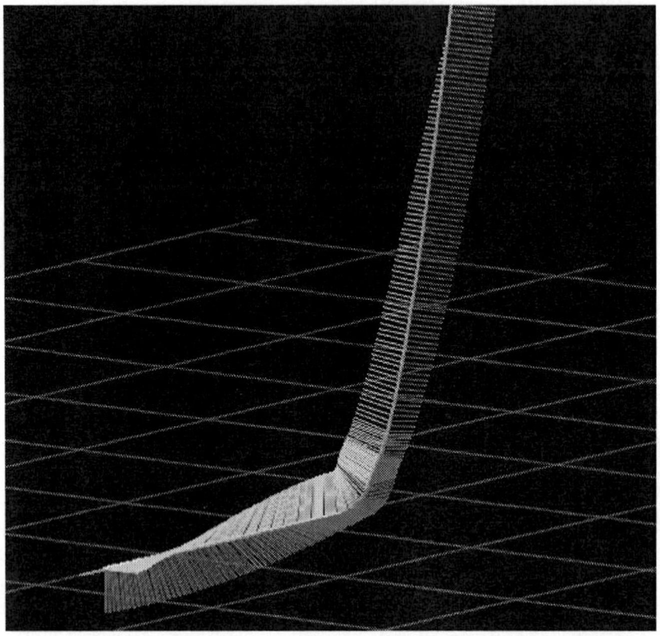

Figure 4 Rotation of RC Section in Sagbend Segment

The twist angles (rotation about pipe centerline axis) of an RC section increases with water depth and for deep water, the effect of the out-of-plane bending associated with twist is significant. Rotation of the RC section is beneficial to reduce stress in catenary during installation and to obtain a desired orientation and as-laid shape on seabed. The governing parameters for design of RC sections are: RC strain and length, water depth, pipe torsional stiffness, top-side clamping stiffness, and soil resistance. Typically, shallow water presents a more critical scenario due to the risk that the RC section will not rotate sufficiently out of the vertical plane (least favorable). Depending on the water depth and the cross-section of pipeline, different amounts of rotation are measured, with a maximum of 200 degrees. The twisting of the pipe also creates a threat to the levelness of the in-place pipeline end structures, which may affect structure foundation and jumper designs. Furthermore, for RC section, any unexpected large angles may result in increased stress/strain of pipeline due to lateral buckling in the operating conditions compared to the case that RC section is in the same plane as seabed. During lowering of the RC section, it may be necessary to adjust the orientation of the RC section just prior to touchdown on the seabed by using ROV intervention, changing the ship heading or by increasing lay tension. Thus, RC section orientation may be subject to limited adjustment as it approaches the seabed.

It should be noted that the aforementioned papers do not describe the effect of reel-lay installation on the pipeline in-place behavior, therefore, the in-place behavior is worthy of investigation. This paper considers the design of a hypothetical 12-inch pipeline for deep-water installation on soft Gulf of Mexico (GoM) clays with thermo-mechanical buckling mitigated via RCM. It also reviews the effects of changes to key design parameters (sensitivity cases) associated with RCM and typical deep-water GoM pipelines. Design parameters considered for sensitivity checks include the radial orientation of the residual curvature plane relative to the pipe axis, the residual curvature section length, and strength of the clay at the seabed.

METHODOLOGY

FEA Modeling

The lateral buckling comparison of a full reeling model and a simple monotonic bending model show that the RC section can be introduced to the pipeline through a monotonic bending load for in-place lateral buckling analysis if this adequately reproduces the shape of the pipeline after reeling [8]. The selected residual strain of 0.2% influences in-place behavior and reverse bending of the RC section due to installation tension and is selected for the RC sections based on experience. **Figure 5** shows the pipeline configuration with an RC section after applying and in-plane rotational displacement to both ends of the RC length (L) for an initially straight pipeline. The magnitude of the rotational displacements depends on the residual strain, pipe outside diameter, wall thickness, stress-strain curve, residual curvature length, and the element length. These rotational displacements are then removed from the FE model and the pipeline takes its equilibrium position, resulting in a residual curvature (or RC strain of 0.2%) in the pipe at the selected location. The RC profile will be further flattened by the residual lay tension applied afterwards as shown in **Figure 6**.

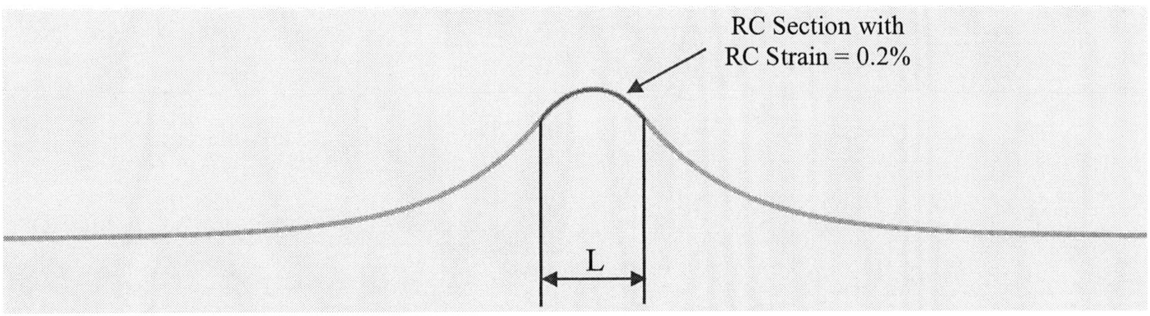

Figure 5 Pipeline Configuration with Residual Curvature Section

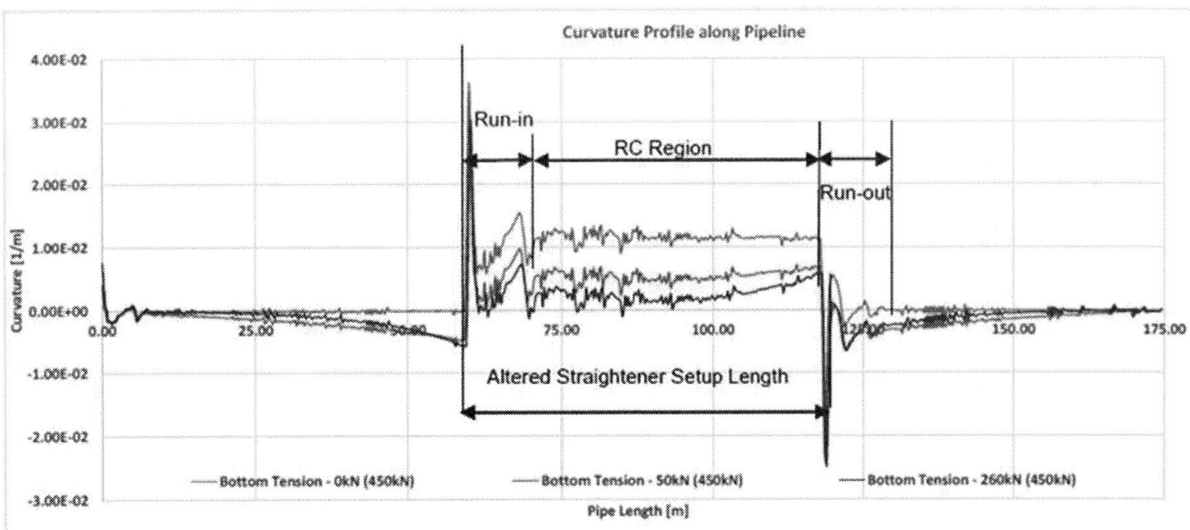

Figure 6 Residual Curvature Distributions under Different Residual Tensions

Axial tension was applied to the pipeline during lay down of the RC section to simulate the installation bottom tensions (343 kN) associated with deep-water (3,500 ft) reeled pipelay. Tension during pipelay results in elastic straightening of the RC section. As the touch down point moves away from the RC section, residual tension remains in the line prior to the hydrotest load step. Additional load steps, consistent with typical operational cycle regimes, are modelled including partial and full cooldowns. The pipeline lateral displacement distributions along the route are plotted in **Figure 7**. The corresponding total nominal longitudinal strain distributions during operation are shown in **Figure 8**.

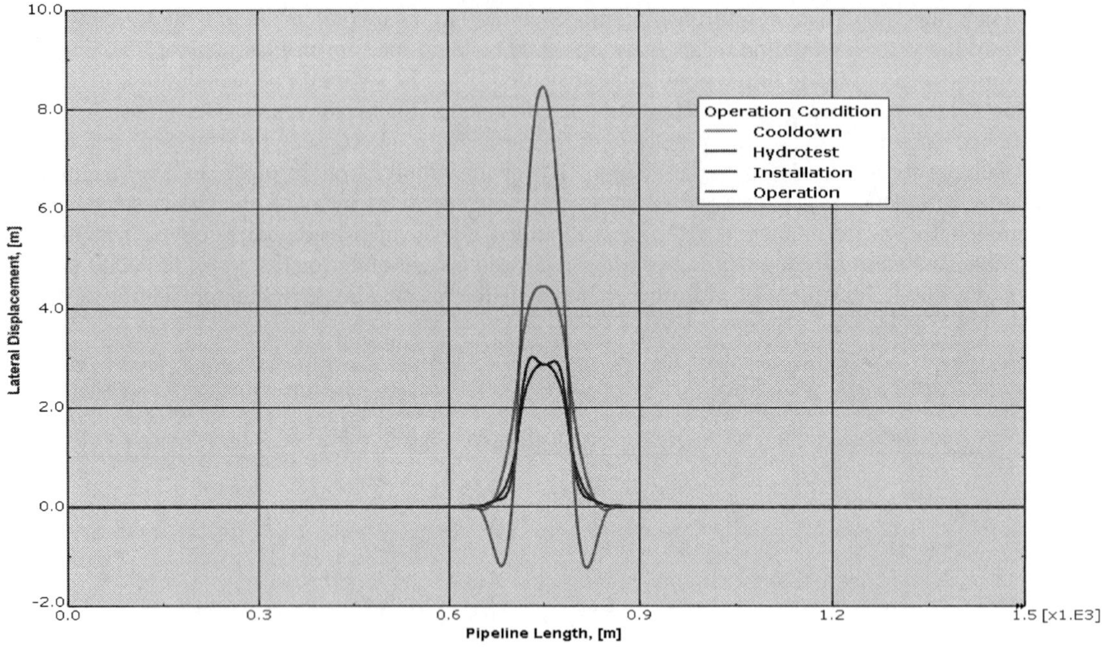

Figure 7 Lateral Displacement Distributions of Flowline with RC Section

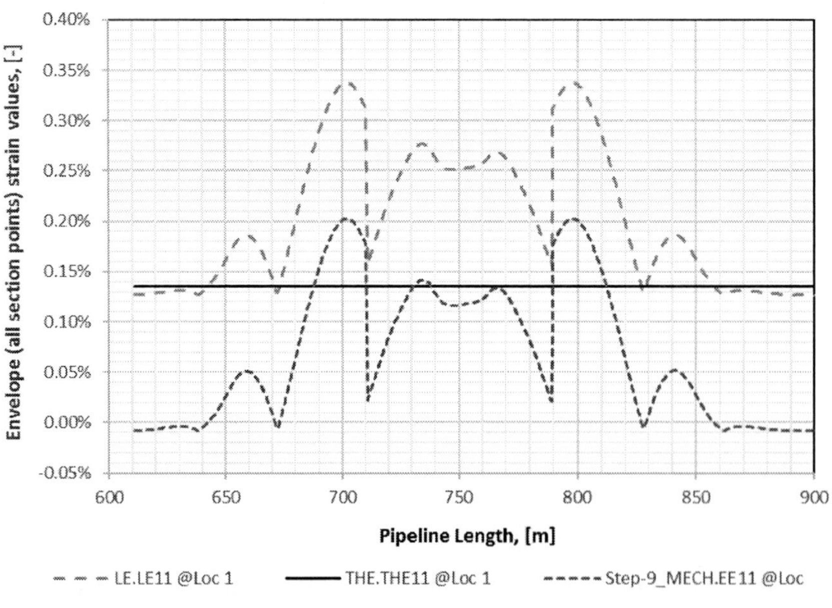

Figure 8 Total Nominal Longitudinal Strain Distribution for Flowline in Operating Condition

Design Criteria

Maximum values of key data, such as stress and nominal longitudinal strain are extracted from the analysis results and DnV OS-F101 [10] based load-controlled checks (LCC) and displacement-controlled checks (DCC) are performed. Fatigue damage is approximated based on typical operational cycle regimes and considering the maximum stress range between load steps. The output data, reviewed below, represent the typical governing criteria of limit states for lateral buckling assessments of rigid pipelines.

- **LE11:** Maximum nominal longitudinal strain in the operating condition, design criteria < 0.4%.
- **DCC:** Displacement controlled local buckling check based on DNV OS-F101, design criteria < 1.0
- **LCC:** Load controlled local buckling check based on DNV OS-F101, design criteria (<1.0) may not be applied for lateral buckling.
- **Fatigue damage:** Cumulated fatigue damage of lateral buckling due to thermal cyclic loads based on Powernap FA data, design criteria < 0.1 (with a safety factor of 10.0).

An extensive set of detailed nonlinear Abaqus FEA analyses were performed using VAS models to evaluate the lateral buckling response of the flowline. The objectives of the analyses include:

- Effect of lateral soil resistance.
- Effect of RC section length.
- Effect of twisted RC section.

ANALYSIS RESULTS

This section summarizes the analysis results and discusses the assessments performed to understand the effects of lateral pipe-soil resistance, RC section length and RC section twist (orientation) on the flowline lateral buckling behavior.

Effect of Pipe Soil Interaction

The effect of lateral soil resistance on the lateral buckling performance of flowline is checked with the design criteria for different lateral resistance using a Virtual Anchor Spacing (VAS) model of the pipeline with RC length of 60 m.

Table 1 Effects of Lateral Soil Resistance on Key Design Parameters

Laterial Soil Resistance	LE11 [-]	DCC [-]	LCC [-]	Fatigue damage [-]
BE	0.36%	0.055	0.95	0.057
UB	0.37%	0.058	1.15	0.072
LB	0.34%	0.050	0.76	0.047

Table *1* summarizes the key limit-state checks for different lateral soil resistance coefficients. The following input parameters are used in determining the tabulated results:

- BE: Best estimate axial and lateral soil resistance coefficients, with $\mu_a = 0.6$, $\mu_{l,breakout} = 0.92$, $\mu_{l,Residual} = 0.81$
- UB: Best estimate axial and upper-bound lateral resistance coefficients, with $\mu_a = 0.6$, $\mu_{l,breakout} = 1.1$, $\mu_{l,Residual} = 1.1$
- LB: Best-estimate axial and lower-bound lateral resistance coefficients, with $\mu_a = 0.6$, $\mu_{l,breakout} = 0.6$, $\mu_{l,Residual} = 0.6$

As the soil resistance coefficient in lateral direction increases, the chance of a planned buckle exhibiting an asymmetric buckle shape increases. With the increase in lateral soil resistance, the key parameters for the design limit state checks increase close to the design allowables defined in Section Design Criteria.

Effect of RC Section Length

The effect of RC section length on the lateral buckling performance is checked with a sensitivity analysis performed using different RC section lengths [9]. As shown in **Figure *9***, for a 90 m RC length case, a smooth, symmetric buckle shape develops, with a bending moment slightly less than the allowable limit. However, for a 100 m RC length, an asymmetric buckle shape is observed, with a higher bending moment. RC lengths equal to or greater than 100 m result in onerous buckle shapes, with a higher bending moment.

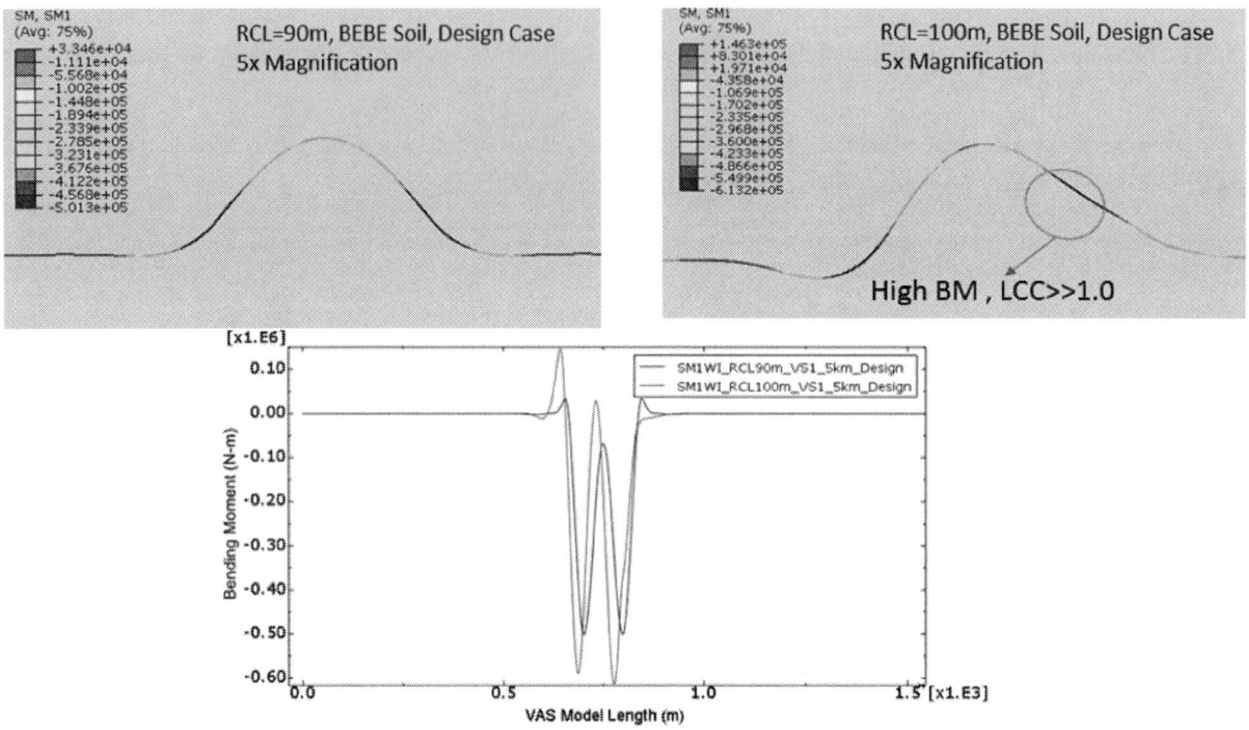

Figure 9 Effect of RC Section Length on Bending Moment

Effect of Twisted RC Section

The design criteria are checked for different installation orientations and the analysis results for RC lengths of 80 m and 60 m for the BE soil condition are summarized herein.

Definitions of terms used in discussion and tabulated results are presented below:

Initial angle: Angle between the RC section plane of bending and flat seabed before the RC section contacts the seabed. -90 degrees for RC section crown downwards (**Figure 10**); +90 degrees for RC section crown upwards.

As-installed angle: Angle between the RC section plane of bending and seabed surface after the RC section is installed on the seabed under residual tension and empty submerged weight.

Figure 10 Example of RC Section with Initial Angle of -90 Degrees

RC length of 80 m

An RC section with a length of 80 m and total bending strain of 0.2% was considered for lateral buckling mitigation. The analysis results of several critical design parameters for several RC section installation orientations are summarized in **Table 2** and **Figure 11**.

Table 2 Effects of Installation Orientation of 80 m RC Section on the Key Design Parameters

Initial Angle	As-installed Angle	LE11	DCC	LCC	Fatigue Damage
[degree]	[degree]	[-]	[-]	[-]	[-]
0.0	0.0	0.34%	0.051	0.98	0.052
22.5	5.9	0.33%	0.051	0.99	
67.5	35.4	0.52%	0.10	1.95	0.186
33.7	21.3	0.31%	0.042	0.78	
101.2	121.0	0.96%	0.208	2.18	0.491

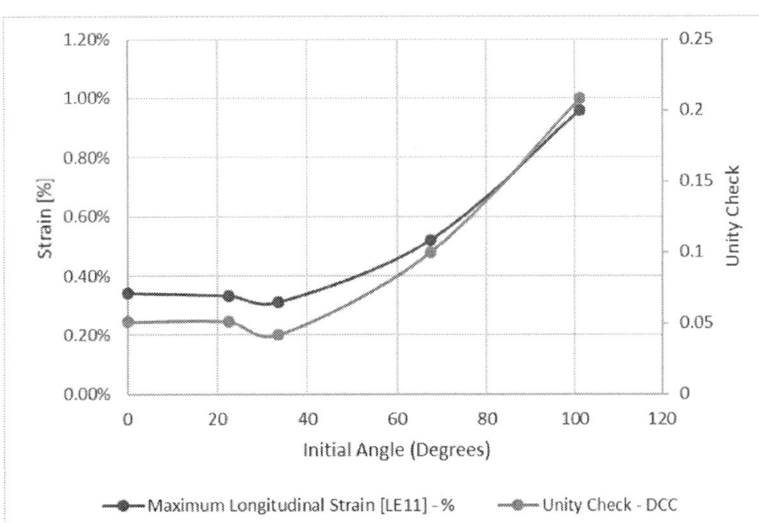

Figure 11 Maximum Longitudinal Strain (LE11) and DCC vs. Initial RCM Angle – 80 m RC Section

The table shows that the critical limit-state parameters are sensitive to the as-installed orientation of the RC section and quickly approach design allowable limits as as-installed angle varies. The design criteria for fatigue damage and maximum nominal strain for initial orientation angles larger than 67.5 degree are higher than the allowable ranges.

RC length of 60 m

Similarly, an RC section with a length of 60 m and total bending strain of 0.2% is considered. The analysis results of several critical design parameters, for several installation orientations, are summarized in **Table 3** and **Figure 12**.

Table 3 Effects of Installation Orientation of 60 m RC Section on Key Design Parameters

Initial Angle [degree]	As-installed Angle [degree]	LE11 [-]	DCC [-]	LCC [-]	Fatigue Damage [-]
0.0	0.0	0.35%	0.0540	0.94	0.053
22.5	6.2	0.36%	0.0540	0.94	
33.7	21.8	0.35%	0.0535	0.98	
67.5	36.1	0.35%	0.0530	1.04	0.059
101.2	121.3	0.34%	0.0513	1.05	0.058
-30.0	-13.2	0.36%	0.0546	0.98	
-90.0	-74.6	0.39%	0.0662	1.40	0.098

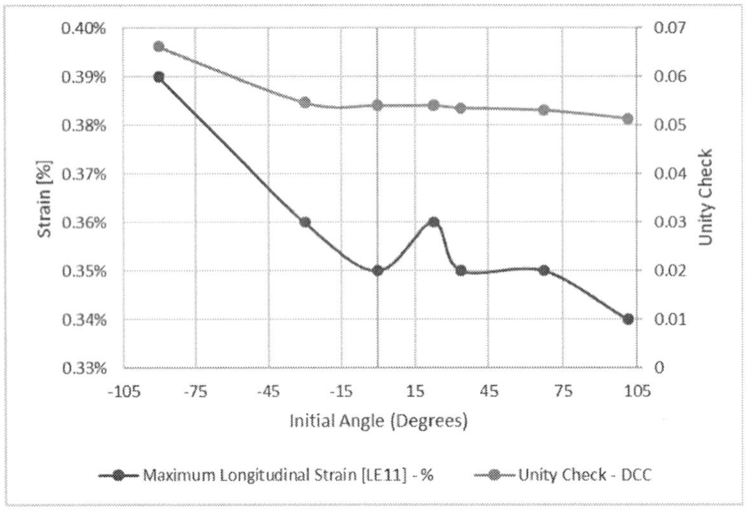

Figure 12 Maximum Longitudinal Strain (LE11) and DCC vs. Initial RCM Angle – 60 m RC Section

The analysis results show that the critical design parameters increase slowly with the as-installed angle. The results are not sensitive to the initial orientation for the 60 m long RC section. The initial orientation angle of -90 degrees (RC section crown downwards) is the worst case for the lateral buckling performance, however all the design parameters are within the design criteria limits for the BE soil condition.

Figure *13* shows the variation of the configurations for the -90 degrees installed RC section for different conditions (load steps).

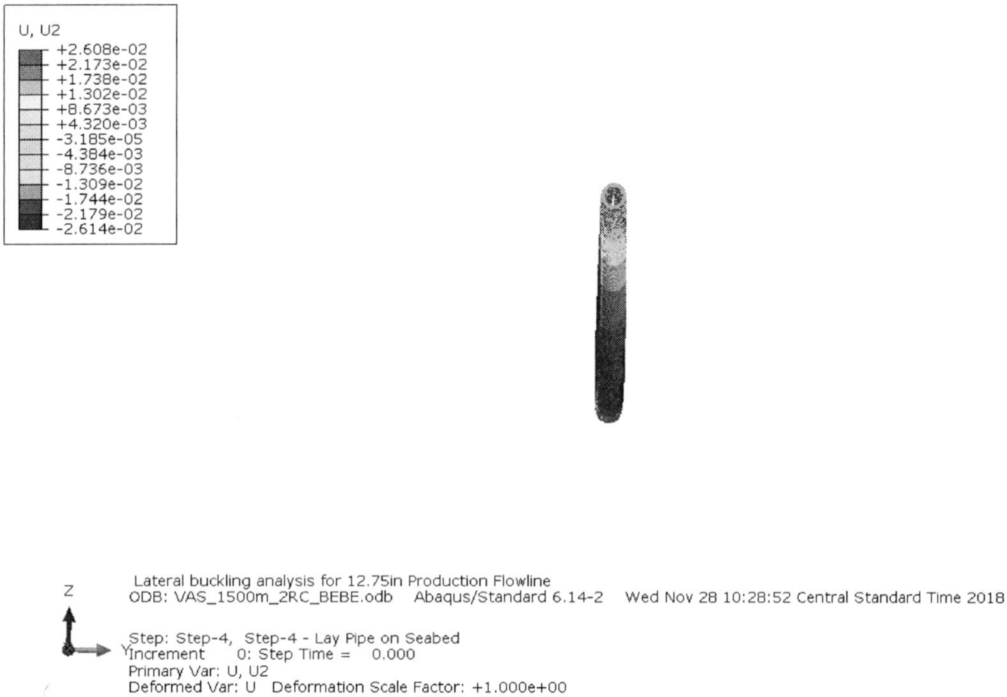

Step-1: Initial Condition- Installation (End View)

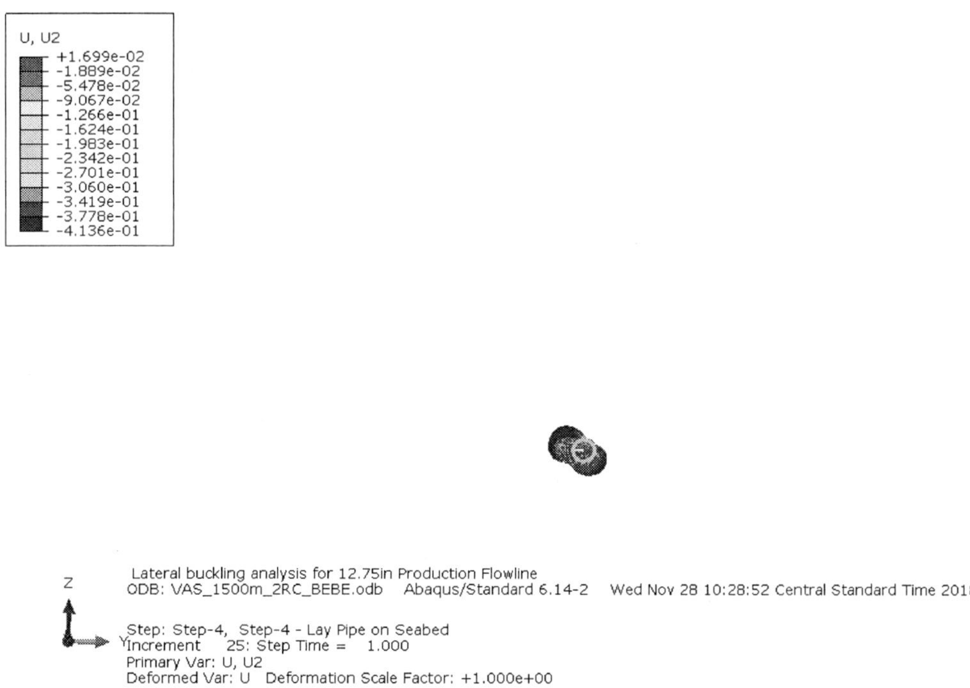

Step-2: Installed Condition (End View)

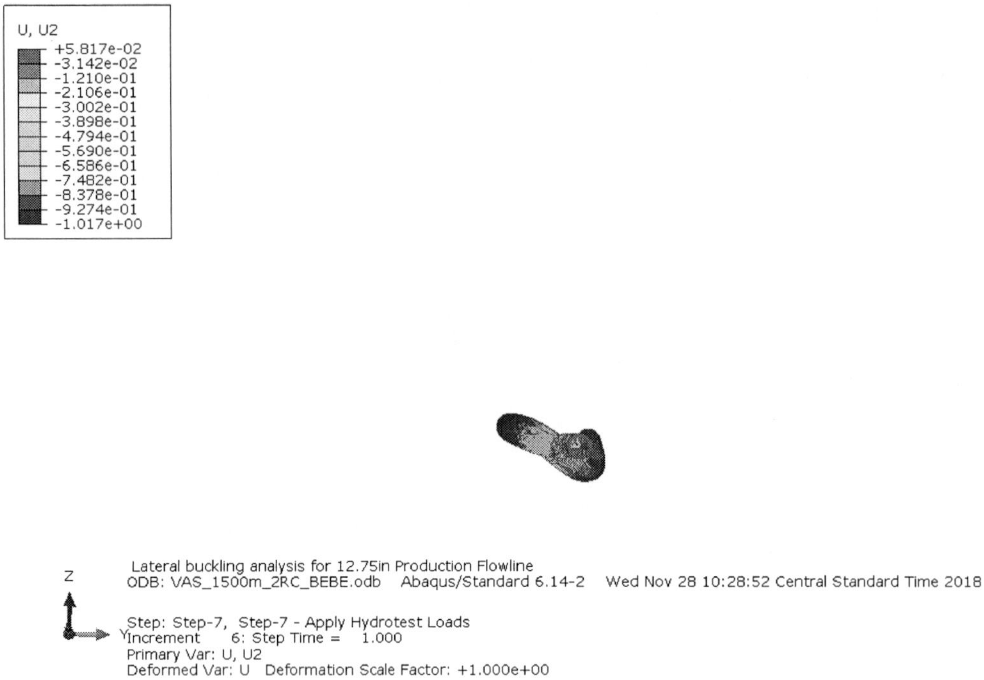

Step-3: Hydrotest Condition (End View)

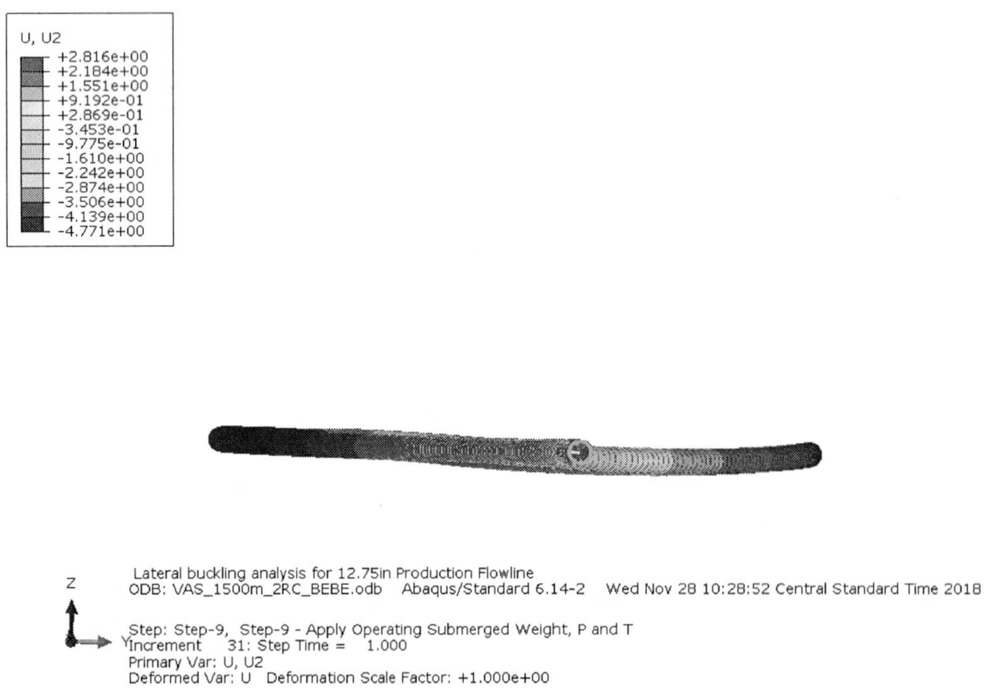

Step-4: Operating Condition (End View)

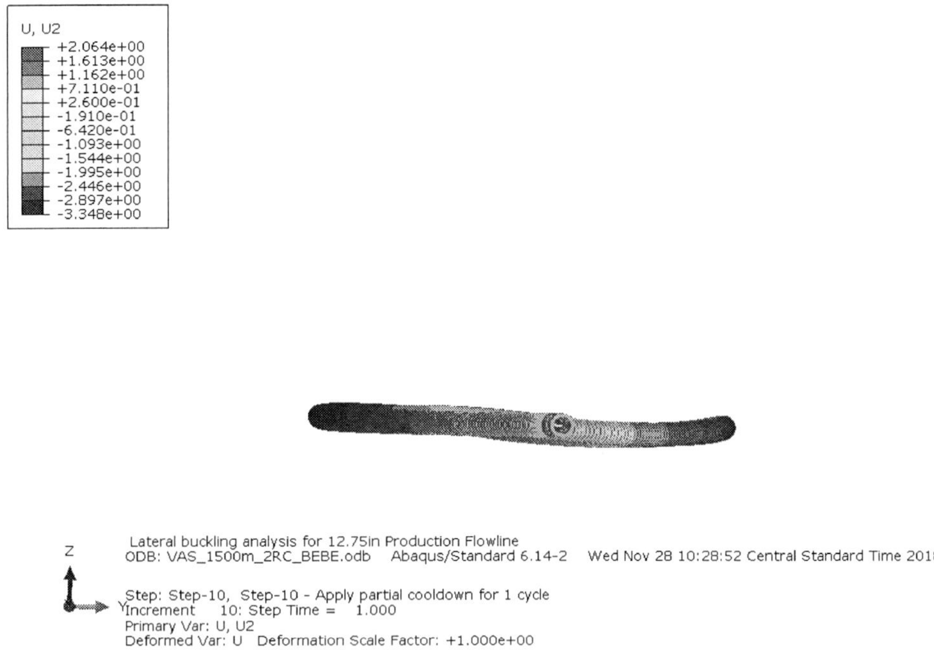

Step-5: Partial Cooldown Condition (End View)

Figure 13 In-service Lateral Responses of Flowline with RC Section Crown Downwards

CONCLUSIONS

This paper studies some of the challenges of using the Residual Curvature Method for lateral buckle mitigation in deep water pipelines, when using a simple monotonic bending model. A series of limit states are checked for the in-service behavior of a pipeline to evaluate the effect of RC section parameters and other input variables on the in-place analyses of the deep-water pipelines. Although RCM has only been installed and operated in shallow water so far, from the observations of this paper it appears that this method is also viable for deep water, so long as certain key parameters are properly managed. Detailed analyses conclusions are summarized below:

- Higher soil lateral resistance increases the chance of causing an asymmetric buckle shape for RC section that showed a symmetric buckle shape in a lower lateral soil resistance (LB) case. Increased soil lateral resistance may result in narrow design margins when considering design allowable limit states. Large lateral resistances may affect the viability of RCM and require consideration of alternate buckle initiation methods.
- Asymmetric buckle shapes occur more easily with longer RC sections which have higher bending moments after lateral buckling. Increasing the RC section length becomes ineffective beyond a certain length, and asymmetric buckle shapes may result in an unacceptable bending strain if the length is too long.
- RC section performance is sensitive to RC section orientation if the RC section length is too long. The critical design limit states are sensitive to the orientation of the in-place RC section and quickly approach the allowable limits as the orientation angle increases for the 80 m RC length. However, the parameters increase slowly with the in-place angle for the 60 m length; and the results are not very sensitive to the orientation angle. Having a shorter RC section length decreases the allowable VAS length, and more RC sections are required for the buckling mitigation. There is also a minimum length limitation for RC section which requires further evaluation to define.
- The initial orientation angle of -90 degrees (RC section crown downwards) is the worst case for the lateral buckling performance of RCM. Even so, all the key limit states stay within the allowable design criteria for the best estimate (BE) soil condition. The suitability of RCM for the deep-water flowlines modelled is confirmed based on these results. However, soil-pipe interaction is a critical aspect of the in-place lateral buckling performance. Furthermore, detailed investigation should be performed for a larger variety of soil-pipe interaction, water depths, etc.

ACKNOWLEDGEMENTS

The authors would like to thank Genesis and TechnipFMC for supporting this paper. Thanks are also expressed to all our colleagues for their fruitful comments and discussions and supporting the production of this paper in various ways.

REFERENCES

1. G. Endal, "Method for Pipelaying from a Coil to the Sea Bed, Controlling Thermal Expansion", US patent, US6910830.
2. R. Bhide, C. Hunter and J. Tainsh, "Wave 1 Project – Deep Energy Residual Curvature Pipelay Lessons Learned", OED, Feb 2018.
3. R. Barnes, A. McRae, R. Rai., and G. Abdelmaged, "Residual curvature method for reeled PiP – a framework for installation", OMAE 2017-61924, 2017.
4. V. Taams, "The Onset of Pipeline Twist during Reel lay Operations", Master Thesis, Deft University of Technology, 2016.
5. M. Abdullayev, "ExxonMobil Liza – Installation the Pipe with RC Section", OED, 072039T001-N-17-0001, 2016.
6. E Karjadi, et al, "Benefits and Deep Water Install Ability Challenges of Residual Curvature Method for Lateral

Buckling Mitigation", OMAE 2017-61387, 2017, Trondheim, Norway.

7. P. Cooper, T. Zhao, anda F Kortekaas, "Residual Curvature Method for Lateral Buckling of Deepwater Flowlines", OPT 2017,
8. M. Gallegillo, et al. "Post-reeled Behaviour of Pipelines with Global Buckling Mitigation by the Residual Curvature Method.", OMAE 2017-62481, 2017.
9. R. Ferdous, "Lateral Buckling Mitigation for Peregrino Phase-II Water Injection Flowline using RCM", 10th Technip-FMC Global Buckling Forum, October 2018, London.
10. Det Norske Veritas, "Submarine Pipeline Systems, DNV-OS-F101", 2013.

SNAME
THE INTERNATIONAL COMMUNITY
FOR MARITIME AND OCEAN PROFESSIONALS

DESIGN AND INSTALLATION OF DYNAMICALLY INSTALLED PILES FOR DEEPWATER MOORING

X.Y. ZHANG
PRINCIPAL ENGINEER, AMERICAN BUREAU OF SHIPPING, USA

F. YONG
RESEARCH FELLOW, NATIONAL UNIVERSITY OF SINGAPORE, SINGAPORE

Y.P. LI
LECTURER, HOHAI UNIVERSITY, P.R.CHINA

J.T. YI
PROFESSOR, CHONGQING UNIVERSITY, P.R.CHINA

F.H. LEE
PROFESSOR, NATIONAL UNIVERSITY OF SINGAPORE, SINGAPORE

X. CHEN
PROFESSOR, BEIJING JIAOTONG UNIVERSITY, P.R.CHINA

S.Q. WANG
MANAGER, AMERICAN BUREAU OF SHIPPING, USA

Proceedings of the 24th Offshore Symposium, February 20th 2019, Houston, Texas
Texas Section of the Society of Naval Architects and Marine Engineers

ABSTRACT

The quest for reliable and cost-effective solution of installing piles in deepwater led to the development of dynamically installed piles that embed themselves into the seabed through free-fall. Several variations of dynamically installed piles have been devised and successfully entered into service at deepwater offshore sites. The most notable one is the torpedo pile patented by Petrobras.

To facilitate the design and installation of the dynamically installed piles, ABS has developed Guidance Notes to provide geotechnical design and structural assessment methods. This paper presents an overview of the guidance and details of the technical development that forms the basis of the recommended methods.

In support of the development of the guidance, finite element analyses and centrifuge tests were conducted to study pile/soil interaction and to verify and further improve the prediction methods for pile pullout capacity. The pile inclination after installation, which has a significant effect on the pile pullout capacity and is of significant concern to the offshore industry, was thoroughly studied. Since the dynamic installation process results in lower short-term pullout capacity of the pile, it is recommended that the piles be installed for a sufficiently long period to allow the development of the pullout capacity. A prediction of the pile capacity restoration over time was developed based on the results of a series of centrifuge tests. A framework on the normalized vertical and horizontal component is proposed to predict the pile pullout capacities subjected to different loading angles.

***Keywords:* dynamically installed piles, pullout capacity, inclination angle, loading angle, finite element analyses, centrifuge tests**

INTRODUCTION

Dynamically installed piles are finned piles designed to be released from a height of 20-100 m above the seabed, reaching velocities of 20-35 m/s at the seabed before self-embedment. They may be stabilized with multiple fins at the trailing edge. The tip penetration is expected to be two to three times the pile length, and pullout capacity after consolidation is three to six times the pile weight. Dynamically installed piles have been widely regarded as the most promising deepwater anchor concept in terms of the cost effectiveness and easy installation (Ehlers at al. 2004). Several different forms of dynamically installed piles have been devised, the most notable of which are the torpedo pile, a concept patented by Petrobras; see Figure 1.

Compared to the high cost of full-scale field trials, physical modeling of pile installation processes, particularly in the centrifuge, provides a cost effective alternative for studies. O' Loughlin et al. (2009) and Richardson (2008) focused on centrifuge tests to evaluate embedment depth and vertical holding capacities. Hossain et al. (2013) conducted centrifuge tests to evaluate holding capacities of torpedo anchor models extracted at different angles.

With regard to anchor capacity, the guidelines recommended by API (2007) for pile axial capacity have typically been used (Lieng et al. 2000; Audibert et al. 2006). The anchor capacity has also been assessed using three dimensional finite element analysis (Lieng et al. 2000; Medeiros 2001; Fu et al. 2017).

To facilitate the design and installation of the dynamically installed piles, ABS (2016) has developed guidance notes to provide geotechnical design and structural assessment methods for the dynamically installed piles. This paper presents an overview of the guidance and details of the technical development that forms the basis of the recommended methods.

Figure 1 Typical torpedo pile geometry (after Bonfim dos Santos et al. 2004)

DESIGN CRITERIA

Factor of Safety for Pile Pullout Capacity

The factor of safety for pullout capacity is defined as the pile's ultimate pullout capacity divided by the maximum load obtained from a dynamic analysis that accounts for mooring line dynamics. Factors of safety for dynamically installed piles should be 2.0 for the intact condition and 1.5 for the broken line (damaged) condition.

For a clustered area, the most critically loaded line should be selected as representative of the entire cluster for geotechnical calculation purposes. The reverse catenary line attached to the pile's padeye below the seabed should be determined, including the mooring line load on the top of pile and its orientation.

Allowable Yielding Stresses

The equivalent von Mises stress σ_{eqv} generated by applied loads should be limited to the following stresses for the *line broken* and *intact* conditions, respectively.

$$\sigma_{eqv} \leq 0.9\,\sigma_{yield} \quad \text{(line broken)} \qquad \text{Eq. (1)}$$

$$\sigma_{eqv} \leq 0.67\,\sigma_{yield} \quad \text{(intact)} \qquad \text{Eq. (2)}$$

where

σ_{yield} = yield stress of the considered structural component

For those stresses of a highly localized nature, local yielding of the structure may be accepted provided it can be demonstrated that such yielding does not lead to progressive collapse of the overall structure and that the general structural stability is maintained.

FINITE ELEMENT METHODOLOGY

The nonlinear finite element analysis (FEA) was conducted using Abaqus 6.14. At the boundary of the FEA model, soil deformation normal to the vertical faces and the base of the boundary are set to zero, while tangential components are allowed. If a plane of symmetry is used, the out of plane displacements are restrained, see Figure 2. The extent of the soil model should be as large as necessary so that boundary conditions do not affect the soil-pile interaction. The Equivalent Pile Diameter (EPD) is the circle diameter that circumscribes the pile shaft and fins. As shown in Figure 2, the vertical extent of the soil model below the pile's lower end is suggested to be not less than 2 times the EPD. The lateral extent of the soil model away from pile's extreme ends is not less than 2.5 times EPD in any position.

The soil was assumed to be Drucker–Prager material with undrained shear strength S_u increasing with depth z following the relation.

$$S_u = 7.4 + 1.8z \qquad \text{Eq. (3)}$$

This is considered to be a reasonable strength profile for overly consolidated clay. Pile structure was modeled by solid elements and its geometry, see Figure 2. Figure 3 shows the pullout load vs displacement curves of piles subjected to the loading directions of 0°, 15°, 30°, 45°, 60°, 75° and 90° from the elevation view. Because of severe mesh distortion, the curves were not complete under the loading direction of 45°, 60° and 75°. Generally the magnitude of the pile pullout capacity is the least in the axial direction.

During the design stage the estimated pile penetration depth for geotechnical analysis can be predicted based on the drivability test in-situ or analytical solution. However, the pile may not penetrate to the exact depth as predicted. Hence, the pile capacity analysis should be conducted for at least three different penetration levels in a reasonable range. Similarly, at least 5 different loading angles between horizontal and axial axis are to be evaluated, including 0°, 45°, 90°, and any other two angles. Second order regression can be applied to provide a pullout capacity vs loading angle curve, see Figure 4. The holding capacity may be obtained by interpolation for the actual penetration depth and loading angles with those from numerical analysis.

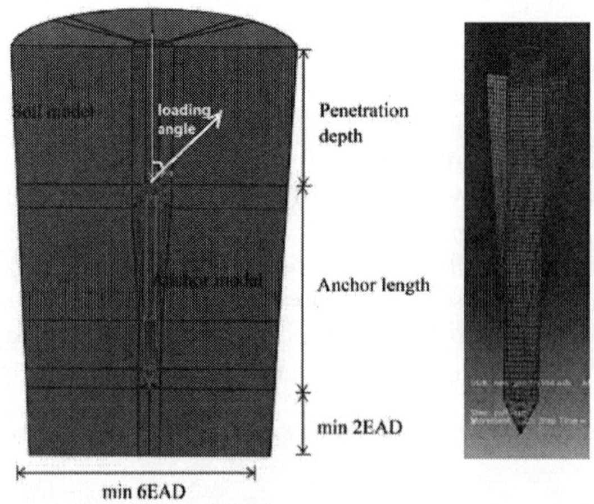

Figure 2 FE Model

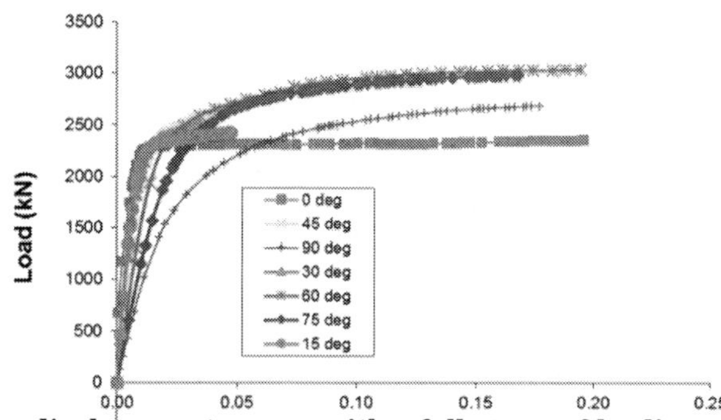

Figure 3 Pullout load vs displacement curves with a full range of loading angles

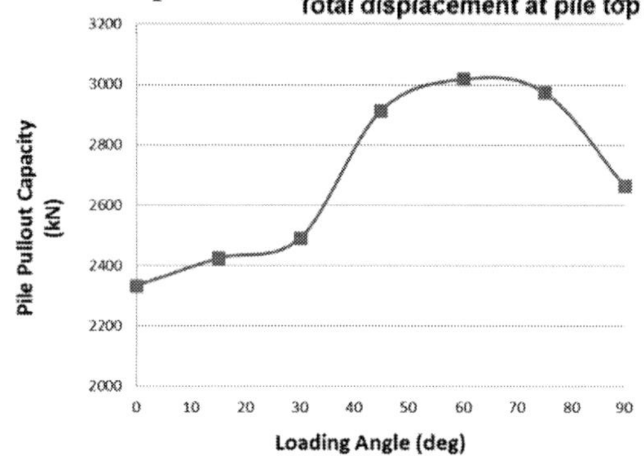

Figure 4 Example of pullout capacity vs loading angle

EFFECT OF PILE INCLINATION ON PULL-OUT CAPACITY

According to field tests almost all piles penetrated into the soil in an inclined position with the inclination angle less than 25°. Therefore a critical case of pile inclination angle of 30° from vertical axis was studied, as shown in Figure 5. As the pile is inclined in the soil and not symmetrical along the vertical axis, the effect of loading direction on pile pullout capacity was studied with the angles of 30° and 45° in the left and right half space along the vertical axis. Load-displacement curves of pile at an inclined position with loading angle in the left half space shows the equivalent or larger capacity when compared to those piles at the same loading angle without inclination; see Figure 6. In the right half space a relatively lower pullout capacity of the pile at an inclined position is found. This is because the right half space is in line with the pile inclination direction and the axial pullout capacity is always the least among the different loading angles. This is consistent with the findings from Brandao et al. (2006) that the pile inclination may significantly change the pile pullout capacity, and it is one of the key parameters to address after the pile installation.

The difference in the pullout capacity in the current studies is within 15%. In the drivability tests the pile inclination angle is usually less than 10° and the difference in capacity could be lower. For simplicity of design the soil surface may be rotated at the inclination angle so that the pile is embedded in the soil vertically, see Figure 5. The loading angle is to be adjusted accordingly and the more conservative pullout capacity should be used for design.

COMPARISON OF NUMERICAL RESULTS WITH SIMPLIFIED SOLUTIONS

The theoretical capacity under axial loading is provided by the frictions from shaft, fin and pile weight. The ultimate pile pull-out capacity under axial loading F_v in cohesive soil may be calculated by the equation:

$$F_v = W_s + F_{shaft}$$

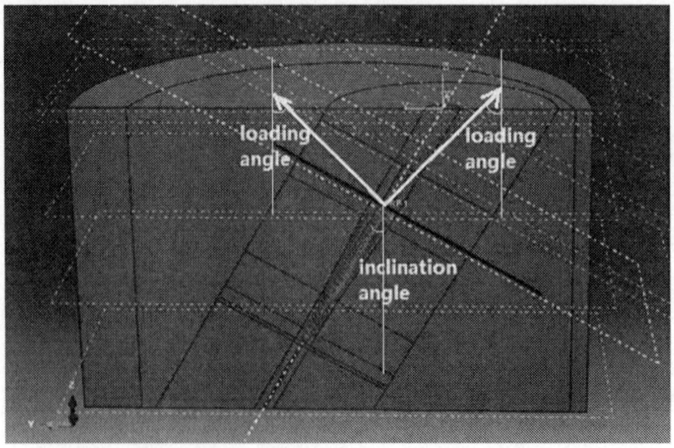

Figure 5 FE model with pile installed with inclination angle

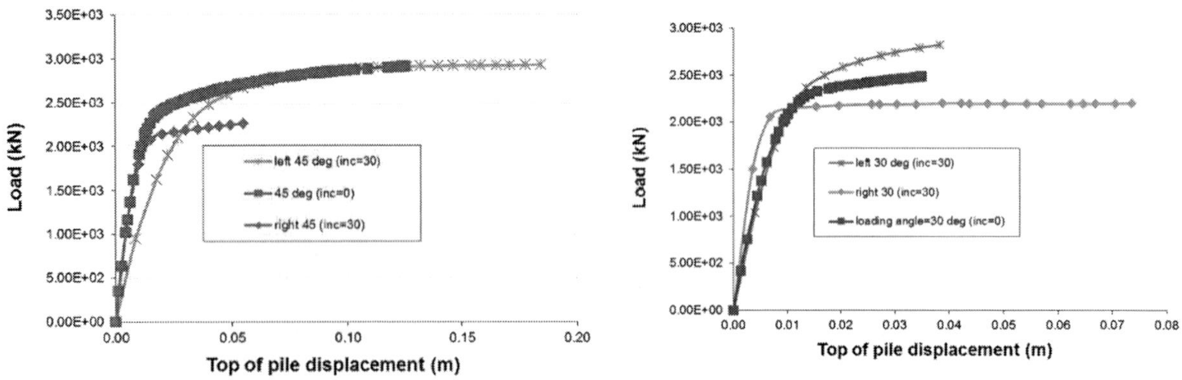

(a) loading angle of 45 degree **(b) loading angle of 30 degree**

Figure 6 Load vs displacement of pile installed with and without inclination angle

where

W_s = the submerged weight of the pile

F_{shaft} = the frictional resistance along the shank and fin walls

The maximum allowable shear stress for cohesive soil, *f*, is obtained per API RP 2A (2007) as follows:

$$f = \alpha \, Su \qquad \text{Eq. (4)}$$

$$a = \begin{cases} 0.5\left(\frac{S_u}{\sigma_v'}\right)^{-0.5} & for\ \frac{S_u}{\sigma_v'} < 1.0 \\ 0.5\left(\frac{S_u}{\sigma_v'}\right)^{-0.25} & for\ \frac{S_u}{\sigma_v'} > 1.0 \end{cases} \qquad \text{Eq. (5)}$$

where σ_v' is the effective overburden pressure and α is the adhesion factor.

Figure 7 shows the comparisons of axial pile capacity from API method with the numerical results. The pile penetration depths are 7.9, 10.8 and 11.3m respectively. It is found that the simplified solutions predict the lower bound of the pullout capacity and the difference is within 15%.

Poulos and Davis (1980) noted that the ultimate soil resistance for piles in purely cohesive soils increases with depth from 2*Su* at the surface to 8*Su* to 12*Su* at a depth of about three pile diameters below the surface. In the absence of more definitive criteria, the following is recommended for a short-term lateral pullout capacity:

$$F_H = 9 S_u A_s \quad \text{Eq. (6)}$$

where

A_s = pile laterally projected area of main body.

To be conservative the side surface area of pile fin is not taken into account. As shown in Figure 8, the simplified solution for lateral capacity coincides well with the numerical results.

ULTIMATE PULLOUT CAPACITY UNDER DIFFERENT LOADING ANGLE

The normalized horizontal component (H/H_0), is defined herein as the horizontal component of the ultimate inclined holding capacity (H) normalized by the ultimate horizontal holding capacity (H_0). The data sets were collected and summarized from the numerical analysis and centrifuge test results [de Sousa et al, 2011; Fu et al., 2017; O'Beirne et al, 2015; Lieng et al, 2000]. As shown in Figure 9(a), H/H_0 is narrowly banded and increases with loading angle θ from 0° to 65° and remains nearly unchanged in the range of around 65° to 90°. As can be seen in Figure 9(b), the normalized vertical component (V/V_0) also follows the same trend, remaining nearly constant when the loading angle increases from 0° to 30°, before decreasing approximately linearly from 30° to 90°. The above discussion indicates that at loading angle below 30°, vertical failure will occur when the vertical component V reaches the ultimate vertical holding capacity, V_0, hence the ultimate inclined holding capacity, P_0, is dictated by the ultimate vertical holding capacity V_0. When the loading angle is between 65° and 90°, the failure mode is controlled by the horizontal pullout and the ultimate holding capacity, P_0, is dictated by the horizontal holding capacity H_0. Coupling between vertical and horizontal capacity is important at the loading angle between 30° and 65°, and the interaction diagram is shown in Figure 10. This leads to the following relationship.

$$0 \le \theta \le 30^\circ \qquad \frac{V}{V_0} = 1; \qquad P_0 = \frac{V_0}{\cos\theta} \quad \text{Eq. 7(a)}$$

$$30^\circ < \theta < 65^\circ \qquad \frac{V}{V_0} = 1.5 - 1.0832 * \frac{H}{H_0}; \qquad P_0 = \frac{V}{\cos\theta} \quad \text{Eq. 7(b)}$$

$$65^\circ \le \theta \le 90^\circ \qquad \frac{H}{H_0} = 1; \qquad P_0 = \frac{H_0}{\sin\theta} \quad \text{Eq. 7(c)}$$

These equations can be used to give a quick evaluation of ultimate inclined pullout capacity of dynamically installed piles once the ultimate axial and lateral pullout capacities have been evaluated using the API's (2007) and the generalized Poulos and Davis (1980) methods.

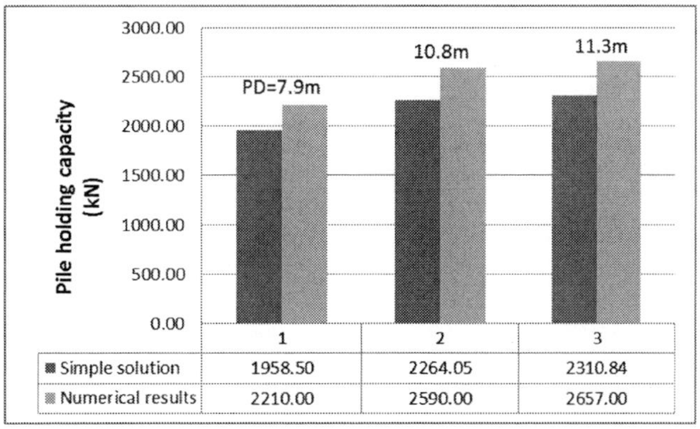

Figure 7 Comparison of pile axial pullout capacity (loading angle=0°) with different penetration depth (PD) from numerical analysis and API method

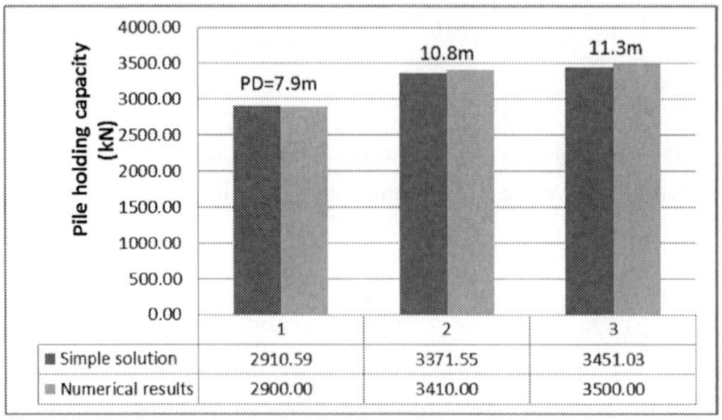

Figure 8 Comparison of pile lateral pullout capacity (loading angle=90°) with different penetration depth (PD) from numerical analysis and simplified solution

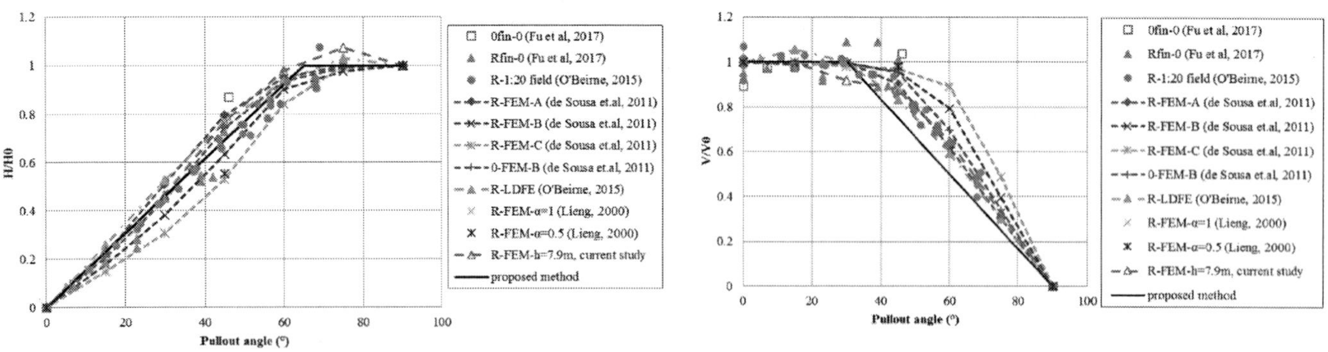

(a) H/H_0 vs pullout angle (b) V/V_0 vs pullout angle

Figure 9 Normalized vertical and horizontal components vs pullout angle

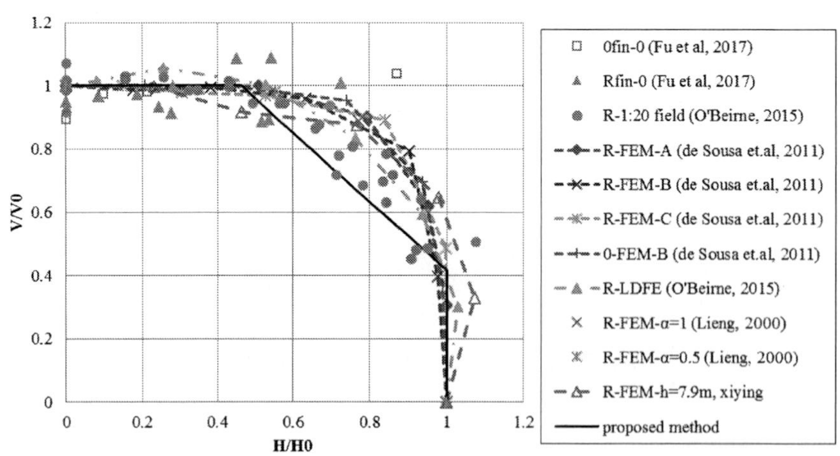

Figure 10 Normalized vertical component vs horizontal component in different studies

CONSOLIDATION TIME

The short-term pile capacity is the capacity immediately after installation, while the long-term one is the capacity after complete dissipation of the excess pore pressure around the pile. The dynamic installation process results in lower short-term capacity than quasi-static installation. Traditional driven pile installation is considered a quasi-static event since it occurs at relatively low penetration velocities when compared to dynamic impact velocities of 20-30 m/s expected in the field. The higher impact velocity leads to the lower short term capacity. This is possibly due to the entrainment of water in the boundary layer close to the pile, resulting in lower effective stress.

A plot for evaluation of pile capacity regain vs non-dimensional time T = $c_h t/d^2$ is presented in Figure 11 with centrifuge data from Richardson et al. (2009). A preliminary assessment of the regain of the pile capacity Fv could be conservatively obtained using the following equation:

$$\frac{F_v - W_s}{F_{\max}} = 1.1 - \frac{1.08}{1+\left(\frac{T}{6.5}\right)^{0.42}} \qquad \text{Eq. (8)}$$

$$T = \frac{c_h t}{d^2} \qquad \text{Eq. (9)}$$

where

W_s	=	submerged weight of the pile
F_{max}	=	long term maximum pile capacity
T	=	non-dimensional time
t	=	consolidation time after installation, in sec
d	=	pile shaft diameter
c_h	=	horizontal coefficient of consolidation, in m²/year

Based on the consolidation data, for typical consolidation values of c_h= 3-30 m^2/year, the time required for 50% consolidation of a prototype dynamic pile with a shaft diameter of 1.2 m ranges from approximately 35–350 days, and 2.4-24 years to achieve 90% consolidation. Therefore, the dynamically installed piles should be installed for a sufficiently long period to allow the development of the long term pullout capacity.

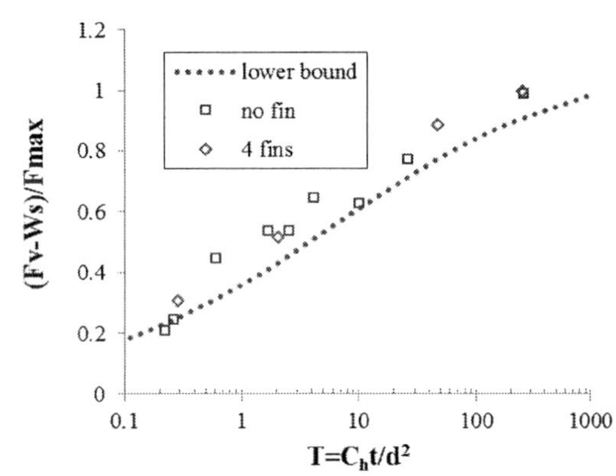

Figure 11 Regain of Normalized Pile Capacity with Consolidation Time

STRUCTURAL ASSESSMENT

In the structural analysis, the pile should be modeled with a suitable non-linear stress-strain curve. The most critical scenarios, including the magnitude of the load at the pile top and mooring loading angle should be determined for the "intact" and "broken line" conditions, respectively.

Table 1 shows examples of structural assessment with different loading angles in plan and elevated views. In general, a mooring load applied in a horizontal direction in the elevation view generates the largest bending moment compared to the loads applied at an inclined angle. The pile load is to be applied in the direction aligned with the fin in the plan view, aiming to produce the highest possible local stresses as shown in Figure 12(b). Von Mises stress distribution indicates the hot spot stresses in the fin connecting to the pile shaft. The results of the finite element analyses of the structural models must be checked for yielding in Eq. (1) and (2).

The fatigue assessment should also be performed to demonstrate the adequacy of the mooring line attachment components for the expected service life of the mooring system.

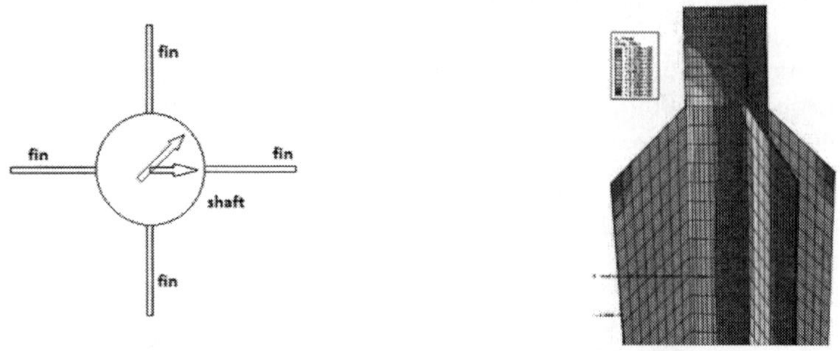

(a) Loading direction **(b) Von Mises Stress Distribution (Model-8)**

Figure 12 Structural assessment

Table 1 Examples of Structural Assessment

Model	Loading direction, Figure 12(a) (Degree)	Load applied (kN)	Mooring line load angle, Figure 2 (Degree)	Von Mises stress σ_{max}(MPa)		Max Von Mises stress (MPa)	Max pile displacement (m)	σ_{max}/σ_y
				Shaft	Fin			
Model-8	0	1400.	0	77600	132900	132900	0.029	0.64
Model-9	45	1400.	0	72930	120300	120300	0.030	0.58
Model-11	0	1400.	45	56790	85440	85440	0.02	0.41

VERIFICATION OF PILE INSTALLATION

The installation of a dynamically installed pile should be monitored to make sure that the installation proceeds as expected and the pile is installed as designed. The measurements should include coordinates of pile after installation, designated penetration depth, acceptable tolerances for pile position, actual penetration depth, inclination angle and azimuth. In case the installed pile position and/or penetration differs from the planned conditions, additional calculations and inspections should be performed to determine the specific ultimate pullout capacity of the pile, and the as-installed mooring system's capability to provide the required ultimate pullout capacity.

CONCLUSIONS

Included in this paper is an overview of the methodologies for the geotechnical design and structural assessment of dynamically installed piles. Recommendations are based on finite element analysis using appropriate soil modeling and simulation of soil-pile interaction. A framework on the normalized vertical and horizontal component is proposed to predict the pile pullout capacities subjected to different loading angles. The parameters required in the framework are the axial and lateral capacities which can be calculated from the simplified solutions. Compared to three-dimensional finite element analysis, this approach is simple and can be used to give a quick and preliminary assessment of pile pullout capacity. The proposed framework has not been verified against more sophisticated pile configurations such as OMNI-Max. The extension of this framework to a wider range of pile configurations is part of an ongoing study.

In the field tests almost all piles penetrated into the soil in an inclined position. For simplicity the soil surface can be rotated at the inclination angle so that the pile is embedded in the soil vertically. The loading angle should be adjusted accordingly to take into account the effect of inclination and the most conservative pullout capacity should be used for design.

Since the dynamic installation process results in the lower short-term capacity of the pile, it is recommended that the piles should be installed for a sufficiently long period to allow for the development of the pull-out capacity. A prediction of the pile capacity restoration over time is developed based on the results of a series of centrifuge tests.

ABBREVIATIONS

ABS: American Bureau of Shipping
API: American Petroleum Institute
FEA: Finite Element Analysis
EPD: Equivalent Pile Diameter
PD: Penetration Depth

REFERENCES

[1] ABS. 2016. Guidance notes on design and installation of dynamically installed piles. American Bureau of Shipping (ABS), Houston, Tex.
[2] API. 2007. Recommended Practice for Planning, Designing and Constructing Fixed Offshore Platforms - Working Stress Design. RP 2A-WSD. American Petroleum Institute (API), Washington, D.C.
[3] Audibert, J.M., Movant, M.N., Jeong-Yun, W., and Gilbert, R.B. 2006. Torpedo piles: laboratory and field research. In Proceedings of the 16th International Offshore and Polar Engineering Conference, San Francisco, Calif., 28 May –2 June 2006, International Society of Offshore and Polar Engineers. (ISOPE), Document ID ISOPE-I-06-339.
[4] Brandao, F.E.N., Henqiques, C.C.D., de Araujo, J.B., Ferreira, O.C.G. and dos Santos Amaral, C. 2006. OTC 18243.
[5] Bonfim dos Santos, A., Henriques, C.C.D., Pimenta, J.M.H.A. (2004). Improvements achieved in the project of FPSO-P-50, Offshore Technology Conference, OTC16705.
[6] de Sousa, J.R.M., de Aguiar, C.S., Ellwanger, G.B., Porto, E.C., Foppa, D., and de Medeiros, C.J. 2011. Undrained load capacity of torpedo anchors embedded in cohesive soils. Journal of Offshore Mechanics and Arctic Engineering, 133(2): 021102. doi:10.1115/1.4001953.
[7] Ehlers, C.J., Young, A.G. and Chen, J.H. 2004. Technology Assessment of Deepwater Anchors. Offshore Technology Conference. OTC 16840.
[8] Fu Y., Zhang X.Y., Li Y.P., Gu H., Sun J., Liu Y. and Lee F.H. Holding capacity of dynamically installed anchors in normally consolidated clay under inclined loading. *Canadian Geotechnical Journal*, 2017, 54(9): 1257-1271.
[9] Hossain, M.S., Kim, Y.H., and Wang, D. 2013. Physical and numerical modelling of installation and pull-out of dynamically penetrating anchors in clay and silt. In Proceedings of the 32nd International Conference on Ocean, Offshore and Arctic Engineering, Nantes, France, 9–14 June 2013. American Society of Mechanical Engineers. Paper No. OMAE2013-10322, pp. V006T10A019. doi:10. 1115/OMAE2013-10322.
[10] Lieng, J.T., Kavli, A., Hove, F., and Tjelta, T.I. 2000. Deep Penetrating Anchor: Further development, optimization and capacity verification. In Proceedings of the 10th International Offshore and Polar Engineering Conference, Seattle, Wash., 28 May – 2 June 2000. International Society of Offshore and Polar Engineers. Document ID. ISOPE-I-00-171.
[11] Medeiros, C.J. (2001). Torpedo anchor for deep water. Proc. Deepwater Offshore Technology Conference. Rio De Janeiro.
[12] O'Beirne, C., O'Loughlin, C.D., Wang, D., and Gaudin, C. 2015. Capacity of Dynamically Installed Anchors as assessed through field testing and three dimensional large-deformation finite element analyses. Canadian Geotechnical Journal, 52(5): 548–562. doi:10.1139/cgj-2014-0209.
[13] O'Loughlin, C.D., Richardson, M.D., and Randolph, M.F. 2009. Centrifuge Tests on Dynamically Installed Anchors. In Proceedings of the 28th International Conference on Ocean, Offshore and Arctic Engineering, Honolulu, Hawaii, 31 May – 5 June 2009. American Society of Mechanical Engineers. pp. 391–399. doi:10.1115/OMAE2009-80238.
[14] Poulos, H.G., and Davis, E.H. 1980. Pile Foundation Analysis and Design. Series in Geotechnical Engineering. Wiley, New York.
[15] Richardson, M.D. 2008. Dynamically installed anchors for floating offshore structures. Ph.D. thesis, The

University of Western Australia, Crawley, Australia.

[16] Richardson, M.D., O'Loughlin, C.D., Randolph, M.F. and Gaudin, C. 2009. Setup Following Installation of Dynamic Anchors in Normally consolidated clay. Journal of Geotechnical and Geoenvironmental Engineering. ASCE. pp.487-496

SNAME
THE INTERNATIONAL COMMUNITY
FOR MARITIME AND OCEAN PROFESSIONALS

NUMERICAL STUDY ON OPB/IPB INTERLINK ANGLE WITH UNDERWATER CHAIN STOPPER SYSTEM

WOO CHUL CHUNG
TEXAS A&M UNIVERSITY (PH.D. STUDENT)

HEON YONG KANG
TEXAS A&M UNIVERSITY (RESEARCH ASSISTANT PROFESSOR)

MOO HYUN KIM
TEXAS A&M UNIVERSITY (PROFESSOR)

Proceedings of the 24th Offshore Symposium, February 20th 2019, Houston, Texas
Texas Section of the Society of Naval Architects and Marine Engineers

ABSTRACT

Station-keeping capability of offshore oil rig is important so that mooring integrity become a critical design factor. Several unexpected accidents of chain mooring failure were related to new mechanism called Out-of-Plane Bending (OPB) and In-Plane Bending (IPB). The main factors causing OPB-induced fatigue are tension, friction, and interlink angle. To consider this mechanism during the design stage, chain interlink angles from global-system simulation should be computed accurately in time domain. The results are sensitive to the degree of accurate modeling of fairlead connection.

In this study, time-varying chain interlink angles with underwater chain stopper system (chain-hawse) are investigated with hull-mooring-riser coupled time-domain simulation program CHARM3D, which has been developed in TAMU for the past two decades. Several scenarios of chain-hawse system and environment are investigated to observe the corresponding effects on interlink angles.

Keywords: Chain Mooring, Interlink Angle, OPB/IPB, Chain-hawse, Global Performance, underwater chain stopper, Fully-Coupled

INTRODUCTION

Many floating-production-platforms use chain-mooring connected to fairlead with chain stopper for their station keeping. It is well known that chain mooring may have OPB fatigue problem depending on how the chain-mooring is connected to the platform. Since the report of Girasol SPMB (single-point-mooring-buoy) chain-mooring failure in 2002, there have been several cases of similar chain-mooring fatigue-induced failures near fairlead during the past fifteen years despite assigning sufficient safety factor based on tension-fatigue-failure estimation in the design stage [Amir. et al., 2018]. In this regard, offshore industry formed several JIPs (joint-industry projects) and reported that the failures are most likely caused by OPB-induced fatigue, which was also supported by experimental investigations. After those studies, it is now commonly required in many design guidelines and rules that the OPB-induced fatigue of chain-mooring needs to be checked during the design stage.

Before the OPB problem of chain-mooring was introduced, chain-mooring legs have been designed with the consideration of only tension-induced fatigue under the assumption that the chain-mooring lines have no bending stiffness. However, the high tension with unexpectedly large friction can produce appreciable cyclic bending moment to the chains near fairlead depending on the platform movement and level of restriction at the connection. This bending-induced fatigues are called "out-of-plane bending (OPB)" fatigue or "in-plane bending (IPB)" fatigue according to the relative rotational direction of two adjoining links. It is also known that there may be appreciable residual stress near the fatigue hot spot after proof-loading test to worsen the resilience against fatigue.

[Jean, et al., 2005; Lim, et al., 2010; Melis, et al., 2005; Rampi, et al., 2006] have worked on the nonlinear stress analyses under extremely high-tension during proof-loading test and its effects on fatigue failure. Generally, the bending stiffness between chain links has been neglected in the previous chain-mooring designs and the OPB-induced fatigue has not been extensively documented. After the fatigue-failure incident of the Girassol SPMB chain-mooring, many JIPs were formed and both experimental and computational approaches have been researched. As a result, several useful guidelines were documented. It was revealed that the fatigue failure due to OPB/IPB occurred mainly at the interlink corresponding to the 1st link out of chain-hawse [Jean, et al., 2005]. Many experiments and computations also showed that the bending stiffness of the mooring chain changes in time according to the interlink angle and its tension [Brake, et al., 2007; Lim, et al., 2010; Hwang, 2012; Kim,S.J. et al., 2017].

In this regard, to accurately estimate the additional OPB-induced stress in the time-domain global-performance simulation for fatigue assessment, local dynamic FEM analysis near fairlead should be coupled and conducted simultaneously with the global-performance time-domain simulation. In other words, global-performance simulation provides instantaneous tension and OPB/IPB interlink angles and they are inputted to the nonlinear local FEM program to produce the corresponding chain-link bending stiffness, with which the global-performance simulation of the next time step is conducted. However, this kind of multi-scale full-coupling method may be extremely time consuming and thus not practical especially for long-term fatigue-damage estimation requiring many long time-domain simulations. More practical approach is to pre-calculate the time-varying bending stiffness for a reasonable range of tension magnitudes and interlink angles for a given system so that the corresponding values (instantaneous bending stiffness and principal stresses at hot spots) can be picked up at each time step from the pre-calculated table by using nonlinear FEM program. Authors are developing this approach to more accurately estimate the OPB/IPB induced internal stresses for chain mooring lines.

In this paper, the corresponding global-system simulation program is applied to produce the instantaneous interlink angles for the given platform, chain mooring, chain-hawse, and chain stopper. A chain-polyester-chain mooring with underwater chain stopper system is taken as an example. Several sensitivity studies with relevant design parameters are also conducted. For the purpose, hull-mooring-riser fully-coupled global performance in-house software CHARM3D [Kim M.H. et al., 2005; Yang et al., 2010; Kang et al., 2014] is used.

An example case includes a drilling semi-submersible with 12 mooring lines (3 lines at each corner with 5-degree spreading angle) and 4 steel catenary risers. In this paper, we selected only one mooring line for systematic comparisons.

MODELING OF UNDERWATER CHAIN STOPPER

As previously mentioned, the detailed modeling of the fairlead-chain-hawse is very crucial to analyze the OPB/IPB problems for chain mooring near fairlead. In this study, it is assumed that the chain-hawse is a circular tube with 10 mm wall thickness that has a chain stopper at its outlet. The chain stopper holds the chains tightly inside. Each moori ng line has one chain-hawse of 3.1m length at its top. Polyester rope are used between the top (near fairlead) chain bl ocks and bottom (near anchor) chain blocks. The chain-hawse is connected to the hull by freely-rotating ball-joint. F igure 1 and Figure 2 are shown modeling of chain-hawse and mooring line near fairlead and cross-section of chain-h awse. Mooring system material properties are also shown in Table 1.

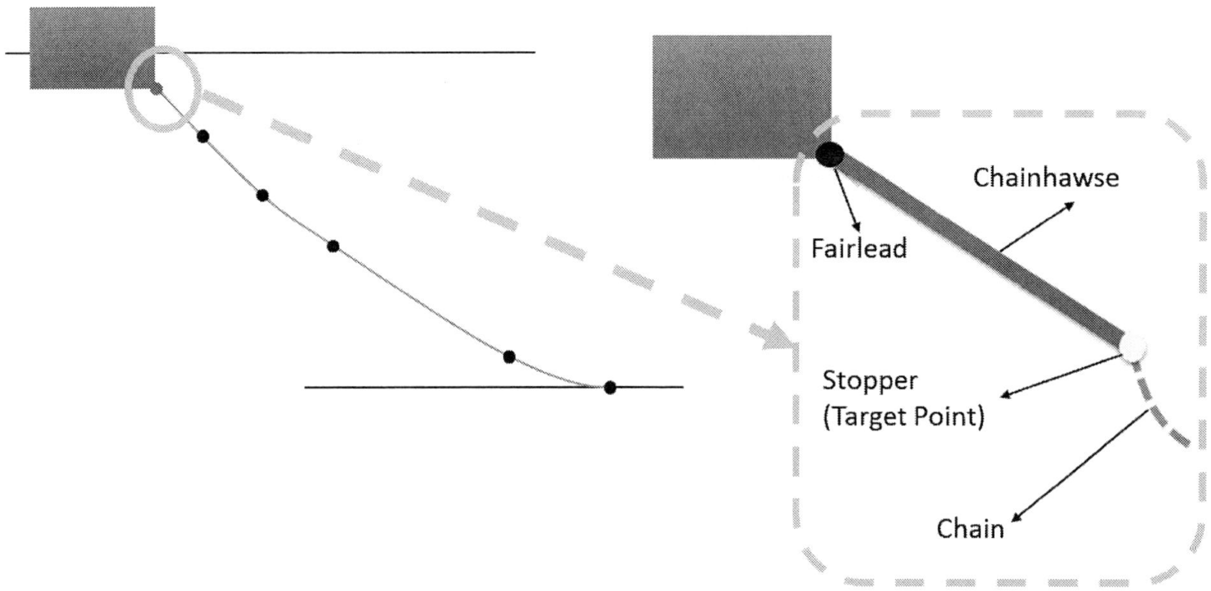

Figure 1 The modeling of chain-hawse and mooring line near fairlead (semi-submersible is simplified by a box in the following figures)

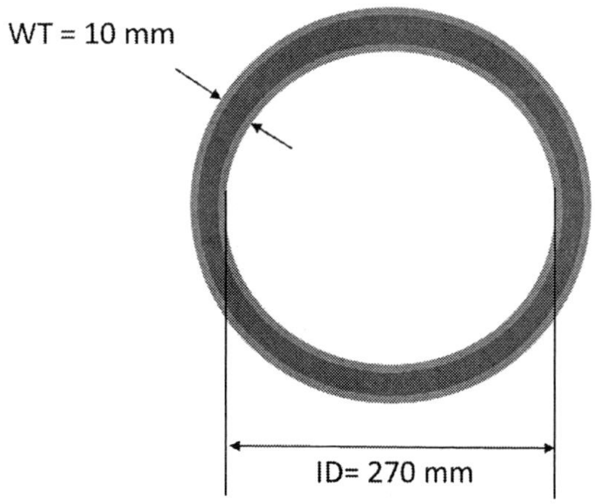

Figure 2 Cross-section of chain-hawse

Table 1 Mooring System Material Properties

Mooring Material Properties				
	Unit	Chain	Rope	Chain-hawse
Type	[-]	R3 Studless	Polyester	Circular Tube
Diameter	[mm]	147	220	290
Mass in Air	[kg/m]	430	39	499
Displaced Mass	[kg/m]	56	29.2	124
MBL(Min. Braking Load)	[kN]	14700	8250	14700
Axial Stiffness (EA)	[MN]	1845.4	52.76	1845.4
Bend Stiffness (EI)	[MN-m^2]	0	0	18.3

DESCRIPTION OF NUMERICAL TOOL (CHARM3D)

CHARM3D is TAMU-in-house computer program that has been developed by the 3rd author's lab during the past 20 years for multi-hull-mooring-riser fully-coupled nonlinear dynamic response analysis in time domain [Kim, M.H., 2000;]. In CHARM3D, the slender-elastic-rod theory developed by Garrett [Garrett, 1982] is used for risers and mooring lines. Their governing equations are derived with respect to the position vector r(s, t) which is function of arc length (s) in general coordinate and time (t).

$$(\mathrm{P}\mathrm{r}'')'' - (\lambda \mathrm{r}')' - \mathrm{q} - \rho\ddot{\mathrm{r}} = 0 \quad (1)$$

$$\mathrm{P} = \mathrm{EI}$$

$$\mathrm{r}' \cdot \mathrm{r}' = (1 + \frac{\mathrm{T}}{\mathrm{A_I E}})^2 \approx 1 + 2\frac{\lambda}{\mathrm{A_I E}} \quad (2)$$

In Eq. (1) and (2), E=Young's modulus, I=Sectional Moment of Inertia, q =distributed load, ρ = rod density, λ = Lagrangian multiplier, T = tension, A_I=effective cross sectional area, primes represent derivative with respect to arc length (s) and dot represent differentiation with respect to time (t). The distributed load (q) is composed of rod element weight (w), hydrostatic load (F^s), and hydrodynamic load (F^d). Those loads are based on the unit length of the rod. Also, hydrostatic load (F^s) can be estimated using Eq. (3).

$$\mathrm{F^s} - \mathrm{B} + (\mathrm{P A_I r'})' = 0 \ (3)$$

where B = buoyancy, P = hydrostatic pressure at position of r point

Also, hydrodynamic load can be computed following Eq. (4).

$$F^d = -C_A \rho A_I \ddot{r}^n + \left[(C_A + 1)\rho A_I \dot{V}^n\right] + \frac{1}{2} C_D \rho A_D |V^n - \dot{r}^n|(V^n - \dot{r}^n) \quad (4)$$

In Eq. (4), V^n , $\dot{V}^n$ are velocity and acceleration vectors of fluid which are normal to the center line of the rod. And $\dot{r}^n$, $\ddot{r}^n$ are velocity and acceleration vectors of rod element which are normal to the center line of the rod. Also, C_A, C_D, A_D, ρ represent added mass coefficient, drag coefficient, projected area to plane normal to the center line of rod, seawater density, respectively. Through Eq. (1) ~ (4), Eq. (5), the final equation of motion for rod element, can be derived as below.

$$m\ddot{r} + C_A \rho A_I \ddot{r}^n + (EIr'')'' + (\tilde{\lambda} r')' = \tilde{w} + \tilde{F}^d \quad (5)$$

m = mass per unit length of the rod element

$\tilde{\lambda} = \tilde{T} - EIk^2$

$\tilde{T}$ = effective tension of the rod element

k = local curvature of the rod element

$\tilde{w}$ = net weight of the rod element

$\tilde{F}^d = \left[(C_A + 1)\rho A_I \dot{V}^n\right] + \frac{1}{2} C_D \rho A_D |V^n - \dot{r}^n|(V^n - \dot{r}^n)$

The external dynamic loads are evaluated at the lines instantaneous position at each time step. Once this final non-linear equation of motion is solved with respect to r and λ, the displacements and forces of the rod element can be found respectively. To solve the line dynamics in time, high-order FEM was used. The banded matrix for line dynamics is combined with the full matrix of 6DOF platform dynamics through proper connection conditions. The combined full-matrix equation is solved at each time step and the time-marching continues.

METHODOLOGY OF CHAIN INTERLINK ANGLE MEASURING

The definition of chain interlink angle in the fully-coupled global performance simulation is similar to that of high-fidelity non-linear FEM analysis [Lim, et al., 2010; Hwang, 2012; Choung, et al., 2016; Kim MS, et al., 2017; Kim,S.J. et al., 2017]. It can be measured at each time step during the hull-mooring-riser fully-coupled global performance simulation. The OPB chain interlink angle is the relative angle of two adjacent vectors according to the localized horizontal plane x-y. The same concept can be applied to IPB angle with respect to the localized vertical x-z plane. The target reference point is at the end of chain-hawse, as shown in Figure 3, first free-to-rotate chain link which is the weakness point against fatigue-induced failure.

To evaluate chain interlink angle near the target point (the discontinuity caused by the chain-hawse), the local directional-cosine vectors of mooring lines which is direct output in CHARM3D is required. To observe the effects of chain-hawse on OPB/IPB interlink angles, the chain mooring lines with and without the chain-hawse were modeled and compared. In the case of with chain-hawse, it is assumed that the chain-mooring inside chain-hawse moves with chain-hawse due to chain stopper inside. In the initial chain-mooring global-performance simulation, it is also assumed that the bending stiffness of the chain-mooring can be neglected, which is actually true for the entire length of long chain-mooring.

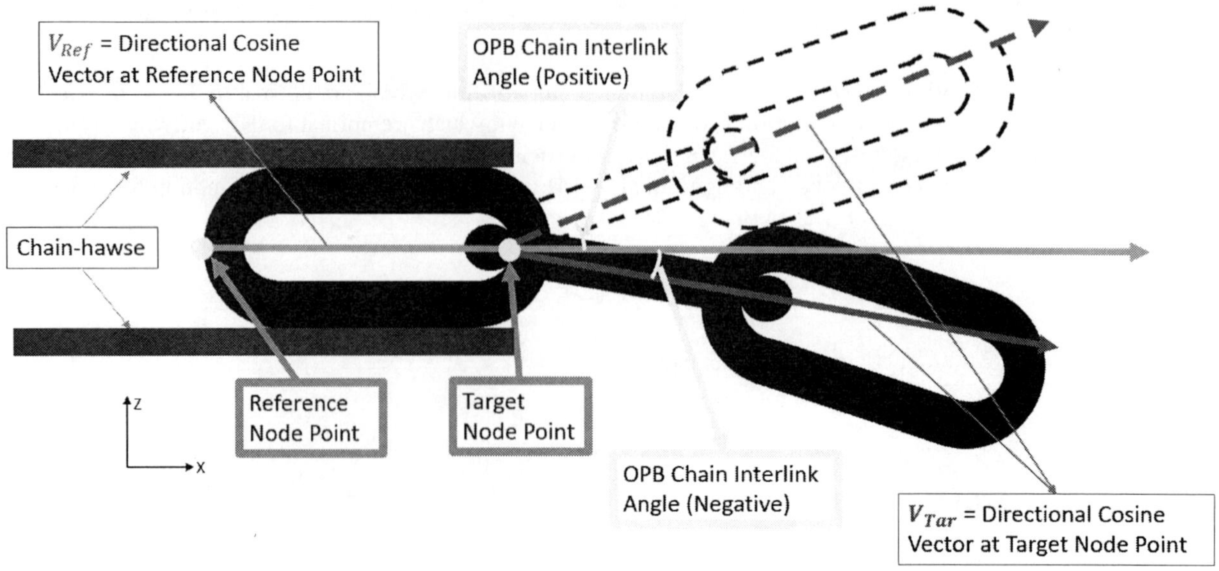

Figure 3 Definition of OPB Chain Interlink Angle – Global Performance Simulation

EFFECTS OF UNDERWATER CHAIN-HAWSE MODELLING

Figure 4 shows a schematic view of chain mooring with and without chain-hawse. For comparison, Figure 5 shows the results at the target point after including chain-hawse. The OPB chain interlink angles with chain-hawse are significantly increased compared to the case without it because of line discontinuity. In this case, the IPB interlink angles are very small as expected in head wave (H=5m, T=10s, Heading =180º)). In case of OPB chain interlink angles, the amplitude differences between with and without chain-hawse case are around 55~60% based on with chain-hawse case. There is no phase shift between with chain-hawse and without chain-hawse. For both OPB/IPB case, the chain-hawse is assumed to be connected to the platform by free-to-rotate hinge. If the restriction at the fairlead is imposed, the resulting interlink angles are expected to become larger. Based on this comparison, if the global dynamic analysis is done without more accurate modeling including chain-hawse, the OPB effects are to be significantly underestimated.

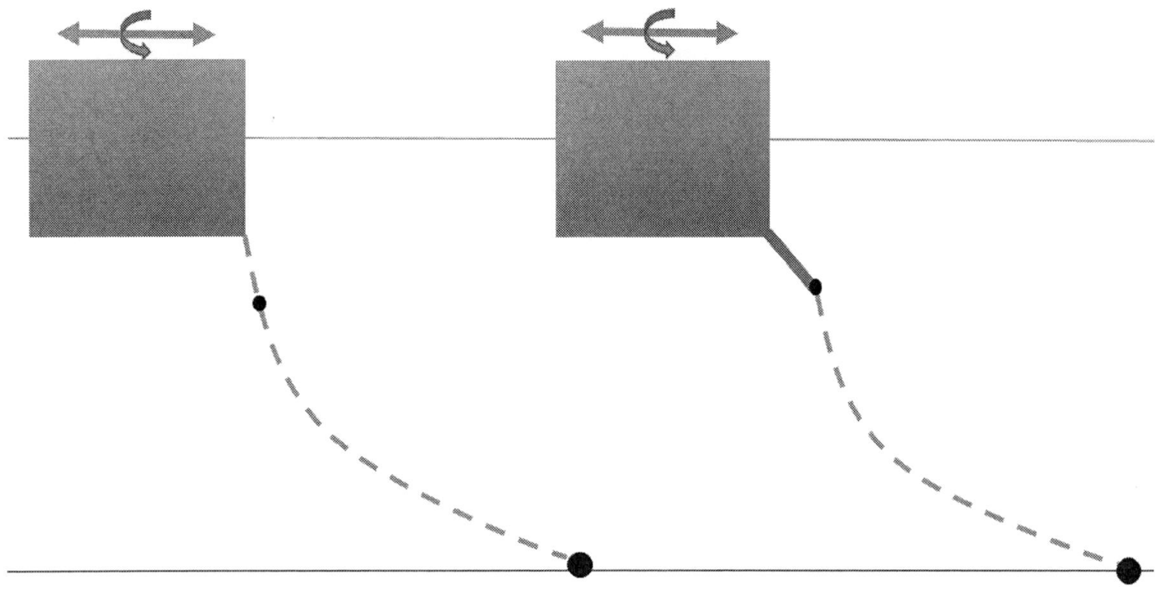

Figure 4 Schematic View of chain mooring without Chain-Hawse vs with Chain-Hawse

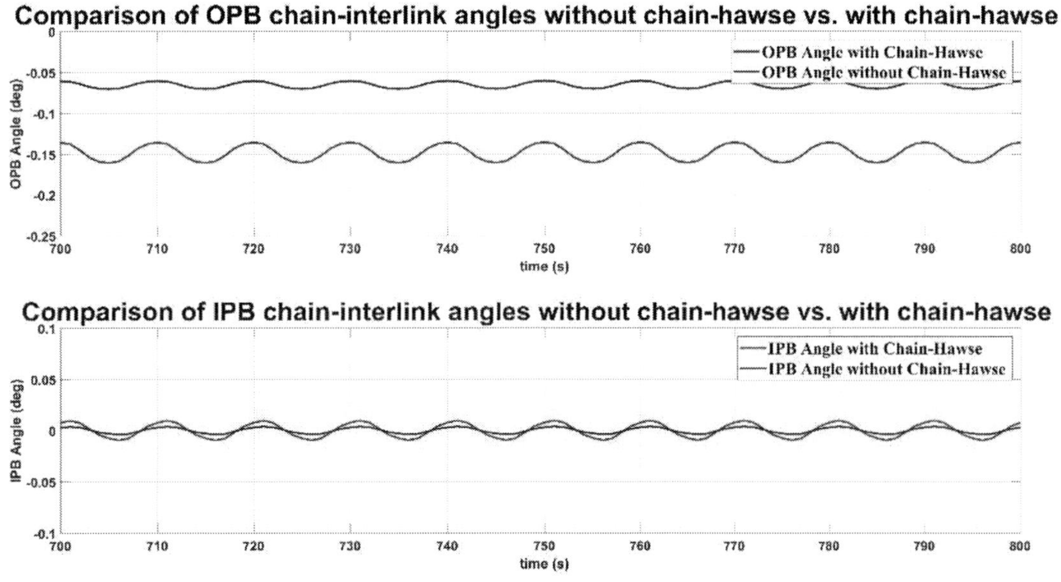

Figure 5 Comparisons of interlink angles without Chain-Hawse vs with Chain-Hawse, Regular Wave (H=5m, T=10s, Heading =180°)

EFFECTS OF UNDERWATER CHAIN-HAWSE LENGTH

It can be expected that underwater chain-hawse length will affect interlink angles. Figure 6 shows the schematic view of chain-hawse-length sensitivity test. Figure 7 shows the comparisons of the resulting interlink angles for three chain-hawse lengths (2.1m, 3.1m, 4.1m, and 5.1m). As chain-hawse length becomes longer, the interlink angles are increased due to larger initial discontinuity angle, as shown in Figure 6. As shown in 7, the corresponding change in IPB interlink angles is small in head waves.

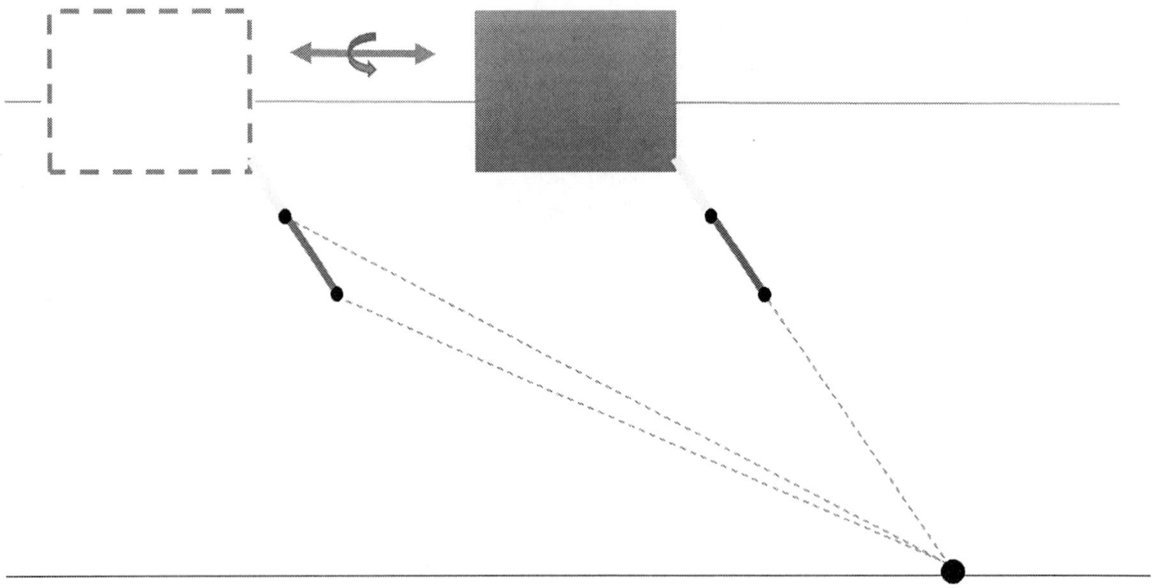

Figure 6 Schematic View of Chain-Hawse-Length Sensitivity Test

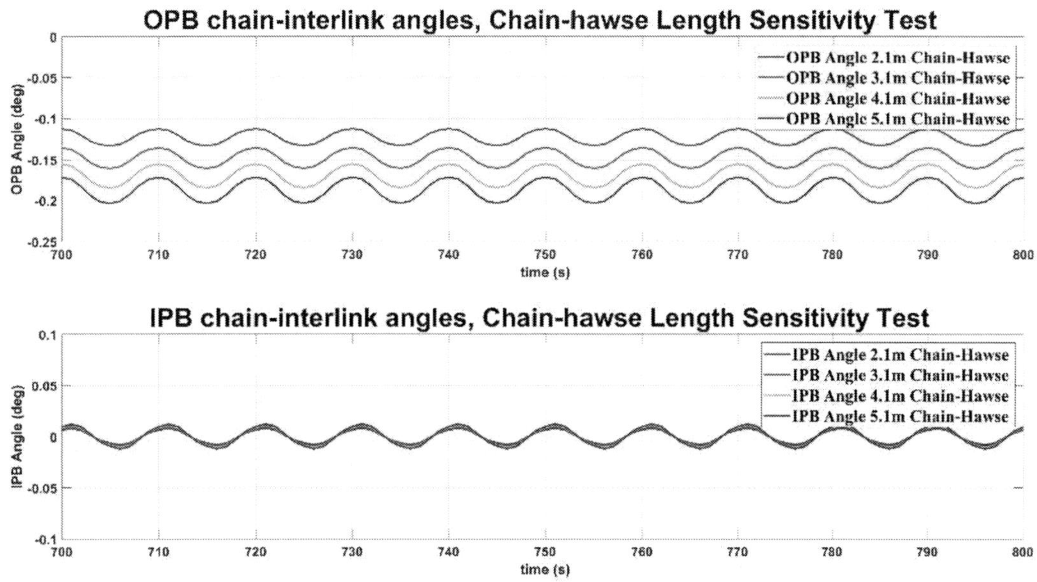

Figure 7 Chain-Hawse-Length Sensitivity Test Result, Regular Wave (H=5m, T=10s, Heading =180°)

EFFECTS OF UNDERWATER CHAIN-HAWSE BENDING STIFFNESS

Next, let us consider the effects of varying chain-hawse bending stiffness. The chain-hawse bending stiffness will affect the rate of line discontinuity between chain-hawse and chain mooring. To check the pattern of change, three bending flexural rigidities (50%, 100%, 150% based on 10mm wall thickness bending stiffness) are simulated and the results are presented in Figure 9. As chain-hawse bending stiffness increases, larger line discontinuity occurs and the corresponding OPB angles increase. Since the chain weight and chain-hawse bending occur in the vertical plane, only OPB angles are affected i.e. the corresponding change in IPB angles is small. Also, it is found that once chain-hawse bending stiffness reaches sufficient strength, there is little change by further increasing bending stiffness. It is also proved that if the bending stiffness of chain-hawse is infinitely small, it does not influence the top portion of chain-mooring lines at all, and thus the resulting OPB interlink angles there will also become small similar to the case without chain-hawse as in Figure 5. The chain-hawse bending-stiffness effects are expected to increase if the restriction of fairlead joint is increased. For quantity comparison, it is checked that relative difference in OPB interlink angle case are around 5% in 50%_EI case, 2% in 150%_EI case based on 100% EI case.

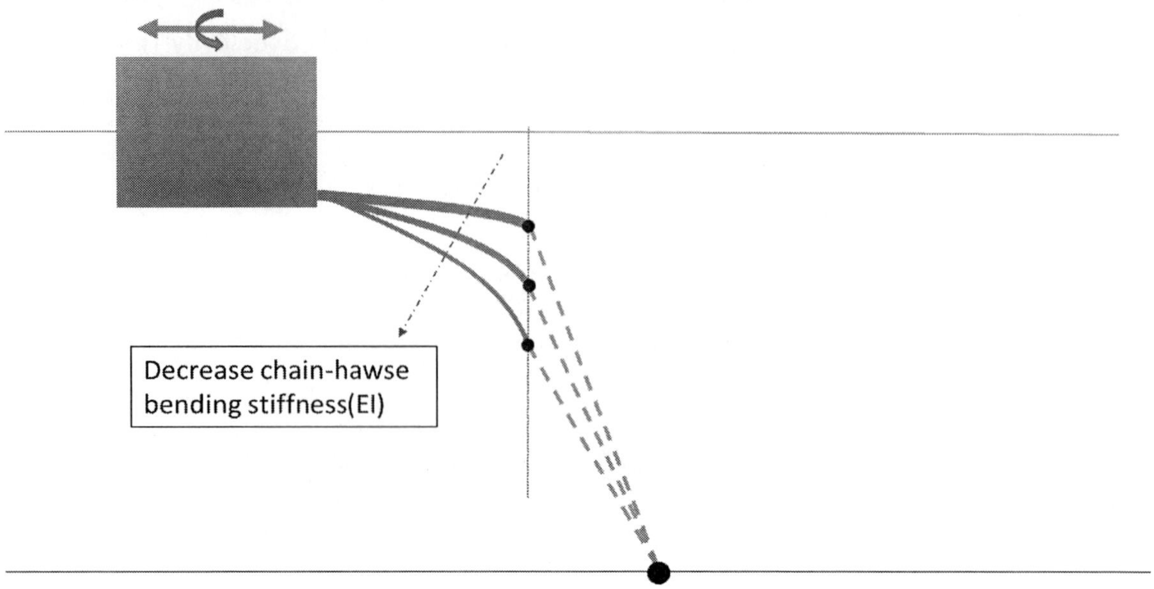

Figure 8 Schematic View of Chain-Hawse bending-stiffness Sensitivity Test

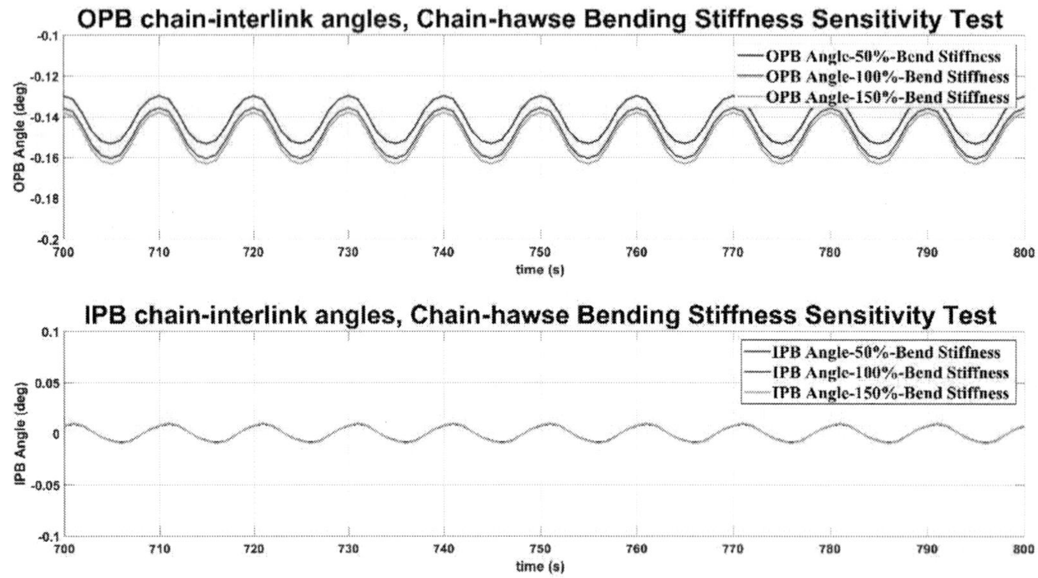

Figure 9 Chain-Hawse bending-stiffness Sensitivity Test Result, Regular Wave (H=5m, T=10s, Heading =180°)

EFFECTS OF WAVE HEIGHT

It can intuitively be expected that larger platform motions in larger waves cause more OPB problems. To observe the anticipated trend, various wave heights (H=4m, 6m, 8m, 10m) are inputted while keeping the same wave period of 10s. The platform motion time histories are presented in Figure 10. It is seen that the motion amplitudes monotonically increase with increasing wave heights. The downstream mean surge offsets caused by the second-order wave effect are also increased quadratically with increasing wave heights. Also, the corresponding OPB/IPB chain interlink angles according to the increase of wave heights are presented in Figure 11. The dynamic OPB interlink angles for mooring lines also monotonically increase with the increase of wave heights and platform motions. Interestingly in this case, the IPB angles are appreciable and nontrivial even in head waves. In oblique incident waves, both sway-roll-yaw motions and the corresponding IPB angles are expected to increase.

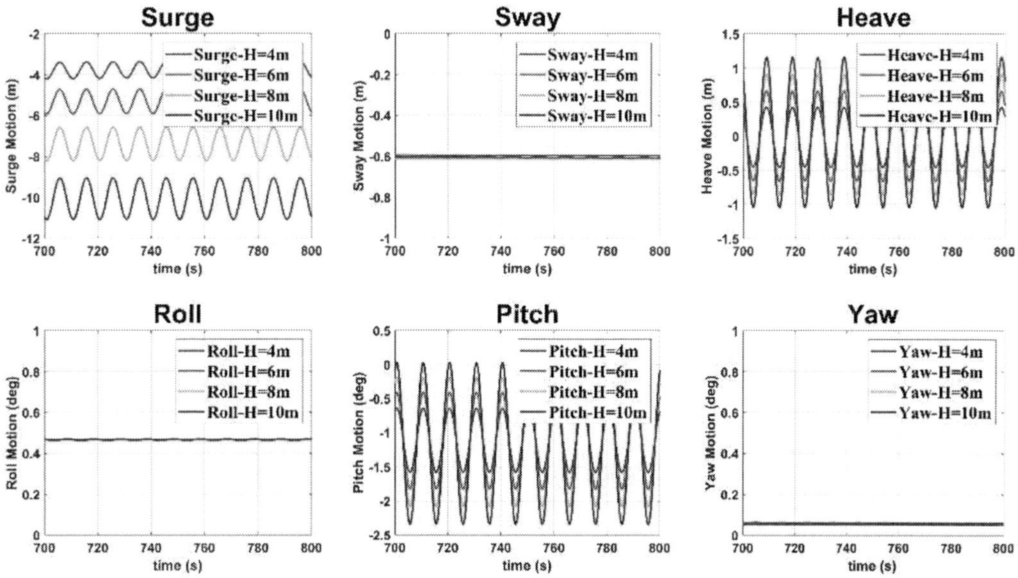

Figure 10 6DOF Platform-Motion Time Histories with increasing wave heights

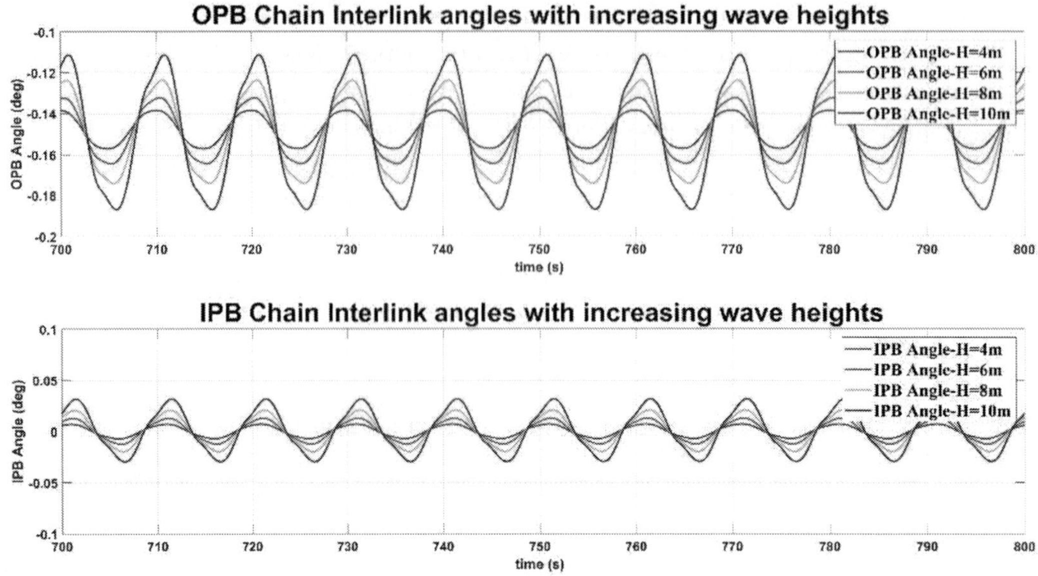

Figure 11 Wave-Height Sensitivity Test Result

CONCLUSIONS

A multi-scale local-global approach is suggested for the accurate OPB/IPB fatigue estimation of chain-mooring system including chain-hawse and chain-stopper. In the present paper, the examples of global time-domain simulations including chain-hawses are given. From the present study, it was found that the accurate modeling of the chain-mooring connection to the platform during the global-system-dynamics simulation is very critical for the reliable OPB/IPB interlink angles and the corresponding bending-stress estimation for fatigue analysis. From the sensitivity analysis, several findings are summarized as follows.

- Line discontinuity due to underwater chain stopper (or chain-hawse) leads to the increase in OPB/IPB chain interlink angles
- Larger chain-hawse length and bending stiffness tend to increase OPB/IPB chain interlink angles
- The greater wave heights, the larger OPB/IPB chain interlink angles

REFERENCES

1. Jean, P. Goessens, K. L'Hostis, D., 2005. Failure of Chains by Bending on Deepwater Mooring Systems, Offshore Technology Conference 17238
2. Kim, M.H., 2000, "Charm-3D User's Manual", 3rd ed., Texas A&M University, pp.5-23.
3. Lim, Y.C. Kim, K.S. Choung, J.M. Kang, C.H., 2010. A Study on Out-of-Plane Bending Mechanism of Mooring Chains for Floating Offshore Plants. Journal of the Society of Naval Architects of Korea Vol.47 No.4
4. Melis, C. Jean, P. Vargas, P.M., 2005. Out-of-Plane Bending Tests of Chain Links, International Conference on Offshore Mechanics and Arctic Engineering
5. Rampi, L. Vargas, P., 2006. Fatigue Testing of Out-of-Plane Bending Mechanism of Chain Links, International Conference on Offshore Mechanics and Arctic Engineering
6. Hwang Oe Ju, 2012. Fatigue Assessment for Mooring Chain Link of Spread-Moored FPSO, Korea Maritime University, Dissertation
7. Choung JM, Han SO, 2016. A Novel Procedure for Mooring Chain Fatigue Prediction based on Maximum Principal Stress Considering Out-of-Plane and In-Plane Bending Effects, Journal of the Society of Naval Architects of Korea
8. Kim MS, Kim YI, 2017. Nonlinear Finite Element Analysis for Mooring Chain Considering OPBIPB, Journal of Ocean Engineering and Technology
9. Garrett D.L., 1982. Dynamic analysis of Slender Rods, Journal of Energy Resources Technology
10. Brake E. ter, Cammen J. van der, Uittenbogaard R., 2007, Calculation Methodology of Out of Plane Bending of Mooring Chains, International Conference on Offshore Mechanics and Arctic Engineering.
11. Amir Izadparast, Caspar Heyl, Kai-tung Ma, Pedro Vargas, Jun Zou, 2018, Guidance For Assessing Out-Of-Plane Bending Fatigue On Chain Used In Permanent Mooring Systems, the 23rd Offshore Symposium of The Society of Naval Architects and Marine Engineers.
12. Kim, M.H., Koo, B.J., Mercier, R.M., and Ward, E.G., 2005, Vessel/mooring/riser coupled dynamic analysis of a turret-moored FPSO compared with OTRC experiment, Journal of Ocean Engineering Vol.32, 1780-1802.
13. Yang, C.K. and Kim, M.H., 2010, Transient effects of tendon disconnection of a TLP by hull-tendon-riser coupled dynamic analysis, Ocean Engineering, Vol.37 #9, 678-687
14. Kang, H.Y. and Kim, M.H., 2014, Safety assessment of Caisson transport on a floating dock by frequency- and time-domain calculations., Ocean Systems Engineering, International Journal, Vol.4, No.2,.
15. Kim, S.J. and Won, D.H., 2017, Bending Behavior of the Mooring Chain Links Subjected to High Tensile Forces, Journal of Korean Society of Steel Construction, Vol.29, No.2,pp.99-110.

SNAME
THE INTERNATIONAL COMMUNITY
FOR MARITIME AND OCEAN PROFESSIONALS

HIGH STRENGTH AND HIGH STIFFNESS ROPE FOR PERMANENT DEEPWATER MOORING: TECHNOLOGY DEVELOPMENT AND CHALLENGES

XIAOHONG CHEN
SENIOR PRINCIPAL ENGINEER, ABS

SUE WANG
MANAGER, ABS

MARK AUSBORN
ADVANCED MARINE ENGINEER, EXXONMOBIL UPSTREAM RESEARCH COMPANY

ZHEN LI
MARINE TEAM LEAD, EXXONMOBIL UPSTREAM RESEARCH COMPANY

Proceedings of the 24th Offshore Symposium, February 20th 2019, Houston, Texas
Texas Section of the Society of Naval Architects and Marine Engineers

ABSTRACT

Recent mooring research indicates that fiber ropes with higher strength and higher stiffness would benefit floating offshore platforms in water depths beyond 2000 meters in terms of reduced offset and reduced weight in comparison with polyester rope mooring. More advanced fibers with high strength and high stiffness are also entering into market. The industry has used high strength and high stiffness ropes for temporary moorings and mobile offshore drilling unit moorings. However, high strength and high stiffness fiber ropes have not yet been used for permanent moorings. This paper summarizes studies conducted by the industry on the high strength and high stiffness fiber ropes. An overview is provided for the existing research results, testing conducted, application guidelines, rope qualification processes, project experience, lessons learnt and the challenges of using high strength and high stiffness ropes for permanent moorings. Based on the industry experience of using polyester rope for permanent mooring and knowledge gained on high strength and high stiffness rope, this paper provides recommended assessments that could facilitate the application of high strength and high stiffness ropes for permanent deepwater moorings.

Keywords: Synthetic Fiber Rope, Polyester, HMPE, Dyneema, Deepwater Mooring System

INTRODUCTION

Offshore oil and gas exploration and production have been going to deeper water depths. In order to keep offshore floating platforms on station in deep water, mooring system technologies have advanced over the years and new technologies keep emerging. The mooring systems have evolved from steel catenary mooring systems, semi-taut mooring systems to taut leg mooring (TLM) systems for deep water applications. In the TLM systems, synthetic fiber ropes are used as the main components of the mooring lines and the mooring system restoring forces are mainly provided with the stiffness of the synthetic fiber ropes. The industry has more than 20 years' experience with the use of polyester rope moorings for offshore drilling units and production facilities. In the current industry deepwater projects, synthetic fiber ropes used for the permanent mooring system are all polyester ropes. The main technical advantages of polyester rope are weight reduction and lower axial stiffness compared to steel systems. These two factors lead to substantial reductions in operating tensions, maximum line tensions, vessel offsets, anchor radius, and installation loads and costs (Fulton, et.al., 2002). However, for water depths greater than 2000m, mooring systems using polyester rope may experience large offsets, which can be a challenge for riser system design.

Many studies have been conducted on high strength and high stiffness fiber ropes. Different fiber types, such as aramid (aromatic polyamide), liquid crystal polymer (LCAP or LCP) and high-modulus polyethylene (HMPE), have been studied. Based on the literature review, it is found that offshore applications of aramid and LCP rope are very limited. Most of the studies and developments of high strength and high stiffness rope are related to HMPE ropes.

HMPE fibers are characterized by high strength and high modulus. They produce lighter and smaller diameter high stiffness rope and provide both technical and operational advantages over traditional polyester mooring lines. A stiffer HMPE mooring system is potentially more riser-friendly than polyester. HMPE rope typically has an extension at break of 2%-2.5% elongation. HMPE fiber ropes have shown a longer fatigue life compared to polyester ropes for the same rope construction and are not vulnerable to axial compression fatigue compared to aramid fiber (Vlasblom, et.al., 2012). However, their unique properties and influence on the mooring system performance need to be properly accounted for in the design and verification process.

In the present study, a comprehensive literature review was conducted to collect and evaluate results of past studies, test procedures and data, code requirements, and fiber rope mooring analysis methods. A gap analysis was performed for the use of HMPE rope in permanent mooring systems in comparison with polyester rope and industry expectations. Recommendations and suggestions on the path forward for the application and qualification of synthetic fiber rope, especially on HMPE ropes for permanent moorings were developed. The following results of this study are discussed and summarized in this paper:

- Overview of synthetic fiber mooring ropes
- Detailed review of HMPE ropes status
- Gap analysis for using HMPE ropes for offshore permanent moorings
- Recommendations and suggestions for further studies

OVERVIEW OF SYNTHETIC FIBER MOORING ROPE

Polyester, nylon, and polypropylene rope has been used for decades as mooring lines for vessels alongside of piers, as towing hawsers, and as single point mooring (SPM) hawsers. (Flory, et.al., 2007). Serious consideration of synthetic mooring lines began in the 1980s with the trial use of aramid fiber ropes. The highly publicized failure of the Ocean Builder 1 construction barge mooring lines was a major setback for aramid, attributed to axial compression fatigue, a previously unknown failure mode of the fiber. This led to extensive research into synthetic fiber ropes and eventually to the wide use of polyester (PET) mooring ropes for permanent floating production systems pioneered by Petrobras. (Corbetta and Sloan, 2001). By 2010 (Rossi, et.al., 2010), 24 permanent mooring systems had been installed for platforms in Brazil with polyester ropes.

Prior to 2001, most of the synthetic fiber rope mooring applications were installed in Brazil by Petrobras (Petruska, et.al. 2010). In 2001 both Shell and BP successfully performed a full scale field trial of polyester mooring systems from a Mobile Offshore Drilling Unit (MODU). Since then, such systems have become more commonly used. In 2004, BP was the first to install a polyester mooring system in the Gulf of Mexico on its Mad Dog spar. Anadarko followed shortly with an installation of a polyester mooring system on its Red Hawk cell spar. Polyester rope mooring has since been used by several other projects. Table 1 provides a summary of the projects using synthetic fiber ropes as permanent mooring in the Gulf of Mexico, with the information of platforms, rope diameters, minimum breaking strength (MBS), rope types and rope manufacturers (Flory, et.al., 2007, Haslum, et.al., 2005, Tule, et.al., 2005, Jatar and Dove, 2005, Kindel, et.al., 2007, Couch, et.al., 2007, Herman, et al., 2010, Lohr and Smith, 2010, Miller, et.al., 2011, Mullett, et. al., 2012, Otten and Leite, 2013, Ganguly, et.al., 2013, Davidson, 2015, Fowler, 2015, www.bsee.gov, www.offshore-mag.com, www.oedigital.com, http://www.lankhorstoffshore.com, www.bexco.be, and Pallanich, 2012). Additionally, polyester rope has been employed for the Kikeh spar moorings in Malaysia as well as several CALM (Catenary Anchor Leg Mooring) buoy moorings and turret moorings throughout the world.

Table 1 Gulf of Mexico Permanent Polyester Mooring Projects

Year	Water Depth	Platform	Rope			
	m (ft)	name type	Diameter (mm)	MBS (tons)	Manufacturer	Type
2004	1,348 (4,420)	Mad Dog Spar	270	1,932	Marlow Ropes	Superline
2004	1,615 (5,300)	Red Hawk Spar	215	1,300	Whitehill	VETS 379
2006	915 (3,000)	Rowan Midland (MODU convert)	165	820	Whitehill	Torque-matched parallel subropes
2007	2,414 (7,920)	Independence Hub Semi	235	1,588	Marlow Ropes	Superline
2008	2,200 (7,215)	Blind Faith Semi	240	1,700	Whitehill	parallel subropes
2008	1,200 (3,940)	Tahiti Spar	248	1,905	Lankhorst Ropes (Quintas)	Gama 98
2009	1,800 (6,000)	Thunder Hawk Semi	254	1,930	Lankhorst Ropes	Gama 98
2009	2,438 (8,000)	Perdido Spar	246	1,800	ScanRope	MoorLine
2010	1,300 (4,300)	Mirage/Titan MinDoc	240	1,600	Whitehill	parallel subropes
2011	945 (3,100)	Who Dat/OPTI-EX semi	205	1,100	Whitehill	VETS 370
2011	2,500 (8,200)	Chinook and Cascade FPSO	170		Lankhorst Ropes	Gama 98
2014	2,100 (7,000)	Jack/St. Malo Semi	275	2,301	Parker ScanRope	MoorLine
2014	1,402 (4,600)	Tubular Bell/Gulfstar 1 Spar	222			
2014	2,165 (7,100)	Lucius Spar		1,907	Lankhorst Ropes	Gama 98
2014	1,372 (4,500)	Delta House Semi		1,700	Bexco	DeepRope
2016	1,615 (5,300)	Heidelberg Spar	254	1,905	Lankhorst Ropes	Gama 98
2016	2,900 (9,500)	Stones/Turritella FPSO		1,250	Bexco	DeepRope

Significant research and development work has contributed to the industry's successful use of fiber rope mooring for more than 20 years. The research and development work includes:

- Fiber, yarn and rope property testing for breaking strength, creep and creep rupture, rope load-elongation, stiffness, and other testing
- Tension-tension fatigue and axial compression fatigue testing
- Rope insert and trial rope study
- Installation study
- Study on the rope residual strength after damage
- Rope contact seabed study (for seabed parking only)
- Failure analysis

- Operational consideration
- Mooring performance analysis methodology
- Industry guidelines on rope qualification
- Industry guidelines on fiber mooring design assessment

A number of Joint Industry Projects (JIPs) have been conducted to facilitate the safe design and safe operation of the fiber rope moorings. A list of the related JIPs is given in Table 2.

Table 2 JIPs Related to Fiber Rope

JIP	Year	Comment
Fiber Tethers 2000	1994	By Tension Technology International (TTI), Noble Denton and National Engineering Laboratory (NEL)
Synthetic Mooring Lines	1994	DeepStar project
Polyester Taut Leg Mooring (TLM) Test – Program Definition	1995	DeepStar project
Polyester Taut Leg Mooring (TLM) Test – Failure Analysis"	1996	DeepStar project
Operational Considerations for Polyester Mooring Systems	1998	DeepStar project
Polyester Taut Leg Mooring (TLM) – Installation Report	1998	DeepStar project
An Evaluation of Deepwater Polyester Rope Mooring Technology	1999	DeepStar project
Polyester Taut Leg Mooring Test: Post Retrieval Testing	2000	DeepStar project
Durability of Polyester Fiber Ropes	2000	by NEL/TTI
Polyester Rope Stiffness Modeling	2009	Managed by ABS
Fiber Rope Damage Assessment and Acceptance Criteria		Managed by DNV
Improving Fiber- Mooring Design Practices		Managed by DNV
Managing the Safe Service Life of Fiber Ropes for Mooring		Managed by DNV
Safe Installation with Ropes in Ultradeep Sea		Managed by DNV

Industry guidelines on fiber ropes have been developed through the research process, most notably by the following organizations.

- Oil Companies International Marine Forum (OCIMF): Mooring Equipment Guidelines
- API: Recommended Practice (RP) 2SM "Design, Manufacture, Installation, and Maintenance of Synthetic Fiber Ropes for Offshore Mooring
- API: API RP 2I "In-service Inspection of Mooring Hardware for Floating Structures"
- ISO: Fibre Ropes for Offshore Stationkeeping-Polyester
- Rules and Guides by Classification Societies, such as ABS, BV, DNV GL, LR and others

Some fiber rope mooring incidents have been reported. An early incident in the 1980s involved axial compression fatigue failure of the aramid ropes used to moor the construction barge Ocean Builder I. The study (Chi, et.al. 2009) indicates that the factors contributing to the failures included rope damage and bending fatigue in the instance in which one rope was too long for the water depth. Torque generated under low tension in the aramid rope induced rotation of the rope, causing shear and compression strains and fiber kinking. Damaged fibers subsequently suffered torsional (shear) and compressive fatigue due to prolonged exposure of the lightly constrained aramid ropes to the constant loading of the open-ocean environment, resulting in loss of 40% to 50% of rope strength. The incident may be one of the primary reasons that the polyester ropes became the research focus for permanent moorings. The study found that the resistance of aramid ropes to axial compression fatigue can be improved through rope design and can be controlled while in use.

Two other fiber rope-related incidents reported (Tule, et.al. 2005, Ma et al. 2013) are related to operations. During the installation of polyester mooring lines for the Red Hawk cell spar in the Gulf of Mexico, one segment accidentally contacted the seabed. Concerns regarding the effect of possible soil ingress on the durability and performance of the polyester rope led to a two-phase test plan. The tests demonstrated that the polyester rope retained 100% of the new rope break strength, modulus, and fatigue life after contact with the seabed. The polyester rope that contacted the seabed was agreed by regulatory agencies to remain in the service. The other one was a mooring line failure of the Girassol buoy in a polyester segment at a pre-existing cut which went undetected by ROV (Remotely Operated Vehicle) inspection/survey.

Two recent rope failures due to axial compression fatigue in fiber ropes were reported by Flory, et.al. (2017). One of these was an HMPE mooring line on a gas carrier and the other was a nylon SPM hawser. In both incidents, the unusual rope design, with very long rope and strand lay lengths, was the root cause of the failure based on the investigation. The investigation indicated that HMPE rope has less axial compression resistance than polyester rope, but much better than aramid rope. It is suggested that unusual rope design may cause axial compression fatigue of HMPE rope. For new rope design with no past experience, an axial compression test may be conducted to qualify the rope for axial compression resistance.

REVIEW OF HMPE ROPE STATUS

While polyester rope is commonly used for deepwater mooring, research and development into the use of high modulus fibers in deepwater applications has been reported. The main driver for deepwater mooring system design is seen to be the vessel's offset requirement of the riser system. HMPE rope can provide the required stiffness with higher strength and reduced weight for deepwater mooring applications. Use of HMPE rope adds the flexibility to reach an optimized design of the mooring system in deepwater.

An HMPE mooring insert line was installed on the Scarabeo III MODU (Corbetta and Sloan, 2001) and an industry first use of HMPE synthetic fiber rope to moor the Sedco 601 MODU in 2,049 m water depth was reported by Berryman, et.al. (2002). In 2003, the Dyneema SK78 fiber grade, a low creep HMPE, was introduced for semi-permanent MODU moorings. Further research on creep rate reduction has resulted in the development of the Dyneema DM20 fiber grade, specifically designed for deepwater permanent mooring. (Vlasblom, et.al., 2012).

A detailed review has been performed on HMPE rope status on the following aspects:

- Testing for fiber and rope properties
- Projects and Joint Industry Projects (JIPs) related to testing and qualification of HMPE ropes
- Industry standards and classes requirements on HMPE testing and certification
- Latest status on the offshore application of HMPE ropes

There is limited information on individual research projects and JIPs related to testing and qualification of HMPE rope. Some of the JIPs noted earlier cover HMPE rope and fiber rope in general. Table 3 provides further information related to testing and qualification of HMPE ropes and testing on fiber and rope properties.

Table 3 Testing for HMPE Yarns and Ropes

Projects/JIPs	Scope	Participants	Results
Fiber Tethers 2000 JIP	Aramid, HMPE, LCAP (LCP), polyester yarns were tested	Tension Technology International LLC (TTI), Noble Denton, National Engineering Laboratory (NEL)	HSE (1999) Flory, et.al. (2007) Flory, et.al. (2017)
French Mooring Line Project (1996-2007)	Polyester, improved polyester, PEN, aramid, HMPE, LCP fiber ropes from yarns to 800 ton break load ropes	A group of French institutes and companies	Davies, et.al. (2002) Davies, et.al. (2008a) Davies, et.al. (2008b) Francois, et.al. (2010)
Sedco 601 MODU Mooring HMPE rope insert test JIP (2002)	Residual strength, residual tension-torque fatigue performance, creep	Unocal Indonesia, Transocean Sedco Forex, Marlow Ropes and DSM High Performance Fibres	Berryman, et.al. (2002)
Creep properties and model Dyneema SK75, SK78	Creep test on yarn	DSM Dyneema	Vlasblom and Bosman (2006)
Dyneema SK78 for MODU Mooring	Stiffness, creep rate, elongation, TN curve	Lankhorst Ropes, DSM Dyneema	Leite and Boesten (2011)
HMPE Mud Rope development for MODU Mooring	Rope static stiffness, breaking strength Yarn test on strength, elongation, creep	Delmar Systems, Inc. Samson Rope	Pasternak, et.al. (2011)
SIRIUS (Safe Installation with Ropes In Ultradeep Sea) JIP	Cyclic Bend Over Sheave (CBOS) fatigue tests Ø100mm fiber rope made out of Dyneema HMPE and ePTFE (expanded polytetrafluoroethylene) fibers	DNV, Technip, Subsea7, DSM Dyneema, Samson Rope	Tornqvist, et.al. (2011)
Dyneema DM20 for Permanent Mooring	Break strengths in subrope Yarn and rope stiffness Fiber creep rate and creep lifetime experiments, rope creep rate tests, yarn-on-yarn abrasion performance, fatigue life on rope samples, Ultra Violet (UV) resistance and thermal aging on fibers	Lankhorst Ropes, DSM Dyneema, IFREMER (Institut français de recherche pour l'exploitation de la mer)	Vlasblom, et.al. (2012)
HMPE rope failure investigation (2017)	Axial compression fatigue new test on HMPE, nylon yarns	Tension Technology International LLC (TTI)	Flory, et.al. (2017)

The industry guidelines and classification Rules and Guides for fiber ropes, in general, cover the requirements for HMPE rope. In order to qualify the high strength and high stiffness rope used in permanent fiber mooring systems, the design, manufacture, installation and inspection should comply with the industry standards and class requirements. Yarn and rope tests are also required to qualify the ropes. The design assessment should be performed to meet the requirements of mooring performance for the following aspects:

- Maximum line tension
- Maximum vessel offset
- Clearance to fairlead and seabed
- Axial compression fatigue analysis
- Tension-Tension (T-T) fatigue analysis
- Creep and creep rupture analysis (HMPE ropes)
- Torque compatibility

In comparison with steel wire rope, the change-in-length properties of fiber rope mooring lines are more complex. These properties are not the same for all fiber rope, even for the same material, but vary with material grade and rope construction. Stiffness is non-linear and dependent on mean load, load amplitude, and frequency. Rope elongation and stiffness should be appropriately modeled in the analysis.

The required yarn and rope tests for HMPE rope applications by ABS and API are given in Table 4. Test procedures and measurement should follow the guidelines given in ABS (2014) or API RP 2SM (2014).

Table 4 Required Yarn and Rope Tests

	ABS	API
Rope	Minimum Breaking Strength	Break Test
	Elongation and Stiffness	Static Load-Elongation Test
	Splice Qualification	
	Particle Ingress Resistance 1)	Filter Barrier Particle Ingress Test5)
	Torque Match with Steel Wire Rope 2)	Torque and Rotation Test 3)
	HMPE Creep Rate Verification	Creep and Delayed Elastic Stretch and Recovery Test
		Dynamic (Low Frequency and Wave Frequency) Load-Elongation Test
		Tension-Tension Fatigue Qualification Test 4)
Yarn	Yarn Dry Breaking Strength and Elongation	Dry break strength / Dry elongation to break
	Yarn Dry Creep for HMPE	
	Yarn-on-Yarn Abrasion Performance	Wet yarn-on-yarn abrasion property
	Yarn Testing during Production	Yarn size (linear density)
		Twist level

Notes:

1) Required for rope preset on seabed and reuse of an accidentally dropped rope
2) Required for fiber rope connected with torque wire rope (6- or 8-strand) for permanent mooring
3) Where torque and rotation properties are of concern, rope torque tests shall be conducted
4) For rope types for which no fatigue data exist, qualification testing shall be performed
5) Required for the rope designed for seabed contact or high turbidity environments, and reuse of an accidentally dropped rope

Rope and seabed contact may occur in the following situations:

- Accidental contact with seabed during installation
- Planned seabed pre-installation

To qualify as particle ingress-resistant, the jacket or filter barrier shall be tested. The test condition of the rope should reflect the condition of the rope when exposed to particles.

In order to gain approval for the use of synthetic fiber for permanent mooring in the Gulf of Mexico, the regulations as set out by the Minerals Management Service (MMS, now BSEE - Bureau of Safety and Environmental Enforcement), and United States Coast Guard (USCG), as well as the delegations of responsibility had to be satisfied or equivalencies established. Class societies may also be involved in the process, with their role being to satisfy Classification requirements and provide third party verification. (Bugg, et.al., 2004). Based on MMS NTL No. 2009-G03 (MMS NTL, 2009), the technical requirements of using synthetic fiber rope in deepwater mooring systems are mainly set out in API RP 2SM, API RP 2I and API RP 2SK.

It has been reported that HMPE rope has been used for several MODU moorings and offshore deepwater installation lines (Corbetta and Sloan 2001, Berryman, et.al. 2002, Gilmore, et.al. 2009, Leite and Boesten 2011, and Vlasblom, et.al. 2012, Torben, et.al., 2007, http://www.samsonrope.com, Crawford, et.al. 2011). There has no application of HMPE rope for permanent moorings as of yet. The applications of HMPE rope are summarized in Table 5 and Table 6.

Table 5 Offshore Applications of HMPE Ropes – MODU Application

Year	Mooring Type	Rope Type	Rope Manufactures
1997	Scarabeo III MODU, rope insert and anchor pendant lines	Plasma 12x12-strand braided rope 80mm diameter 430 tonnes MBS	Cortland / Puget Sound Rope
1999	Sedco 601 MODU temporary mooring	Superline Steelite Xtra SKX75 Dyneema fibre ropes 74mm diameter 357 tonnes (3502 kN) MBS	Marlow Ropes
2006	Anchor mudline and retrieval line for OMNI-Max anchor	M-8 8×3 construction Dyneema SK75 400 tonnes retrieval line 800 tonnes mudline	Samson Rope
2011	Pertobras MODU Deepwater Mooring	Gama 98 Dyneema SK78 630 tonnes MBL	Lankhorst Ropes

Table 6 Offshore Applications of HMPE Ropes – Deepwater Installation Line

Year	Mooring Type	Rope Type	Rope Manufactures
2006	Lifting line for FDRS	12x12 braided construction HMPE and LCP combined fiber	Cortland / Puget Sound Rope
2007	Installation Winch Line for Perdido Spar	Quantum-12 ropes combination of Dyneema-75 and Samson's DPX fiber 63.5mm	Samson Rope
2011	Installation Winch Lift Line	80mm, 12 x 12 strand HMPE	Cortland/Puget Sound Rope

GAP ANALYSIS FOR USING HMPE ROPES FOR OFFSHORE PERMANENT MOORINGS

Based on the current industry practices on using synthetic fiber ropes for offshore permanent moorings, a gap analysis of using HMPE ropes for offshore permanent moorings was performed. Additional considerations of using HMEP ropes with consideration of the industry readiness are summarized and challenges were identified for further study.

In the current industry practices for permanent moorings, synthetic fiber rope is only used as middle sections. The synthetic ropes are submerged and freestanding. The guidelines in ABS (2014), API RP 2SM (2014) and API RP 2I (2015) are provided for synthetic rope as mooring line middle sections which are fully submerged and not intended for seabed contact.

The main differences in mechanical properties between HMPE rope and polyester rope are load-elongation behavior and creep and creep rupture. In addition to the requirements for polyester rope, such differences need special considerations for HMPE rope. More specifically the considerations should be provided to the following aspects:

- Line length management, re-tensioning requirement
- Quasi-static stiffness
- Dynamic stiffness
- Creep and creep rupture analysis
- Creep test and creep model verification
- Tension-Tension fatigue test

Axial compression fatigue is normally not a concern for polyester and HMPE fiber ropes. Based on a recent investigation into an HMPE rope failure, unusual rope design may cause axial compression fatigue of HMPE rope (Flory, et.al., 2017).

Polyester rope has been used for permanent mooring for many years while HMPE has not been applied to permanent moorings. The industry experience of polyester ropes and HMPE ropes is summarized in Table 7 which provides some indications for the areas where more studies may be needed for HMPE ropes.

Table 7 Industry Experiences of Polyester and HMPE Rope

	Polyester Rope	**HMPE Rope**
Rope Design	Around 10 rope designs	Applicable to HMPE
Fiber Testing	Two JIPs, testing for deepwater mooring projects	Two JIPs and seven research projects
Rope Testing	DeepStar JIP, JIPs by NEL/TTI, ABS and DNV, testing for deepwater mooring projects	Two JIPs and five research projects
Installation Study	DeepStar JIP studies, Trail and real project experiences	Applicable to HMPE Rope will float due to positive buoyancy
Damage Study	JIP studies	Applicable to HMPE for the same rope design
Marine Ship Mooring	Around 50 years application	Around 20 years application
Other Applications (deepwater installation lines)	No projects found	3 projects found to replace wire rope
Offshore Mobile Mooring	More than 20 years experiences and a number of projects application (middle section)	4 projects found
Deepwater Offshore Permanent Mooring	More than 20 years experiences and a number of projects application worldwide	No projects yet

	(middle section) (more than 24 projects in Brazil, 17 projects in Gulf of Mexico)	
Design Guidelines	Guidelines developed based on JIP works and individual research projects applicable to fiber ropes in general.	HMPE rope is covered
Design Analysis	Design analysis criteria and methods are defined in guidelines	HMPE rope is covered
Stiffness Model	Non-linear stiffness to be considered in different analysis methods	Same analysis methods Creep needs to be considered for quasi-static stiffness
Fatigue Curve	Fatigue curve available	Limited data, no recommended fatigue curve

Currently there is no experience with the long-term application of synthetic ropes as top sections and bottom sections in permanent mooring systems. To qualify synthetic ropes as top sections in permanent moorings, additional considerations could include, for instance, Ultra Violet (UV) light, bend radius, bending fatigue, salt crystallization, temperature and marine growth, and winch and mooring equipment design. To qualify synthetic ropes as bottom sections in permanent moorings, additional considerations could include, for instance, contact and abrasion with seabed, particle filter performance, bend radius, bending fatigue, inspection and effects on anchor installation and design.

FINDINGS AND RECOMMENDATIONS

From the literature review of synthetic fiber mooring ropes, lessons learned, gap analysis, and advantages and disadvantages of HMPE ropes are summarized. Recommendations and suggestions for the path forward for the application and qualification of synthetic fiber ropes, especially on HMPE ropes are provided.

The lessons learned on using synthetic fiber mooring rope include:

- Installation incident: During the installation of polyester mooring lines for the Red Hawk cell spar in the Gulf of Mexico, one segment accidentally contacted the seabed. Particle Ingress Resistance tests were required for reuse of the rope that accidentally contacted the seabed.
- Axial compression fatigue was found to be the cause of the failure of the aramid rope used to moor the construction barge Ocean Builder I.
- One Girassol buoy failure was in a polyester segment at a pre-existing cut which went undetected by ROV (Remotely Operated Vehicle) inspection/survey.
- Care should be taken for possible interference of HMPE rope floating on the surface with workboats during installation.
- Unusual design of ropes may cause axial compression fatigue of HMPE rope. For new rope design with no past experience, an axial compression test may be conducted to qualify the ropes for axial compression resistance.

The main technical advantages of HMPE rope in term of rope properties are:

- Light weight, positive buoyancy in water
- High modulus (stiffness)
- High strength
- Small diameter
- High abrasion resistance
- High cut resistance
- Good T-T fatigue performance
- Good axial compression fatigue performance compared with aramid rope

These rope properties have several advantages for mooring system performance, handling, installation and maintenance, such as:

- Reduced platform offset
- Reduced vertical loads on platform
- Reduced storage space and ease of handling
- Reduced installation loads
- Low maintenance

Due to the properties of HMPE rope, the technical challenges of using HMPE rope for deepwater offshore permanent moorings may include:

- Low heat resistance
- Creep and creep rupture

Challenges for using HMPE rope as middle sections are:

- Lack of experiences on creep and it's influence on line length, quasi-static stiffness and mooring performance.
- Limited test data on creep model verification.
- Lack of experience on line length management plan.
- Limited test data on T-T fatigue performance.

Other challenges may also include:

- Capability of rope testing facility
- Mooring connectors used in TLM with HMPE rope
- Wire rope and chain fatigue in a TLM with HMPE rope

The benefits of using HMPE rope as middle sections are obvious especially in deepwater where the middle sections are very long. Without using HMPE rope, polyester rope may not provide enough stiffness for the desired mooring system performance. There are no obvious consequences if HMPE rope is used as middle sections.

It is promising to use HMPE ropes as middle sections in permanent mooring systems. The following areas should be considered prior to use:

- HMPE rope creep and influence on line length, quasi-static stiffness and mooring performance
- Creep test data and creep model verification
- Line length management plan
- Tension-Tension fatigue qualification test data
- Capability of rope testing facilities
- Mooring connectors used in TLM (Taut Leg Mooring) with HMPE ropes

HMPE rope with new technology such as Dyneema DM20 fiber have very low creep rate and long creep lifetime. These new technologies need to be qualified based on industry standards, class requirements and regulatory requirements. Besides, regulators will also require T-T fatigue qualification testing. Obtaining these test data will be necessary.

Creep and creep rupture analysis is performed using a creep model based on testing data. This analysis and underlying test data are to be verified to the satisfaction of class or regulators. Once the creep can be predicted and verified, the influence of creep elongation on quasi-static stiffness and mooring performance can be studied. Based on this study, line length management plan can be developed. In the line management plan, the need of re-tensioning and its frequency as well as the use of rope inserts can be studied. It is noted that the rope inserts may still be required by the regulators under certain circumstances. This can be considered in the line management plan.

HMPE rope used for deepwater applications may have a high breaking strength that may exceed capacity of the current rope test facility. The test for creep properties and T-T fatigue can be extensive. A survey of capacities of current rope test facilities is considered necessary.

The mooring connectors connecting the HMPE ropes to the top and bottom sections and to each other should be studied. Since the sizes of HMPE ropes are smaller than polyester ropes with the same strength, the mooring connectors should also be smaller than those used for polyester ropes. However, the mooring connectors are important and will also affect rope properties, such as breaking strength and fatigue performance. The mooring connectors need to be qualified for long-term use. Thus a survey and studies of the mooring connectors are considered necessary.

Higher line tensions due to higher dynamic stiffness can be resolved for a specific design in a specific water depth in the design process. For example, hybrid mooring system can be an option. Thus this issue is not considered as a challenging item. However, this still need to be considered in the design process.

Selection of anchor should be addressed in the design process. Conventional anchors used for the permanent mooring system using polyester ropes should also be suitable for the permanent mooring system with HMPE ropes.

REFERENCES

[1]. American Bureau of Shipping (ABS), 2014. "Guidance Notes on the Application of Fiber Rope for Offshore Mooring", August 2011 (Updated February 2014).

[2]. American Petroleum Institute (API), 2015. API RP (Recommended Practice) 2SK "Design and Analysis of Stationkeeping Systems for Floating Structures", Third Edition, 2005. Addendum, May 2008. Reaffirmed 2015.

[3]. American Petroleum Institute (API), 2014. API RP 2SM "Design, Manufacture, Installation, and Maintenance of Synthetic Fiber Ropes for Offshore Mooring", Second Edition, 2014.

[4]. American Petroleum Institute (API), 2015. API RP 2I "In-service Inspection of Mooring Hardware for Floating Structures", 3rd Edition (2008). Reaffirmed 2015.

[5]. Berryman, C.T., Dupin, R.M. and Gerrits, N.S., 2002. "Laboratory Study of Used HMPE MODU Mooring Lines". 2002 Offshore Technology Conference, Houston, Texas, Paper No. OTC 14245.

[6]. Bugg, D.L., Vickers, D.T. and Dorchak, Jr., C.J., 2004. "Mad Dog Project: Regulatory Approval Process for the New Technology of Synthetic (Polyester) Moorings in the Gulf of Mexico". 2004 Offshore Technology Conference, Houston, Texas, Paper No. OTC 16089.

[7]. Chi, C.H., Lundhild, E.M., Veselis, T. and Huntley, M., 2009. "Enabling Ultra-Deepwater Mooring with Aramid Fiber Rope Technology". 2009 Offshore Technology Conference, Houston, Texas, Paper No. OTC 20074.

[8]. Corbetta, I. and Sloan, F., 2001. "HMPE Mooring Line Trial for Scarabeo III". 2001 Offshore Technology Conference, Houston, Texas, Paper No. OTC 13272.

[9]. Couch, A., Guion, J., Rieth, D., Rager, B. and Bouwman, J., 2007. "Independence Installation". 2007 Offshore Technology Conference, Houston, Texas. Paper No. OTC 18585.

[10]. Crawford, M., Crawford, P. and Shand, D.M., 2011. "Application and Evolution: 17.5Te, 85Te and 250Te to 3000 meters using a Drum Winch Approach, HMPE Fiber Rope and Efficient Offshore Vessels". 2011 Offshore Technology Conference, Houston, Texas, Paper No. OTC 21780.

[11]. Davidson, N., 2015. "Gulfstar One FPS". February 26, 2015.

http://www.mtshouston.org/pdfs/2015/FebruaryDavidson2015.pdf

[12]. Davies, P., François, M., Grosjean, F., Baron, P., Salomon, K. and Trassoudaine, D., 2002. "Synthetic Mooring Lines for Depths to 3000 Meters". 2002 Offshore Technology Conference, Houston, Texas, Paper No. OTC 14246.

[13]. Davies, P., Baron, P., Salomon, K., Bideaud, C., Labbé, J.-P., Toumit, S. Francois, M., Grosjean, F., Bunsell, T. and Moysan, A.-G., 2008a. "Influence of Fibre Stiffness on Deepwater Mooring Line Response". 27th International Conference on Offshore Mechanics and Arctic Engineering, Estoril, Portugal, Paper No. OMAE2008-57147.

[14]. Davies, P., LeChat. C., Bunsell, A., Piant, A., Francois, M., Grosjean, F. Baron, P., Salomon, K., Bideaud, C. Labbe, J.P., and Moysan, A.G., 2008b. "Deepwater Mooring With High Stiffness Polyester and PEN Fiber Ropes". 2008 Offshore Technology Conference, Houston, Texas, Paper No. OTC 19315.

[15]. Det Norske Veritas Germanischer Lloyd (DNV GL), 2018. DNVGL-OS-E303 "Offshore Fibre Ropes", July, 2018.

[16]. Det Norske Veritas Germanischer Lloyd (DNV GL), 2017. DNVGL-RP-E305 "Design, Testing and Analysis of Offshore Fibre Ropes", July, 2017.

[17]. Duggal, A.S., Izadparast, A.H. and Pothuganti, V.K., 2015. "Stationkeeping Technology for Frontier Deepwater Floating Systems" 2015 Offshore Technology Conference, Houston, Texas, Paper No. OTC 26058.

[18]. Flory, J.F., Banfield, S.J. and Berryman, C., 2007. "Polyester Mooring Lines on Platforms and MODUs in Deep Water". 2007 Offshore Technology Conference, Houston, Texas, Paper No. OTC 18768.

[19]. Flory, J.F., Banfield, S.J., Ridge, I.M.L. and Carr, R., 2017. "Axial Compression Fatigue in Long-Lay-Length Fiber Mooring Ropes". 2017 Offshore Technology Conference, Houston, Texas, Paper No. OTC 27664.

[20]. Fowler, R., 2015. "Delta House Project". June 25, 2015. http://www.mtshouston.org/pdfs/2015/JuneDeltaHouse2015.pdf

[21]. Francois, M., Davies, P., Grosjean, F. and Legerstee, F., 2010. "Modelling Fiber Rope Load-Elongation Properties - Polyester and Other Fibers". 2010 Offshore Technology Conference, Houston, Texas, Paper No. OTC 20846.

[22]. Fulton, T.M., Veselis, T., Dove, P.G.S., Bowles, H.T. and Petruska, D.J., 2002. "Introduction of Polyester Taut Leg Mooring Into the Gulf of Mexico". 2002 Offshore Technology Conference, Houston, Texas, Paper No. OTC 14244.

[23]. Ganguly, P., Mastrangelo, C.F. and Daniel, J., 2013. "First Floating, Production, Storage and Offloading Vessel in U.S. Gulf of Mexico". 2013 Offshore Technology Conference, Houston, Texas, Paper No. OTC 24112.

[24]. Gilmore, J., Chou, R. and Zimmerman, E.H., 2009. "Hurricane-Proof Mooring System: Enabling Technology Through the Use of Synthetic Ropes". 2009 Offshore Technology Conference, Houston, Texas, Paper No. OTC 20155.

[25]. Haslum, H.A., Tule, J., Huntley, M. and Jatar, S., 2005. "Red Hawk Polyester Mooring System Design and Verification". Offshore Technology Conference, Houston, Texas, Paper No. OTC 17247.

[26]. Herman, D., Garrett, D., Shivers, R. and Veselis, T., 2010. Performance Benefits of the MinDOC Hull at the Mirage Field in a Post- Katrina/Rita Environment. 2010 Offshore Technology Conference, Houston, Texas, Paper No. OTC 21029.

[27]. Health and Safety Executive (HSE), 1999. "Review of Fiber rope for Offshore Application". Offshore Technology Report, OTO 1995 005. Prepared by National Energy Laboratory.

[28]. Jatar, S. and Dove, P., 2005. "The Planning and Installation of the Red Hawk Spar and Polyester Mooring System". 2005 Offshore Technology Conference, Houston, Texas, Paper No. OTC 17294.

[29]. Kindel, C., Rijken, O., Khodr, R., Cao, P., Galvin, C., van Hofslot, T. and Barnett, P., 2007. "Independence Hub—Turnkey Delivery of Ultra Deepwater Hull and Mooring System". 2007 Offshore Technology Conference, Houston, Texas, Paper No. OTC 18587.

[30]. Leite, S. and Boesten, J., 2011. "HMPE Mooring Lines for Deepwater MODUs". 2011 Offshore Technology Conference Brasil, Rio de Janeiro, Brazil, Paper No. OTC 22486.

[31]. Lohr, C. and Smith, K., 2010. "Perdido Development Project - Spar & Moorings". 2010 Offshore Technology Conference, Houston, Texas, Paper No. OTC 20885.

[32]. Ma. K.T., Duggal, A., Smedley, P., L'Hostis. D. and Shu, H.B., 2013. "A Historical Review on Integrity Issues of Permanent Mooring Systems". 2013 Offshore Technology Conference, Houston, Texas, Paper No. OTC 24025.

[33]. Miller, J.D., Warren, B.J., Chabot, L.G. and Jenkins, J.F., 2011. "Murphy Thunder Hawk: New Findings on Polyester and Chain Mooring Systems". 30th International Conference on Offshore Mechanics and Arctic Engineering, Rotterdam, The Netherlands, Paper No. OMAE2011-50298

[34]. Mineral Management Service (MMS), 2009. NTL No. 2009-G03 "Notice To Lessees and Operators of Federal Oil, and Gas Leases in the Outer Continental Shelf (OCS), Gulf of Mexico OCS Region – Synthetic Mooring Systems".

[35]. Mullett, C., Phillips, C., Huntley, M., Garrity, R., Orson, J. and Pasternak, J., 2012. "WHO DAT Project - Polyester Mooring System Design, Analysis, and Installation". 2012 Offshore Technology Conference, Houston, Texas, Paper No. OTC 23083.

[36]. Otten, J. and Leite, S., 2013. "Thunder Hawk Mooring Rope Particle Ingress, Break, and Fatigue Testing". 2013 Offshore Technology Conference, Houston, Texas, Paper No. OTC 24080.

[37]. Pallanich, J. 2012. "Rolling in the deep". Offshore Engineer, April 1, 2012. https://www.oedigital.com/energy/item/769-rolling-in-the-deep

[38]. Pasternak, J.D., Shelton, J and Gilmore, J. 2011. "Synthetic 'Mud Ropes' for Offshore Mooring Applications - Field History and Testing Data". OCEANS 2011.

[39]. Petruska, D., Kelly, P., Stone, B., Ahjem, V., Zimmerman, E., Garrity, R. and Veselis, T., 2010. "Fiber Moorings, Recent Experiences and Research: Updating API RP 2SM on Synthetic Fiber Rope for Offshore Moorings". 2010 Offshore Technology Conference, Houston, Texas, Paper No. OTC 20836.

[40]. Rossi, R., Del Vesshio, C.J.M. and Goncalves, R.C.F., 2010. "Moorings with Polyester Ropes in Petrobras: Experience and the Evolution of Life Cycle Management". 2010 Offshore Technology Conference, Houston, Texas, Paper No. OTC 20845.

[41]. Torben, S.R., Bunes, Ø., Bull, S., Paterson, J and Davidson, D., 2007. "Fiber Rope Deployment System For Ultradeepwater Installations". 2007 Offshore Technology Conference, Houston, Texas, Paper No. OTC 18932.

[42]. Tornqvist, R., Strande, M., Cannell, D., Gledhill, P., Smeets, P. and Gilmore, J., 2011. "Deployment of Subsea Equipment: Qualification of Large Diameter Fibre Rope for Deepwater Construction Applications". 2011 Offshore Technology Conference, Houston, Texas, Paper No. OTC 21588.

[43]. Tule, J., Huntley, E., Phillips, C. and Haslum, H., 2005. "Red Hawk Project Polyester Soil Ingress Testing". 2005 Offshore Technology Conference, Houston, Texas, Paper No. OTC 17259.

[44]. Vlasblom, M.P., Bosman, R.L.M., 2006. "Predicting the Creep Lifetime of HMPE Mooring Rope Applications". OCEANS 2006.

[45]. Vlasblom, M., Boesten, J., Leite, S., and Davies, P., 2012. "Development of HMPE Fiber for Permanent Deepwater Offshore Mooring". 2012 Offshore Technology Conference, Houston, Texas, Paper No. OTC 23333.

DISSECTION, MEASUREMENT AND ANALYSIS OF A RECOVERED DAMAGED FLEXIBLE RISER

AUTHOR (COLIN C. DUFFETT)
AFFILIATION (RISER LEAD, WOOD PLC)

Proceedings of the 24th Offshore Symposium, February 20th 2019, Houston, Texas
Texas Section of the Society of Naval Architects and Marine Engineers

ABSTRACT

This study summarizes a review of the dissection and post-dissection material testing of a recovered flexible riser with a damaged outer sheath and localized wire corrosion, and the subsequent analytical fatigue assessment of the riser based on findings from that dissection. Flexible pipe technology and experience is still relatively young when compared to the operational experience of rigid pipeline systems. This drives the continued effort to understand performance characteristics and to maintain integrity management and good practice guidance, particularly as ongoing projects continue to extend the environmental, structural, and service life requirements of such flexible pipe systems

.

Incidents relating to outer sheath damage and annulus flooding account for most flexible pipe damage and failure statistics. The ultimate consequence of annulus flooding is principally a failure resulting from either reduction in armour capacity through corrosion or accelerated corrosion-fatigue.

The fatigue of armour wires is in many cases the limiting factor for the design life of flexible risers. Fatigue design of armour wires is, in general, based on full scale tests in which a given pipe design is subjected to a fatigue load history simulating the accumulated damage of the design life. These methods are costly and time consuming, thus this study aims to quantify the reduction in fatigue life due to metal loss/corrosion in the armour wires using advanced numerical analysis.

To evaluate the damage, Wood led a dissection of the recovered flexible pipe and collected wire samples for material and dimensional analysis. Wood observed an increase in pipe circumference around breach location; noteworthy as the sections of the outer sheath around the breach location were thinned out or missing completely, indicating a likely bird-caging effect of the tensile armour wires.

A layer-by-layer dissection was then performed such that each of the layers constituting the makeup of the flexible riser was removed individually and a condition assessment of each layer recorded. All tensile wires were retained following dissection for examination. Selections of wires were sampled for mechanical properties (yield strength and UTS) and metallography, dimensional measurement and testing. Once the amount of material loss in the armour wires was measured, a local analysis of the flexible pipe was performed to quantify any increased wire stress and annualized fatigue damage

between the original design wire size and the measured wire size.
Finally, integrity management and life extension recommendations for the current risers connected to the facility were made for future risers, all identical in size and service type to the recovered pipe.

Keywords: SNAME, Offshore, Symposium, Flexible Pipe, Flexible Risers, Life Extension, Integrity Management

INTRODUCTION

A client had recently replaced a water injection riser and upon removal of the existing riser, suspicions of an outer sheath breach were confirmed. Damage was noted in the area under the bend stiffener/bend stiffener connector interface (Figure 1). This was consistent with pressure test results that indicated the riser annulus had become flooded sometime in the last two years.

The riser was originally wet stored upon removal; currently the top 22m of riser has been cut and recovered for inspection, dissection and material analysis.

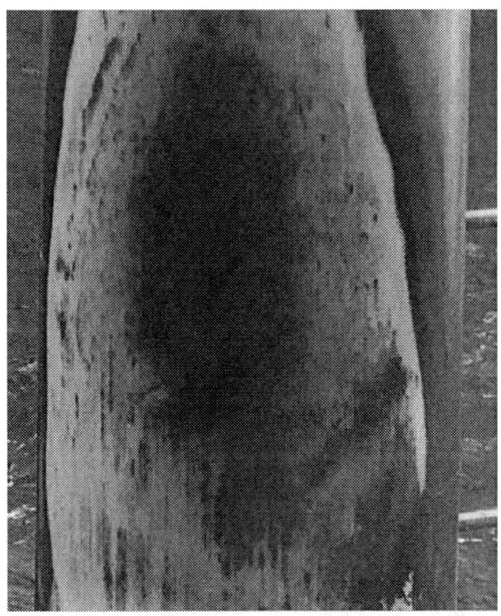

Figure 1 Riser Damage

The main objective of the study is the data gathering, dissection and material analysis and mechanical testing of the recovered riser layers. The findings from the flexible pipe dissection and post-dissection material testing were then focused on the fatigue assessment of the riser based on findings from that dissection.

Fatigue of armour wires is in many cases the limiting factor for the design life of flexible risers. Fatigue design of armour wires is in general based on full scale tests, in which a given pipe design is subjected to a fatigue load history simulating the accumulated damage of the design life. These methods are costly and time consuming, thus this study aims to quantify the reduction in fatigue life due to metal loss/corrosion in the armour wires using advanced numerical analysis.

METHODOLOGY (RISER DISSECTION)

Personnel from Wood and a local R&D lab performed the dissection and sample extractions for material analysis of the 10-meter section of pipe. To aid with the data recording for each layer, four axial reference lines were marked at 90° intervals around the circumference of the Riser. The dissection occurred at the site of the outer sheath breach as well as 3 meters on either side of the breach location. Illustrations of the marking locations are presented in Figure 2 and Figure 3.

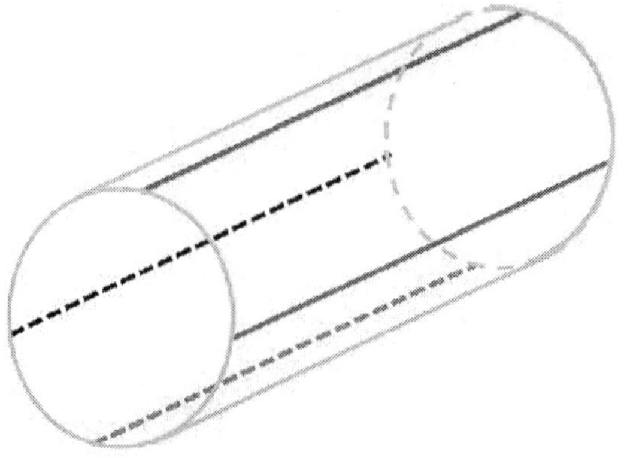

Figure 2 Identification Lines

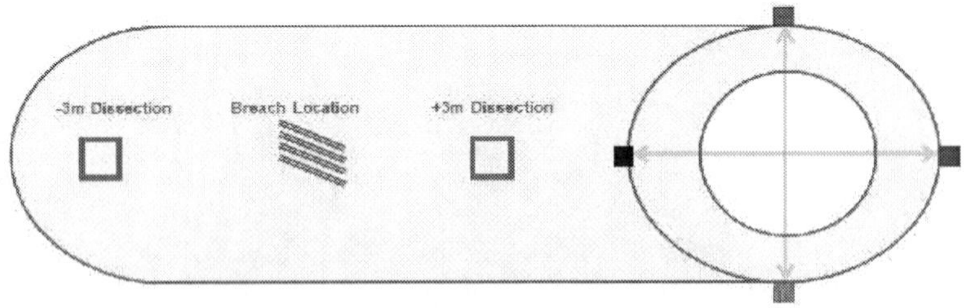

Figure 3 Dissection Locations

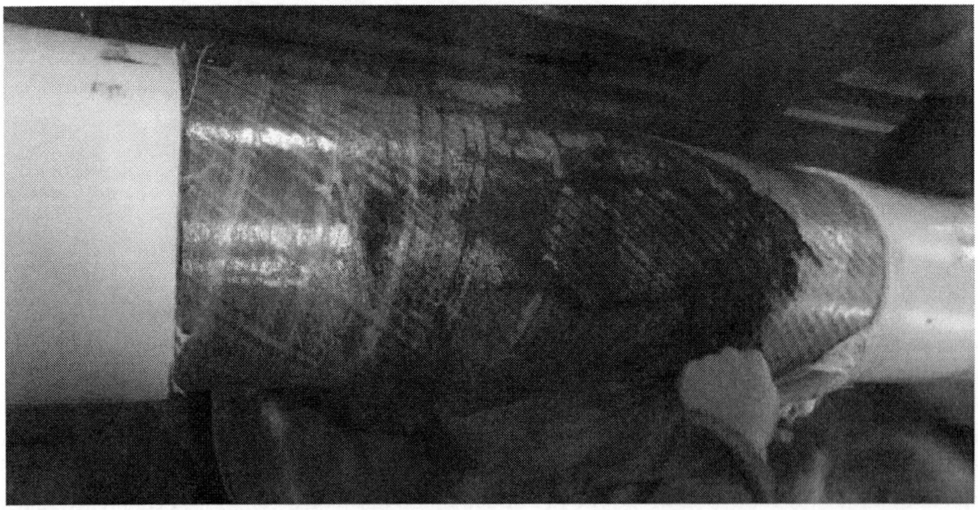

Figure 4 Riser Breach Location

Once the riser section was prepared for inspection, the outer circumference of the pipe was measured at various sections. This was due partly to a visible bulge in the outer sheath near the breach location and to verify the nominal size at sections thought to be intact and free of any defects.

Other than the breach location, the external protective sheath was coherent along the length of the pipe, though showed multiple minor superficial scratches and shallow indentations. There were no visible through thickness damages to the external sheaths at these areas.

Evidence of marine growth was not visibly present anywhere along the length.

As can be seen in the table below, the outer circumference at the breach location and +/- 1 meter away were different to the undamaged portions at the topsides and subsea ends. Additionally, the increase in circumference around breach location is compounded when it is observed that sections of the outer sheath are thinning or missing completely at the measurement point.

Table 1 Pipe Circumference (Outer Sheath)

Designation	**Location**	**Circumference (mm)**
A	Breach Location	966
B	Breach + 3m (towards subsea end)	938
C	Breach + 3m (towards topsides end)	938
X	Breach + 1m (towards subsea end)	941
Y	Breach + 1m (towards topsides end)	935
Nominal Circumference = 937 mm		

Once the outer sheath sections were removed, the outer circumference of the pipe was measured at various sections on the second tensile armour layer. This was to confirm the measurements and anomalies from the previous measurement.

As can be seen in the table below, the outer circumference at the breach location was much greater than the undamaged portions at the topsides and subsea ends (+/- 3m).

Table 2 Pipe Circumference (Tensile Armour)

Designation	**Location**	**Circumference (mm)**
A	Breach Location	926
B	Breach + 3m (towards subsea end)	893
C	Breach + 3m (towards topsides end)	892
Nominal Circumference = 881 mm		

It was also observed that the corrosion and degradation of the tensile armour wires was localized to the breach location and did not extend past the exposed area in the lateral or radial directions to any great extent. Shiny, undamaged metal could be observed in these areas as shown in Figure 5 and Figure 6. It should be noted that the orange coloured fabric in between the individual wires is residue of the tape layer directly beneath it.

Figure 5 Tensile Armour Wires (adjacent to breach location)

Figure 6 Tensile Armour Wires (opposite of breach location)

While not part of the original dissection process, some outer armour wires were cut away to a suitable length to allow safe observation of underlying layers wires that had failed.

A number of these inner wires were able to be manually moved or shifted, indicating their fracture in the vicinity of the breach location. Because of the number of outer wires still intact, it was not possible to trace the fracture location with exact certainty.

METHODOLOGY (MATERIAL ANALYSIS)

A layer-by-layer assessment was performed such that each of the layers constituting the make-up of the flexible riser were removed individually and a condition assessment of each layer recorded.

All the tensile wires were retained following dissection for examination. Selections of wires were sampled for mechanical properties and metallography. Wood personnel recovered short samples of wires for its internal measurement and long samples to be furnished to a local laboratory (Element Labs) for their measurement and testing. The long samples were sent to ensure there was enough material to fit in any specific testing machines or sampling apparatus.

A chemical analysis (SEM-EDX) was completed on the four wire samples from the three locations. The metallography of all tested wires did not reveal any anomaly that would be associated with premature failure of the wires. The analysis showed that the chemical compositions of the wires are in line with typical tensile armour metals from the flexible pipe manufacturers.

Some shallow localized corrosion has occurred; it is impossible to gauge when this corrosion initiated, but it is probable that moisture ingress following rupture of the external sheaths has increased the rate of corrosion.

A mechanical tensile test was then performed on all 12 samples of tensile armour wire that were recovered. The mechanical testing of the wires showed no non-compliance with respect to OEM data sheet requirements for yield strength and UTS for typical tensile armour wires on past Wood riser projects. Wood referenced datasheets from various analogous projects and while none of those risers had wires of the exact dimensions of the dissected flexible pipe (12mm x 4mm), data from similar wires (12mm x 3mm) and (12mm x 5mm) showed scalable, consistent results.

Outer sheath samples from locations 'B' and 'C' underwent tensile testing (ASTM D638) to assess the mechanical properties of the material to its standards. The results show that the riser samples had tensile stress, elastic modulus and break strains above the minimum standard for a Rilsan-like material.

METHODOLOGY (DIMENSIONAL ANALYSIS)

The thickness measurements of outer sheath indicate that it has reduced thickness nearer to the breach location. It was thinnest in the vicinity of the initiation and fracture points at the fracture surface. This is consistent with the damage noted in the area under the bend stiffener/bend stiffener connector interface.

A dimensional analysis was performed on the tensile armour wires by Wood and a local laboratory to determine the magnitude of any material losses due to corrosion or other factors.

For the width of the wires, Wood measured its short and long wire samples with digital calipers on the base metal, while a local laboratory used their own proprietary equipment. While the three measurements produced different values for the wire width, in general the differences in width between individual wires remained constant for each technique. The results are present in Table 3 below.

The lengths measurements were performed by Wood on its short and long wire samples with digital calipers on the base metal. Length measurements were not possible for Element to perform as the tensile testing required a reduction in cross section area to fit in the testing equipment. For any future analysis considering the material loss of tensile armour wires, it was recommended to use the minimum values as a conservative assumption

Table 3 Tensile Armour Wire Measurement (width, mm)

Specimen ID	Laboratory	Wood Sample (Short)	Wood Sample (Long)
A1 – 875	3.912	4.053	4.027
A2 – 876	3.886	4.053	4.073
A3 – 877	3.937	4.023	4.043

Specimen ID	Laboratory	Wood Sample (Short)	Wood Sample (Long)
A4 – 878	3.785	3.913	4.043
B1 – 879	3.861	4.057	4.100
B2 – 880	3.912	4.073	4.097
B3 – 881	3.912	4.047	4.047
B4 – 882	3.937	4.067	4.060
C1 – 883	3.886	4.037	4.040
C2 – 884	3.861	4.037	4.017
C3 – 885	3.912	4.047	4.043
C4 – 886	3.912	4.038	4.070
Average	3.893	**4.037**	**4.055**
Minimum	3.785	**3.913**	**4.017**
Nominal Width = 4.00mm			

Table 4 Tensile Armour Wire Measurement (length, mm)

Specimen ID	Wood Sample (Short)	Wood Sample (Long)
A1 – 875	11.990	11.900
A2 – 876	11.937	11.967
A3 – 877	11.930	11.937
A4 – 878	11.840	11.970
B1 – 879	11.967	11.917
B2 – 880	11.977	12.037
B3 – 881	11.937	11.967
B4 – 882	11.957	11.957

Specimen ID	Wood Sample (Short)	Wood Sample (Long)
C1 – 883	11.893	11.893
C2 – 884	11.943	11.973
C3 – 885	11.893	11.920
C4 – 886	11.960	11.960
Average	11.935	11.950
Minimum	11.840	11.893
Nominal Width = 12.000mm		

FATIGUE ANALYSIS METHODOLOGY

The wave fatigue assessment consisted of three main steps.

1. Global analysis to extract tension and curvature time histories at locations of interest
2. Local analysis to compute the corresponding stress time histories experienced at each armour layer
3. Post-process for fatigue damages at each armour layer (rainflow counting method) and identify the governing fatigue damage.

The global analysis uses a nonlinear finite element (FE) model of the water injection riser complete with the FPSO information, clamp details, and boundary conditions. The riser is modelled as fixed to the FPSO buoy through the bend stiffener interface and anchored to the seabed through two clamps. In the FE model, the riser is represented using beam column elements in Flexcom [3] to capture bending and axial responses due to environmental loads and vessel induced motion.

The details of the bend stiffener and dimensions are shown in Figure 7. All dimensions shown in this figure are in millimeters. The material properties of the bend stiffener were assumed based on in-house data. Young's modulus of the Bend Stiffener was assumed to be 67,500,000 N/m2. A linear material model of the bend stiffener is conservatively assumed in this study.

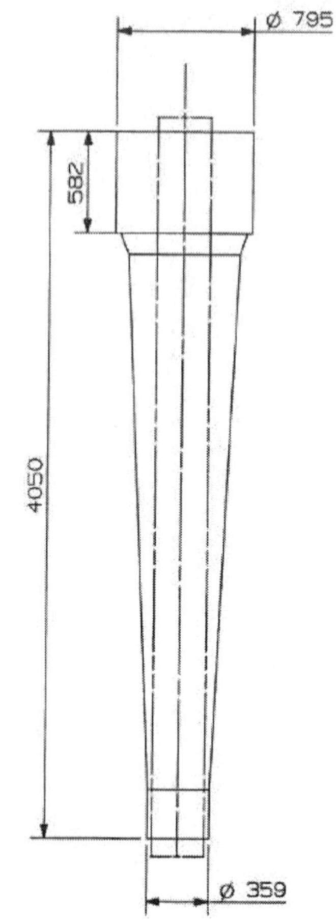

Figure 7 – Bend Stiffener Dimensions

The long-term distribution of sea states is characterized by the fatigue seastate scatter diagram provided by the Client in the riser design basis. The wave scattered diagram was condensed into 132 bins with unique probability distribution. The wave bins were implemented using the JONSWAP spectral wave approach. Wave simulations were run for 3,600 seconds (1 hour), with a 200 second ramp time and a maximum time step of 0.2 second.

For each dynamic simulation, time histories of tension and curvature are extracted at fatigue hot spots of interest and served as input to the Layercom [2] local analysis.

Three fatigue hot spots were selected from the global analysis for further local analysis of the armour layers. These hot spots are at the tip (1), middle (2), and near the base (3) of the stiffener as they are in the areas of high tension and curvature. Local analyses of the tensile armour layers using both the design thickness versus the material testing thickness from the dissection were conducted. The fatigue lives between the two were then compared.

The local analysis of the riser is carried out using Layercom. The local analysis accounts for pipe cross-section details and computes wire stresses across the different armour layers. Tension and curvature time histories at each hot spot are extracted from the global analysis and analyzed in Layercom.

Two tensile armour thickness were considered; 4mm (design value) and 3.785mm (the minimum thickness in the samples taken from the dissection).

The fatigue damage post-processing is also computed within Layercom. The fatigue damage calculations are based on the S-N curve approach using wire stresses on the different armour layers.

The S-N curve used in this study is the DNV B1 [1] curve in seawater. Service experience has shown that a pipe annulus may contain environments that are aggressive with respect to steel and could affect fatigue damage significantly. The B1 curve is more conservative than the curves used by the manufacturer's in-house analysis, as the

B1 data is not primarily based on the proprietary material and wire size.

RESULTS

The unfactored fatigue lives calculated using the S-N approach for the three locations of interest are presented in Table 5. The reduction in fatigue lives due to the loss of metal on the armour wires was computed to be approximately 12%.

It should be noted that these lives are not representative of the actual fatigue life of the flexible riser, as the S-N curve was chosen for comparative purposes.

Table 5 Unfactored Fatigue Life reduction due to Armour Wire Metal Loss (yrs.)

Hot Spot	Min. Fatigue Life – Design wire thickness	Min. Fatigue Life – Metal loss	Percent (%) Reduction
Tip of Bend Stiffener (1)	36,252	32,230	11.1%
Mid Bend Stiffener (2)	14,028	12,320	12.1%
Near Base of Stiffener (3)	13,468	11,822	12.2%

It is also noted that the fatigue lives were decreasing from the tip of the bend stiffener to the base of the bend stiffener. This could be caused by several reasons:

1. The bend stiffener material model used in the analysis was too soft for the water injection riser, causing more bending to travel to the base of the stiffener.
2. The fatigue performance at the bend stiffness is tension dominated. This could be due to high heave motion from the FPSO.
3. The bend stiffener length could be too short for the combination of tension and bending experienced.

Sensitivity studies were then conducted to further examine this finding.

A sensitivity analysis was first conducted to determine the effect of tension only response on the flexible fatigue hot spots. Hot spot (3) – near the base of the stiffener is chosen for this sensitivity. In the Layercom local analysis, only tension time histories were considered, and all curvatures set to zero to determine the corresponding fatigue life.

Table 6 shows the computed fatigue life near the base of the stiffness for both scenarios, tension with or without curvatures. The large increase in fatigue life indicates that tension had negligible effect on the fatigue performance at this location. Bending is the governing component for fatigue performance at this location.

Table 6 Effect of Tension on Fatigue Life

Hot Spot	Min. Fatigue Life – Tension + Curvatures	Min. Fatigue Life – Tension Only
Near Base of Stiffener (3)	13,468	1.51E11

A second sensitivity examines the curvature responses at different locations across the bend stiffener. Figure 8 shows the curvature time histories at three locations (tip, mid, and base) on the bend stiffener for one of the wave bins

It is noted that curvature increases from the tip to the middle of the stiffener and stays relatively constant toward the base of the stiffener. This suggests that the stiffness value used for the bend stiffener modelling could be low or that the bend stiffener length was insufficient, causing bending loads to travel to the stiffener interface and potentially contribute to the tensile armour wires to "birdcage" – resulting in the outer sheath to expand and start to interact with the steelwork under repeated loading.

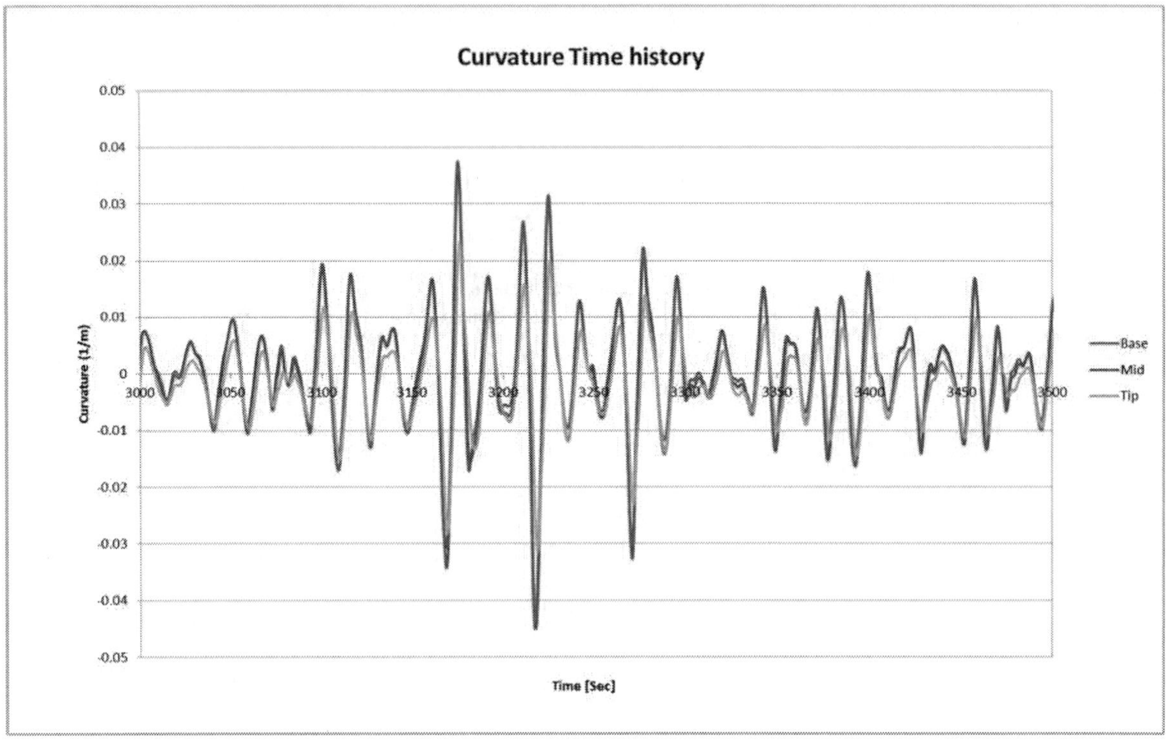

Figure 8 Curvature time histories at 3 locations along the bend stiffener (500 sec)

CONCLUSIONS

In general, the materials used in the wire and outer sheath construction are compliant with the properties stated on the OEM Technical Data Sheet and flexible pipes from Wood project experience.

Apart from the breach location the external protective sheath was coherent along the length of the pipe, though showed multiple minor superficial scratches and shallow indentations however. There were no visible through thickness damages to the external sheaths at these scratches and indentation areas.

The loading dynamics of the riser (as a whole) may have altered the performance of the system near the bend stiffener interface. Specifically, the length of the bend stiffener may not have been optimized for the riser dynamics, and may have cause excessive bending at the interface point. This localized bending may have contributed the tensile armour wires to "birdcage", causing the outer sheath to expand and ultimately start to rub against the steelwork and leading to a breach.

It is impossible to estimate accurately the time scale for the wire corrosion sequence, although the corrosion condition of the observed wires suggests that these may have failed within the last several years. This would be consistent with the timeline supplied from the client.

The armour wire thickness due to the external sheath breach was reduced to 3.785mm from 4.0mm based on the dissection findings in the first phase of this project. The fatigue assessment in this phase showed there is a 12% reduction in fatigue life due to this metal loss when considering a generic, yet conservative S-N curve.

Due to the proximity of the breach location to the bend stiffener, fatigue in this area is highly sensitive to the bend stiffener material performance. A linear material curve was assumed for this study and it was shown that the bend stiffener performance may be insufficient to absorb the bending at the stiffener interface as curvatures were observed to increase from the tip of the bend stiffener to the middle and base of the stiffener. Furthermore, the fatigue damage was also found to increase from the tip of the bend stiffener to the base.

The loading dynamics of the riser (as a whole) may have altered the performance of the system near the bend stiffener interface. Specifically, the length of the bend stiffener and material performance may not have been optimized for the riser dynamics and fatigue and may have cause excessive bending at the interface point. This localized bending may have contributed the tensile armour wires to "birdcage", causing the outer sheath to expand and ultimately start to rub against the steelwork and leading to a breach.

It is possible to further compare the actual fatigue damage accumulation with the predicted damage accumulation pending the availability of historical hindcast sea state data during the operational history. This information can be applied to model the global motions in the FEA program. This would likely result in a less conservative damage but require more computational effort.

S-N curves specific to the water injection riser (provided by the flexible manufacturer) could be used if available, however, based on extensive flexible pipe analysis experience, the SN curves that are publicly available (DNV, JIP's, etc.) are quite accurate, albeit more conservative.

Wood recommended performing a sensitivity analysis to quantify the effect of bend stiffener material and geometry on fatigue performance at the bend stiffener interface. A nonlinear material model of the decommissioned stiffener can be implemented to further assess the suitability of the original bend stiffener design.

A desktop risk assessment study was also recommended to be performed. This study would include a historical and industry-wide review for similar failures, in order to build a probability case for the risk assessment profile and to identify credible failure modes for current and future risers on the same facility.

Results of fatigue analysis and risk assessment can then be used to build a complete Fitness and Risk Profile, including:

- Life Extension Actions.
- A Risk Assessment & Fitness Report.
- A Life Extension Scenario Profile that considers all known risks and uncertainties.

ACKNOWLEDGEMENTS

The authors gratefully acknowledge the support from the Element Labs in Houston, Texas and C-Core in St. John's, Newfoundland, Canada.

REFERENCES

1. DNVGL. (2014). RP-C203: Fatigue design of offshore steel structures
2. MCS Kenny. (2013, June). Layercom Version 1.2 User Manual. Galway, Ireland.
3. Wood Group. (2018). Flexcom. Retrieved from Flexcom: https://www.woodgroup.com/what-we-do/view-by-products-and-services/digital-solutions/flexcom

SNAME
THE INTERNATIONAL COMMUNITY
FOR MARITIME AND OCEAN PROFESSIONALS

FUELING STRATEGY FOR OFFSHORE FACILITIES – A STUDY

ANUP SRIVASTAVA
CHEVRON ASIA PACIFIC SHIPPING PTE. LTD.

Proceedings of the 24th Offshore Symposium, February 20th 2019, Houston, Texas
Texas Section of the Society of Naval Architects and Marine Engineers

ABSTRACT

It is always a challenge to select the "right fuel" for offshore facilities because of the unavailability of well-defined criteria.

*A study was conducted to develop the **fueling strategy** for power generation of offshore facilities and for operation of various offshore vessels. Examples of offshore facilities include central processing plant (CPP), floating production storage and offloading (FPSO) and floating storage and offloading (FSOs). Marine operations include marine support vessels include crew boats, research boats, anchor handling tugs, supply vessels and flotels. These vessels are involved during the hook-up and commissioning (HU&C) and during normal and special offshore operation. Marine support vessels generally bunker fuel from main storage from offshore facilities.*

While suggesting the "fuel type", several points were considered with emphasis on the Original Equipment Maker (OEM) requirements, applicable local & international emission regulation and supply reliability.

*The "**Fueling Strategy**" study evaluated the pros and cons of single versus multi-fuel policy, identified the applicable fuel grade, and identified fuel storage location requirements with associated capacity. The final recommendations also carefully considered long term environmental impact of the suggested fuel.*

The study has also evaluated the economic order quantities (EOQ), while keeping capital stewardship and reliability paramount.

The study provides clear guidance and takes a holistic approach to define the "right fuel" for offshore facilities.

Keywords: fuel storage, offshore structures

OBJECTIVE

The objective is to recommend the appropriate fuel, storage capacity, "economic order quantity" (EOQ) to deliver technical solutions for power generation and support vessels for offshore facilities marine operations. Intent is to strike an appropriate balance between technical, business and operability - reliability criteria.

The recommendations from this study can help offshore projects in selecting the most appropriate fuel and ensuring that the selected fuel meets business objectives. Eventually it will improve project performance.

The study considers an offshore project consisting of a CPP, FSO and several support vessels located in Gulf of Thailand. The FSO provides fuel storage for CPP, FSO and support vessels.

While developing the ***fueling strategy***, the following points were considered -

- marine equipment OEM requirements
- applicable local and international emission regulations
- reliability of supply
- company's directives and long-term economics
- financial calculations basis long term economics and defining EOQ

Scope

- Select the right fuels for electrical power generation for CPP, FSO and main propulsion along with electrical power generations for support vessels
- Evaluate opportunities to see which fuel can be used in all engines involved in above described applications, international and local environment regulations taking into consideration with a futurist's view
- Recommend the right fuel considering reliable supply now and, in the future
- Suggest the EOQ for the recommended fuels

Not In Scope

- Consideration of residual fuels, biofuels and natural gas is not evaluated

FUEL SELECTION CRITERIA

The fuel section criteria depend on relevant Emission Regulations, OEM's recommendations, Company's directives, Reliability in terms of product availability one time and every time, quality etc. and long-term Economics.

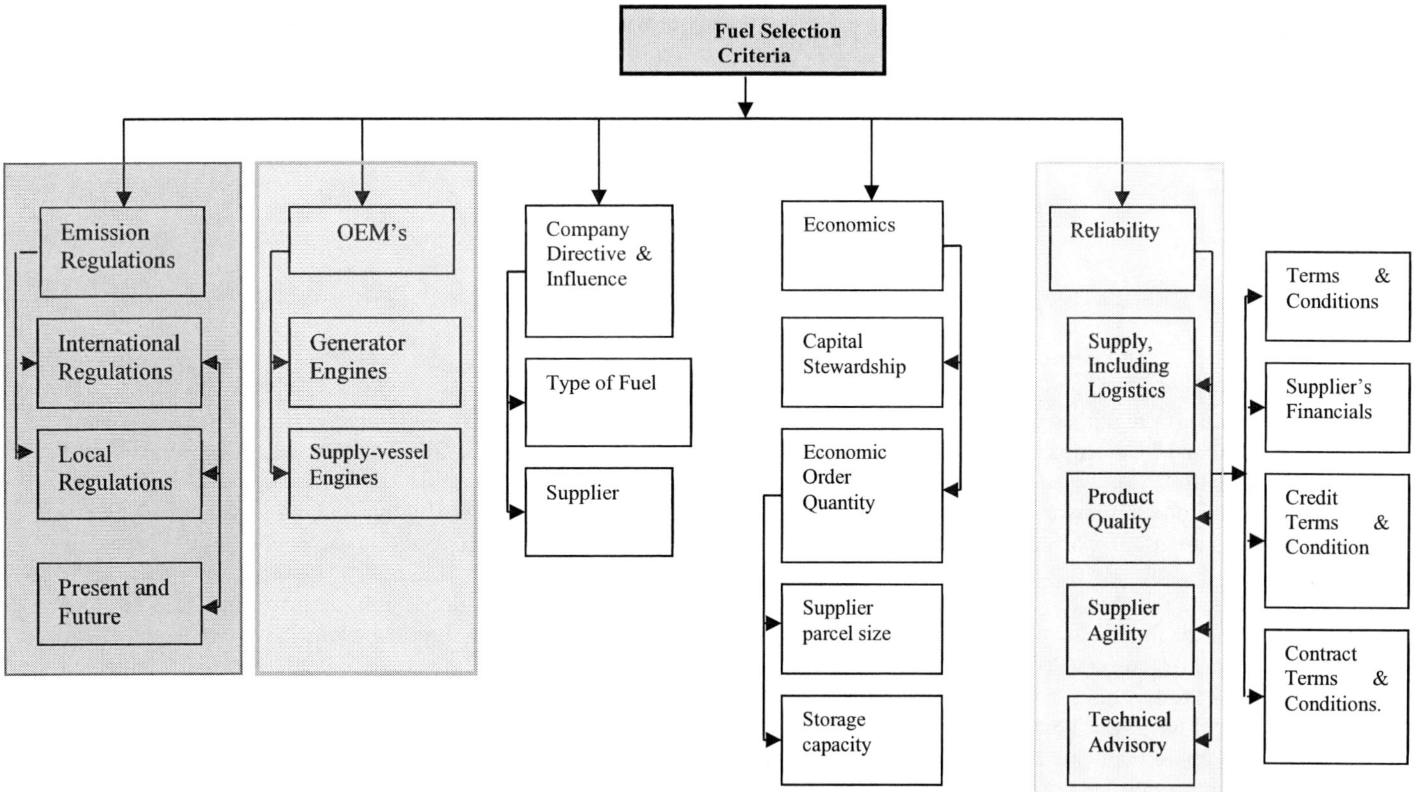

Figure 1: Fuel Selection Concept

The most important selection criteria are:

- Relevant Emission Regulations,
- OEM's Requirements,
- Supply Reliability & Quality.

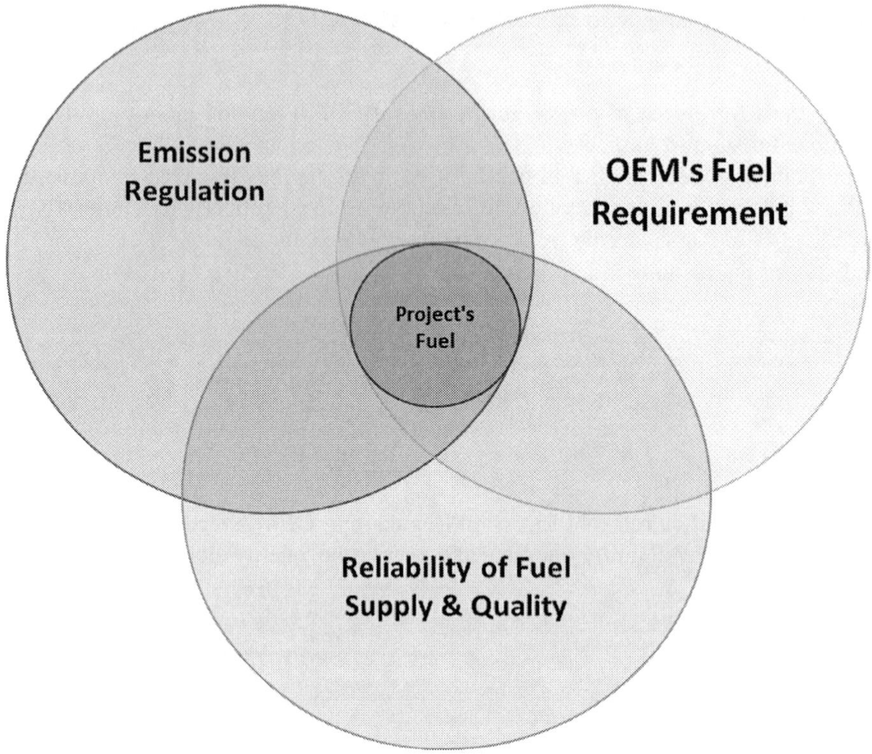

Figure 2 Project's Fuel

EMISSION REGULATION

In 1973, the International Maritime Organization (IMO) agreed to a series of measures to minimize pollution from marine and shipping operations (MARPOL Convention).

The IMO regulates emissions of ODS (Ozone Depleting Substances), NOx, SOx and VOCS (Volatile Organic Compounds) from ships through MARPOL 73/78 Annex VI. **MARPOL defines offshore installations and drilling rigs as ships.**

The Convention was modified in 1997 to address SOx emissions from ships by introducing a global cap on the sulphur content of marine fuel oil and an additional limit in specific waters, referred to as emission control areas, or ECAs.

The sulphur limit in marine fuels has been reduced over time for both the global limit and within the ECAs. The next step change will reduce the global cap on sulphur content for general shipping from 3.50%wt to 0.50%wt beginning 1st January 2020. This is commonly referred to as MARPOL 2020. It is going to be a major game changer that how the vessels are fueled, and strategies used for sulphur compliance. In this paper the discussion is limited to liquid fuel for power generation excluding main propulsion except on support vessels.

The main elements of marine generated pollution are SOx and NOx. Sulphur oxides emission is mainly due to the presence and burning of sulphur compounds in the fuel. The smoke containing sulphur oxides emitted by the combustion of marine fuel will further oxidise and in the presence of catalyst like NO2, will form sulphuric acid which is a major cause of acid rain.

It has also been recognized that the emission of SOx contributes to formation of secondary inorganic aerosol gases, fine particles which are harmful to human health. As of 1st Jan 2020, a maximum Sulphur limit will apply for the Shipping industry as per below graph (for ships not having abatement technology installed).

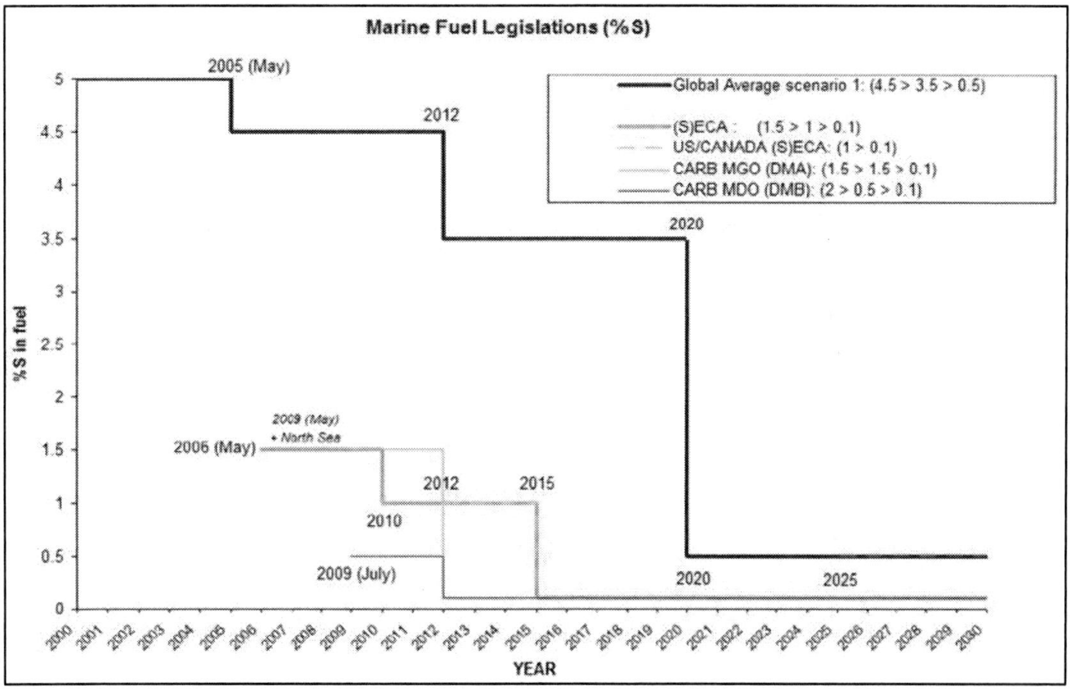

Figure 3: Marine Fuel Legislation (%S) w.r.t. time

The picture below shows the S% limitation on global map.

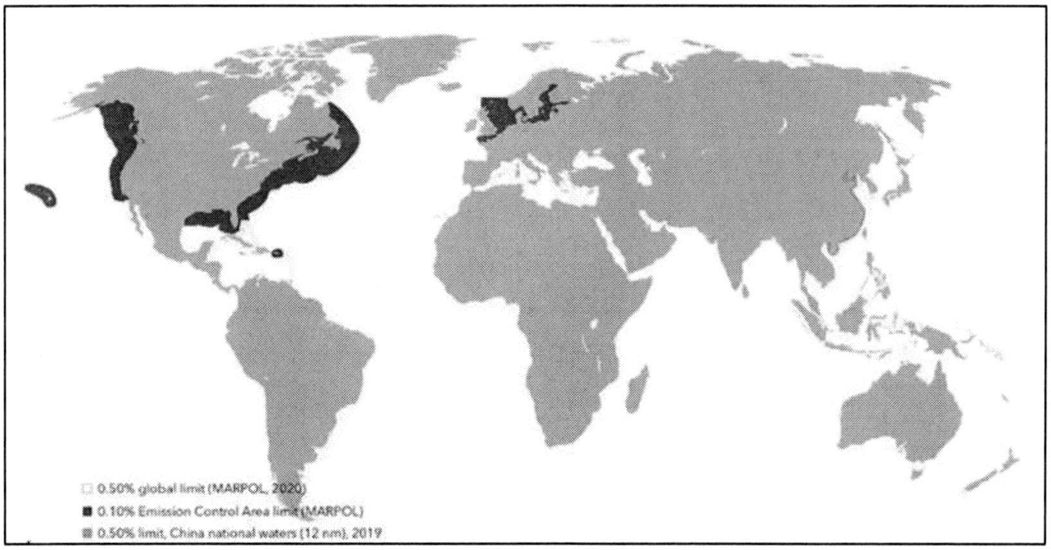

Figure 4: Geographical location for Marine Fuel Legislation (%S)

In accordance with the MARPOL Annex VI Reg.13, NOx emission regulations are applied to diesel engines with a power output of more than 130kW and installed on a ship whose keels are laid on or after 2000, except the engines used solely for emergency such as the emergency generator engine.

There are three levels of NOx emission limits called Tier I, Tier II and Tier III. They are shown in Figure 5.

- Tier I regulation was applied to diesel engines installed on a ship whose keel is laid on or after 2000.
- Tier II was applied to the diesel engines installed on a ship whose keel is laid on or after 1 January 2011.
- Tier III limit will be applied to the diesel engines installed on a ship whose keel is laid on or after 1st January 2016. Tier III emission limits are applied in case where the ship is operating in existing Emission Control Area for NOx.

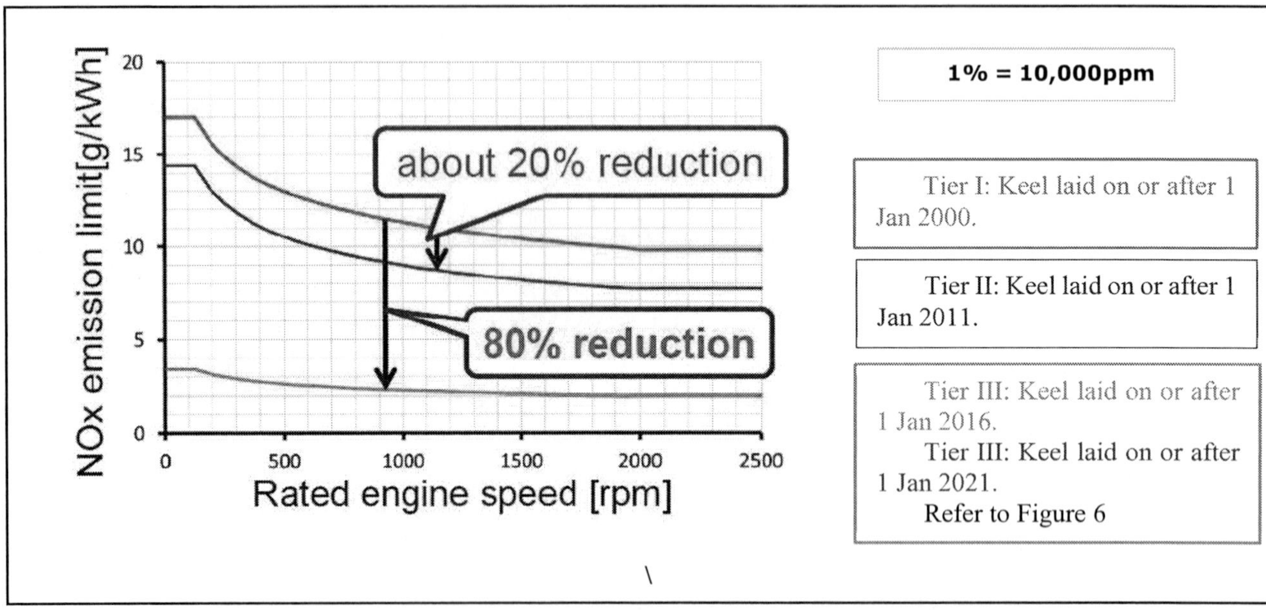

Figure 5: Marine Fuel Legislation (%NOx)

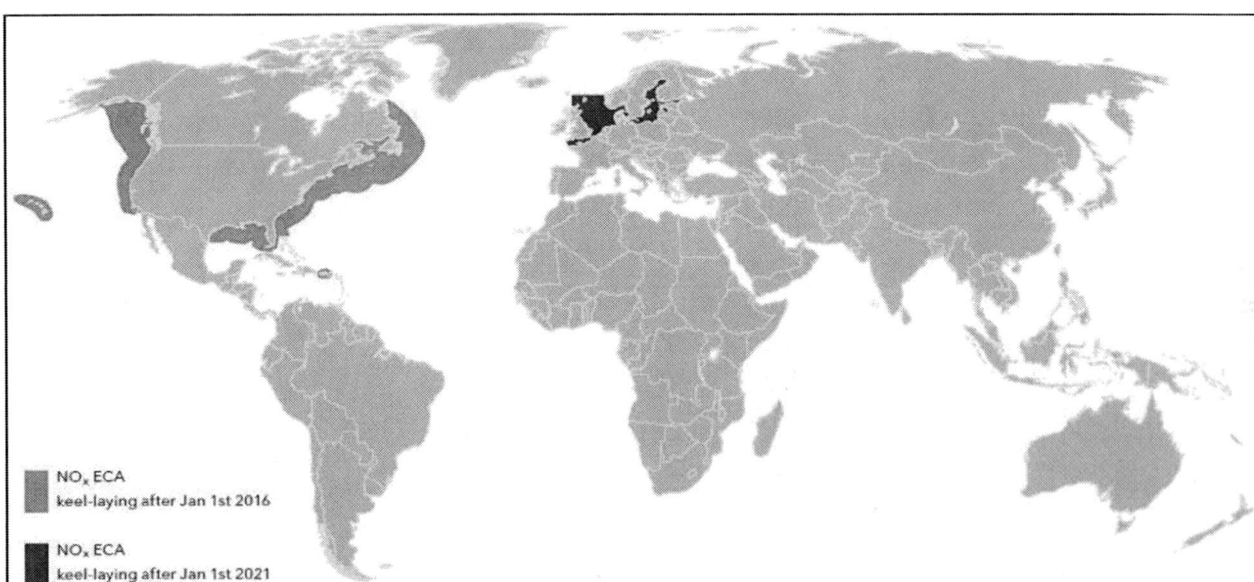

Figure 6: Geographical location for Marine Fuel Legislation (%NOx)

There are various standards used to spec out the fuel when it is purchased such as BS 2869:2017, EN 590 or ISO 8217:2017. However, the most common used standard globally is ISO 8217. Hence, in this study the ISO 8217 standard is used hereafter.

As per ISO 8217:2017, the **upper** Sulphur content limit for these fuel grades is as follows:
Marine Gas Oil (DMA): 1.00% m/m* (10,000ppm)
Marine Diesel Oil (DMB): 1.50% m/m* (15,000ppm)
*m/m: mass to mass percent

When there are no local regulations in place, and as long as the global cap is not exceeded, a higher sulphur distillate product can be burnt, but the availability of these products will be a challenge as they are made towards the commercial specifications.

However, for long term environment care, it is recommended to choose the more stringent fuel specification which does not only supersede the current international and local regulations, but also keeps a futuristic view.

Test	Unit	Test Method			Limits	Grade						
		ASTM	IP	ISO		DMX	DMA	DFA	DMZ	DFZ	DMB	DFB
Viscosity at 40°C	mm²/s (cSt)	D445	71	3104*	max. min.	5.500 1.400	6.000 2.000	6.000 2.000	6.000 3.000	6.000 3.000	11.00 2.000	11.00 2.000
Density at 15°C	kg/m³	D1298	160	3675* or 12185	max.	-	890.0	890.0	890.0	890.0	900.0	900.0
Cetane index	-	4737	380	4264*	min.	45	40	40	40	40	35	35
Sulphur	mass %	D4294	336	8754; 14596	max.	1.00	1.00	1.00	1.00	1.00	1.50	1.50
Flash point	°C	D93	34	2719*	min.	43	60	60	60	60	60	60
Hydrogen sulphide	mg/kg	-	570*	-	max.	2.00	2.00	2.00	2.00	2.00	2.00	2.00
Acid number	mg KOH/g	D664*	-	-	max.	0.5	0.5	0.5	0.5	0.5	0.5	0.5
Total sediment hot filtration	mass %	-	375	10307-1*	max.	-	-	-	-	-	0.10	0.10
Oxidation stability	g/m³	-	388	12205*	max.	25	25	25	25	25	25†	25†
FAME	volume %	D7963*	579		max.	-	-	7.0	-	7.0	-	7.0
Carbon residue, micro	mass %	D4530	398	10370*	max.	0.30	0.30	0.30	0.30	0.30	0.30	0.30
Cloud point	°C	-	219	3015	max.	-16	-	-	-	-	-	-
Pour point												
Winter quality	°C	D97	15	3016*	max.	-	-6	-6	-6	-6	0	0
Summer quality	°C	D97	15	3016*	max.	-	0	0	0	0	6	6
Appearance	–	–	–	–	-	Clear and bright					–	–
Water	volume %	D95	74	3733*	max.	-	-	-	-	-	0.30	0.30
Ash	mass %	D482	4	6245*	max.	0.01	0.01	0.01	0.01	0.01	0.01	0.01
Lubricity‡	µm	-	-	12156-1*	max.	520	520	520	520	520	520†	520†

*Test method reference.
†If the sample is not clear and bright, the test cannot be undertaken and the limit shall not apply.
‡This requirement is applicable to fuels with a sulphur content below 0.05%.
Additional reporting requirements for CFPP and Cloud Point for Winter Grades of DMA, DFA, DMZ.

Table 1: ISO8217:2012 for distillate products

OEM (Original Engine Makers) Requirements

Diesel Fuels – Marine Gas Oil

Diesel fuel quality is an important factor for satisfactory engine life and performance. Fuels must provide adequate combustion without producing excess material which can harm the engine. Additionally, fuel selection involves economic and environmental considerations. The availability of certain grades of diesel fuels may be cost prohibitive or inappropriate for various applications.

In this section, an effort is made to identify appropriate engines / turbines used for marine applications for our example.

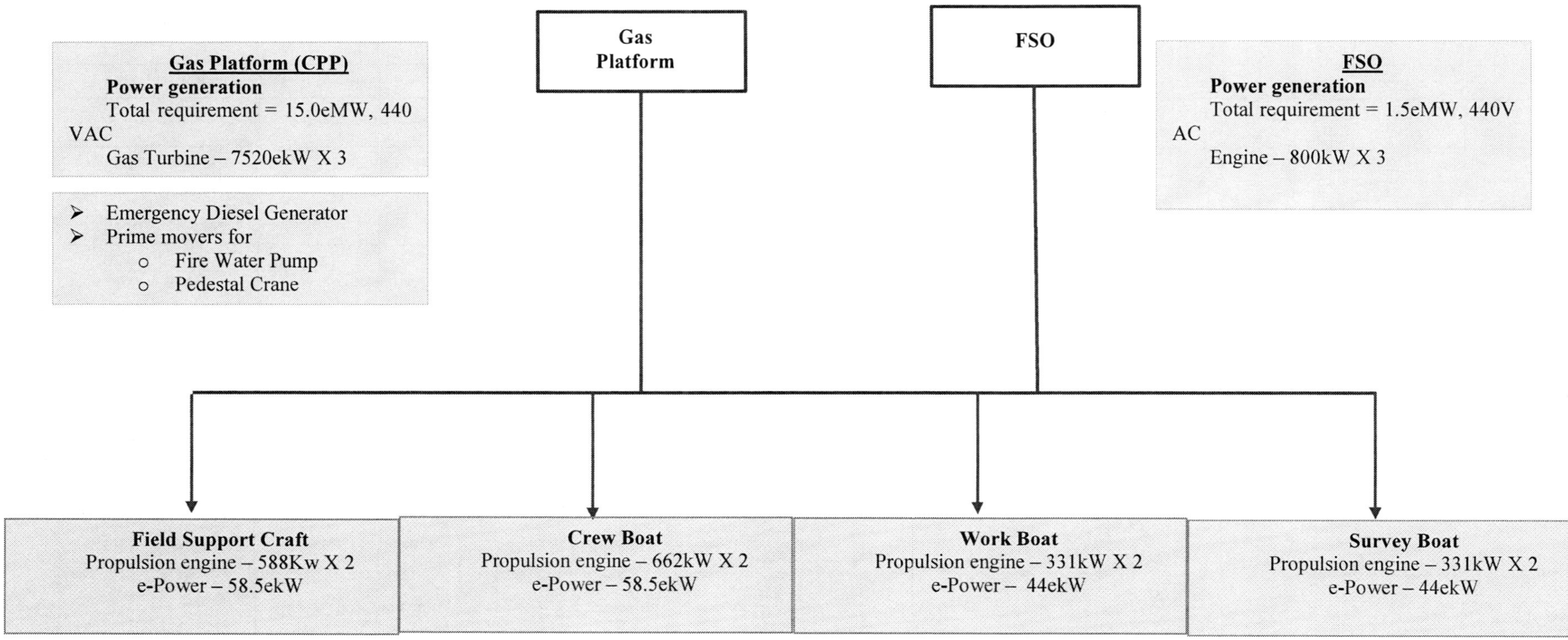

Figure 7: Power prime movers location(s), their purpose, estimated power.

All applications identified in above flow chart are captured in tabular form in the table below. In discussions with various OEMs, the type of fuel which can or can't be used in their application is marked. Various OEMs were consulted. The basis for the information provided by them, fuel standard mentioned are captured. Due to confidentiality issues the engine make and models are not disclosed.

Location	Purpose	Equipment Type	DMX	DMA	DMB
Gas Platform	Power generation	Gas Turbine	Y	Y	N
(CPP)	Emergency Diesel Generator	Reciprocating Engine - Diesel	N	Y	N
	Fire Water Pump engine	Reciprocating Engine - Diesel	Y	Y	N
	Pedestal Crane engine	Reciprocating Engine - Diesel	Y	Y	N
FSO	Power generation	Reciprocating Engine - Diesel	N	Y	Y
	Fire Water Pump engine	Reciprocating Engine - Diesel	N	Y	Y
Field Support Craft	Main propulsion	Reciprocating Engine - Diesel	Y	Y	N
	Power generation	Reciprocating Engine - Diesel	Y	Y	N
Crew Boat	Main propulsion	Reciprocating Engine - Diesel	Y	Y	N
	Power generation	Reciprocating Engine - Diesel	Y	Y	N
Work Boat	Main propulsion	Reciprocating Engine - Diesel	Y	Y	N
	Power generation	Reciprocating Engine - Diesel	Y	Y	N
Survey Boat	Main propulsion	Reciprocating Engine - Diesel	Y	Y	N
	Power generation	Reciprocating Engine - Diesel	Y	Y	N

Table 2: Selection of Prime Mover(s) power and OEM recommendations for fuel type. ***Assumptions have been made while deciding the Engine specification.***

Electrical power generation for gas platforms is considering dual fuel turbines. Purified natural field gas will be used as fuel for these turbines which will improve capital stewardship and fuel reliability. However, during the routine shutdowns liquid distillate fuel will be used. OEM guidelines regarding the selection of liquid fuel for turbine vs suggested fuel for other marine operations and power generation purpose (diesel engines) were evaluated.

During the routine shutdown period (<10days / year), only one turbine has been considered to run on part load basis, ISO 8217:2017 DMA max. 0.10% S (1000ppm) fuel. It was confirmed by the OEM that this operation will not have any adverse impacts on its performance.

As per table 1, it was confirmed that only DMA grade can be used on all engines and turbines without any compromise.

RELIABILITY

Reliability is simply defined as the ability of a company to consistently supply an acceptable product at the required time.

And quality can be defined as the product must be consistently delivered that satisfies the customer's needs. Quality should be guaranteed through documented production practices and defined process controls.

The reliability of a supplier in terms of supply of product is of utmost importance to successfully manage the marine operations.

This is especially true as there are large variations in the fuel characteristics around the world.

It is recommended to choose a company for fuel supply which has reliable track record in terms of supply and maintaining quality and to consider also how the company has resolved disputes if any.

Generally big suppliers offer a variety of fuel and have excellent supply chain setup.

FUEL CONSUMPTION

In The **Original Engine Makers (OEM) Requirements** section, OEM's recommendations were reviewed, and it was concluded that all the selected engines can use ***ISO 8217:2017 DMA grade of Marine Fuel***.

Moving forward with the project, fuel consumption on monthly basis is calculated.

The calculations will assist to design the storage facility at FSO for all off-shore support vessels, Gas Platform.

Please refer to Table 3 for fuel oil consumption on a monthly basis for all selected engines.

Location	Application	Running Hrs per day	SFOC (g/kW-hr)	No. of engines running in parallel	Fuel Consumption per day (MT)	Fuel Consumption per month* (MT)
Gas Platform (CPP)	Power Gen	24	-	1	27.5	For 10days/yr. on liquid fuel basis = **23** / month
	Emergency Diesel Generator	24	-	1	9.08	Neglected
	Fire Water Pump	24		1	4.14	For 10days per year basis = **3.45** / month
	Pedestal Crane	Will run every day for limited time		2	2.9	**87**
FSO	Power Gen	24	185	2	7.104	213
Field Support Craft	Main propulsion	12	222	2	3.13	**94**
	Power Gen	24	-	1	0.45	14
Crew Boat	Main propulsion	12	218	2	3.47	**104**
	Power Gen	24	-	1	0.45	14
Work Boat	Main propulsion	12	226	2	1.39	42
	Power generation	24	-	1	0.37	**11**
Survey Boat	Main propulsion	12	226	2	1.39 / 432	42
	Power generation	24	-	1	0.37 / 115	**11**
					Total	**660 MT**

Table 3: Monthly Fuel Oil Consumption for Marine operations

FUEL TYPE SELECTION AND ITS STORAGE

The fuel type selection and storage strategy has been designed as per **Figure 9**.

There are multiple choices available to choose the fuel. Some of the fuels need heating but most times offshore facilities do not have boilers.

Also, many time offshore facilities use the field gas or condensate and hence, the requirement of additional fuels in limited to run the power generation during the turnaround time or maintain the standby power source and maintain emergencies, e.g. fire water pump, emergency generator, lifeboat engines, etc.

In addition, based on the evaluation of logistics, supply chain and ease of storage, it is recommended to use a single fuel type which does not need heating.

For this case study, the basis of the OEM recommendation for a one fuel strategy (ISO 8217:2017, DMA) is considered prudent.

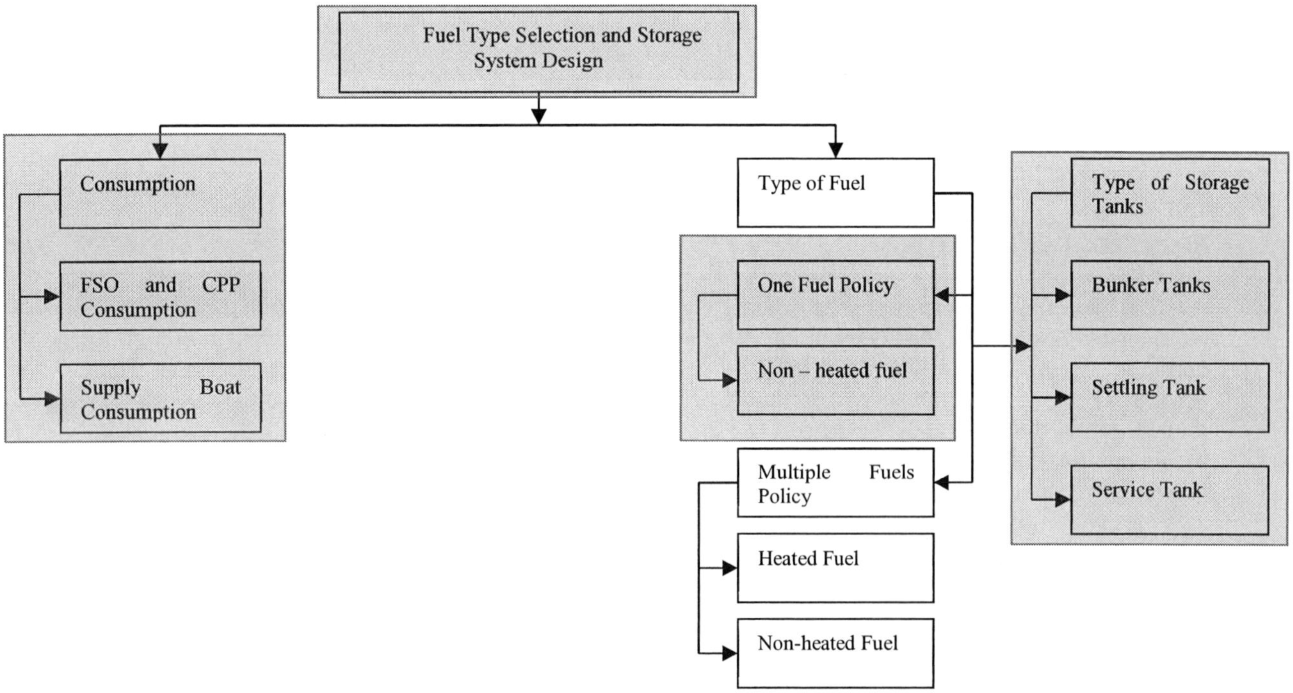

Figure 8: Fuel Storage Strategy

STANDARD PRACTICE FOR STORAGE

Bulk fuel is usually stored in large main storage tanks and transferred to smaller service tanks near engines by electric motor-driven pumps.

If service tanks are not necessary, the main fuel tank must provide a ready fuel supply to the engine-mounted transfer pump.

Another concept of tank in tank has been used recently and introduced on some of the offshore facilities. The arrangement is such that service tanks are located inside the main storage tanks with a weir which overflows fuel to the service tank to main storage tank. This design provides an opportunity to save space.

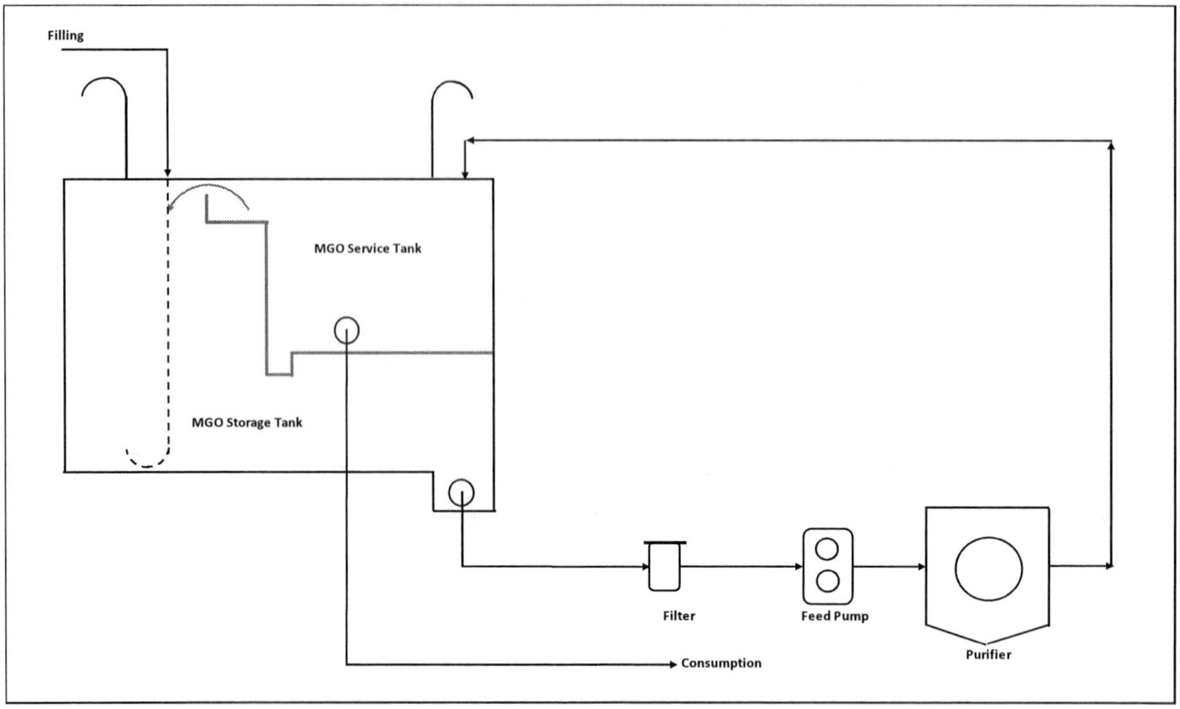

Figure 9 Fuel Storage Option, Service Tank in Side Main Storage Tank

MAIN FUEL TANK SIZING BASIS

The main fuel tank represents the primary fuel reservoir for all applications and must have adequate capacity for the intended application. For determining the tank size, as a rule of thumb, hourly consumption at 100% load is multiplied by number of hours and days between refills. Fuel consumption rates are shown in Table 3 for the considered engines. Additionally, 10% should be added to the total (5% for expansion at the top of the tank and 5% for sediment settlements at the bottom).

In order to maintain reliability and ease of fuel oil storage tank periodic maintenance, it has been advised to make two tanks of equal capacity with an interconnection.

This will allow the oil to be maintained at same level, which will improve the vessels stability as well as allow carrying out of maintenance on any of the tank as and when required.

In addition, it will also allow the safe operations in case one of the tanks is contaminated or has any technical issue.

INVENTORY SIZE AND ECONOMICS

Inventory

The most important question asked is why hold inventory of fuel oil at all?

It is inevitable due to following reasons:

1. Due to an offshore project's location it is not viable to supply the fuel by pipeline or delivery barges on a daily basis.
2. Unexpected change in demand.
3. Rate of consumption may vary and fuel delivery time changes.
4. Even if no uncertainty, the longer the delivery time the higher the storage inventory. It also makes economic sense to stem the bunkers at a discount for large size shipments and capitalize on standard size full load barges.

Economic Order Quantity

Calculation of Economic Order Quantity

One aim of the supply chain manager is to provide operational convenience with minimum possible investment in inventory. Calculation of the right EOQ is essential to minimise the capital tied up in inventory and at the same time mitigating the risk of stock out.

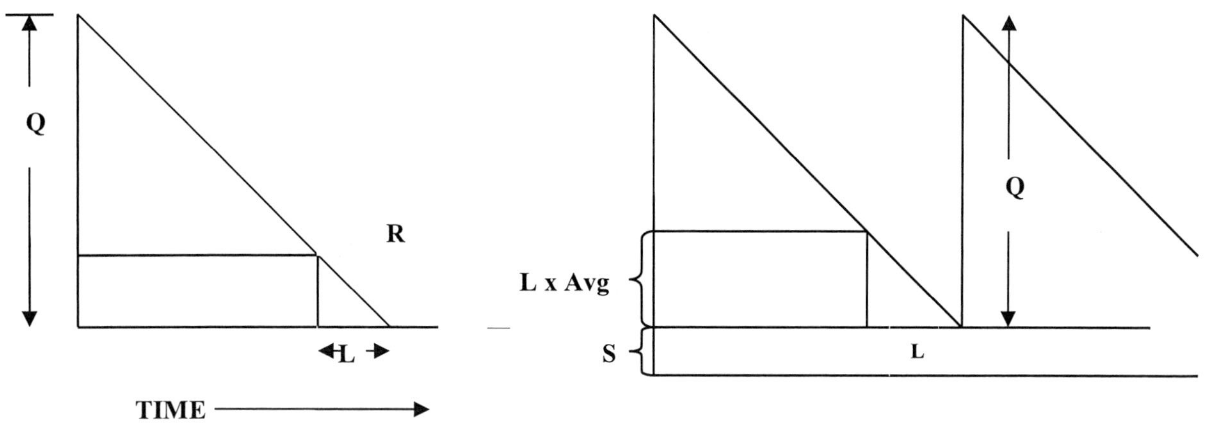

CASE STUDY

Monthly consumption (Q) = 696MT
Lead time for supply (L) = 5 working day.
Safety Stock (S) = Minimum 2 weeks fuel = 348MT
Standard barge supply is 5,000MT and 10,000MT

Order Notification = 15days + 5 Working days = 20 days prior to finish of fuels.
Re-ordering point (R) = 7.5 to 8 months depending on consumption.

It is advised to take precautions for working days. In case there are local holidays, it is recommended to place the order giving sufficient time to supplier.

The selection of bunker stem quantity (parcel size) and ordering frequency will have to be the right evaluation of capital tied up in inventory to risk being out of fuel stock.

From a technical perspective, consideration is to be given to keeping the inventory (Marine Gas Oil) for long time in the main tanks and running the risk for microbiological growth and other related threats.

Water in the gasoil is absolutely unwanted as it will cause damage to the engine and its components. Water appears in the fuel because of condensation, handling and environmental conditions related to the humidity at project's location. Water in the fuel will be more prevalent in humid climates and may cause the microbiological growth.

Taking a balanced view of capital Stewardship and risk against keeping a big inventory for a long time, it is suggested to use the 5,000MT barge and stem the fuel oil every 7.5 to 8.0 months.

EOQ = 5000MT

This will provide the economies of scale as full load barge capacity will be used.

On reviewing Table 2, it is suggested to select the DMA 0.10%S Standard Fuel for the case study basis following benefits:

a) Meeting and exceeding international and local emission regulations now and in future. Depending upon the location of the project, the current most stringent regulations are 0.1%S (1,000ppm).
b) Single fuel for project (marine operations).
c) Saving in engine maintenance and routine maintenance (more lube oil running hrs. between renewals).
d) Optimum fuel S% for engine. In general fuel does not require the use of lubricity additives at this level (lubricity becomes more important when S level is < 500 ppm).
e) Minimal threat of liner lacquering etc. for IC engines.
f) Proactive behavior in terms of emission control on part of Company, **Green image for Company.**
g) Future emission regulations after 2020 are not yet known, however it is likely that Global S% limit will drop. Our recommendation of 0.10% (1,000ppm) S is considered as safe for future for foreseeable time.
h) ISO 8217:2017 DMA grades can be transported and stored in bulk. It is a requirement that this product can be carried only in clean barge tanks. This also ensures the product integrity.

CONCLUSION

Considering the OEM regime (Turbine* and Engines) for selecting the right fuel it was concluded that ISO 8217:2017 DMA (0.10%S, 1,000ppm) grade is accepted with negligible (only on marginal drop in Lube oil renewal hrs.) compromise on offshore vessels with high speed engines.

On considering the present and future local and international emission regulations it has been found that ISO 8217:2017 DMA grade up to 1.0% (10,000ppm) sulphur is accepted. Taking the company's directive on emission regulations, proactive and futuristic approach, it has been suggested to use 0.10% (1,000ppm) sulphur fuel.

Reliable supply was confirmed for select fuel ISO ***8217:2017 DMA 0.10%S*** (1,000ppm) (commonly known as Marine Gas Oil) through a reliable supplier for project's location.

The grade is available as a standard supply and hence will **not** attract any special premium.

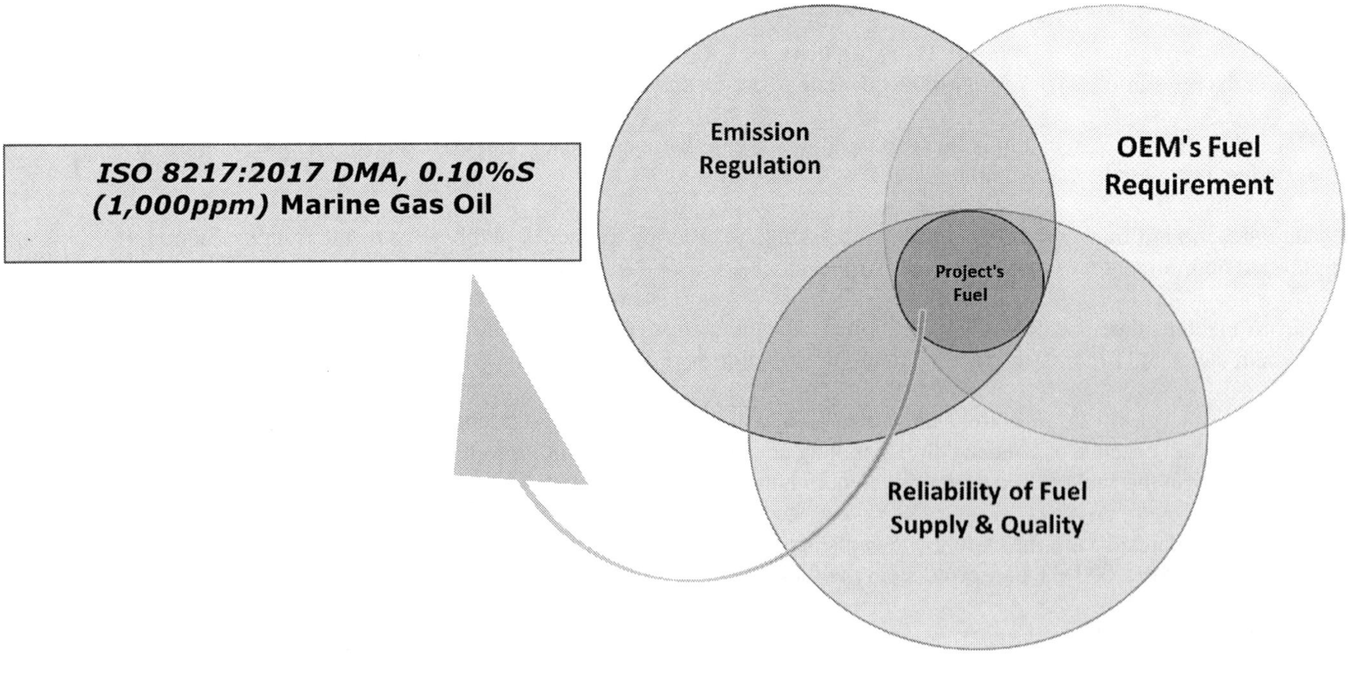

References

1. Caterpillar, Applications and installation guide – Diesel Fuels & Diesel Fuel systems. 2005
2. International Standards Organization (ISO), Fuel Standard for marine distillate fuels. ISO 8217:2017
3. DNV GL. Global Sulphur Cap 2020 Limits

INVESTIGATING OPPORTUNITIES AND CHALLENGES FOR UNMANNED FLOATING PRODUCTION FACILITIES

ANDERS MARTIN MOE
PROJECT DEVELOPMENT MANAGER, WOOD

GAURAV SINGHAL
FLOATING SYSTEMS TEAM LEAD, WOOD

ASTRID BARROS
FLOATING PRODUCTION UNIT MANAGER, WOODSIDE ENERGY LTD

PAUL ERB
SENIOR CONSULTANT, WOOD

Proceedings of the 24th Offshore Symposium, February 20th 2019, Houston, Texas
Texas Section of the Society of Naval Architects and Marine Engineers

ABSTRACT

For many offshore developments, there is a drive to minimize the required manning and enabling remote operations to reduce personnel exposure and to reduce the overall cost (lighter topsides, simpler hull, less maintenance). This paper considers semisubmersible production facilities and investigates the opportunities and challenges of an unmanned hull from an overall field development perspective.

The unmanned hull poses challenges related to typical practice for the frequency of inspections inside the hull and the closed compartments within the hull. In order to achieve a fully unmanned hull, inspections should be minimized, and ideally completely avoided during the intended service life, if possible. The paper highlights key design features required to achieve a hull design with minimal inspections.

Opportunities that will be discussed include, minimizing the hull equipment and system requirements (e.g. ballasting operations), implementing sealed compartments, and enhancing the fatigue life of the hull structure to avoid inspection requirements during the lifetime of the unit.

The paper discusses several design challenges associated with an unmanned hull. These include addressing Class and Flag requirements, (i.e. Class and shelf state regulations that may be a barrier for unmanned operations), risk of entering the unmanned hull, damage tolerance, and damage response capability.

Keywords: SNAME, Offshore, Symposium, unmanned, minimum manning, semisubmersible

INTRODUCTION

Required personnel onboard a production facility is an important parameter for risk exposure (onboard and in crew change), cost of quarters/utilities, and operating expense. This paper focuses on the opportunities available to minimize manpower requirements for the hull of a floating production system. Both normal operation and campaign inspection/maintenance are considered. The latter aspect is a particular challenge in terms of possible interference with normal production operations and maintenance.

The ideal concept of an "unmanned" hull is, admittedly, presented as a stretch objective for conceptual design work. This concept is interpreted as a hull that requires no entry for ordinary marine operations (such as ballast/trim), no routine walk-around inspection of internal equipment/integrity, and (ideally) no inspection of any hull compartments for the life of the field. It can be appreciated that this set of design objectives is very challenging, but similar precedents to some of the features proposed can be found for Spar-type hulls and in a few semisub applications.

A semisubmersible hull form is considered in the discussion. This hull form highlights both benefits and challenges associated with an unmanned hull approach. Semisubmersible production units have evolved from drilling unit design and operating practices. Key design features include manned pump rooms, regular watch "rounds" in the hull, reasonably frequent large load shifts and draft changes (storms, moves, etc.). Semisubmersible hull forms also have stringent structural requirements, particularly for fatigue, and they are sensitive to trim from load shifting or damage. Thus, it can be appreciated that consideration of an "unmanned" concept for this hull form will serve to emphasize merits and limitations of the "unmanned" approach.

This paper provides an overview of the envisioned features of an "unmanned" production semisubmersible hull, based on recent conceptual design work for a realistic field development scenario. Subsequent sections describe changes and impacts on hull systems, stability, hull structures/fatigue, hull/topside interfaces and also a brief discussion of potential simplification of on vessel mooring equipment. Each section addresses strengths and weaknesses of the "unmanned" approach, and provides quantified examples, where possible, based on recent design work. Conclusions and recommendations are then presented for consideration.

OVERVIEW OF THE "UNMANNED" HULL CONCEPT

The fundamental basis of the "Unmanned Hull" concept is to eliminate any requirement for regular entry to perform operational tasks, maintain/inspect equipment, or inspect structure or compartments/coatings within the hull. This change in philosophy for the hull design permits significant simplification in hull systems. Note, however, the hull systems must still provide or permit installation, normal operations, and potential damage response. Structure and stability characteristics must also be shown to meet site metocean and lifespan requirements consistent with Class intent.

The concept of an unmanned hull with no entry for regular maintenance or inspection is outside the normal requirements of Class Rules. However, Class may permit development of new design philosophies and approaches through demonstration of technical equivalence and supported by enhanced requirements and assurance during construction and installation. Flag and local regulatory requirements will need to be addressed concurrent with Class evaluation.

Key design basis considerations for development of the "unmanned" hull concept include:

- Identification of a firm and reasonable field life, applicable for corrosion, fatigue and equipment lifetime considerations.
- Minimizing drivers for inspection and maintenance in hull
- Operational procedures that keep load shifts to a minimum during day-to-day activities in order to limit trim ballast requirements.
- Compartment subdivision and ballast response means to enable appropriate robustness against damage.

HULL MARINE SYSTEMS

This section describes hull systems simplifications that are proposed to support implementation of the "unmanned" hull concept. Simplifications are discussed for each system configuration, together with Class & Regulatory risks & opportunities, and followed by discussions of reliability/operability and robustness/safety. In one design example, overall savings of approximately 50% in weight and cost of both bulks and equipment is estimated. However, the apparent magnitude of this savings must be tempered with the reality that hull equipment and bulks are minimal in weight and cost compared to topside values. Therefore, weight/cost savings in equipment and bulks alone are unlikely to justify the shift to an unmanned philosophy. Reconfiguration and simplification of four key marine systems is discussed below, beginning with the ballast system:

- Permanent Ballast System - As noted above, a key prerequisite for the simplified ballast system is to minimize requirements for active ballasting and thus enable design of most of the ballast tanks for no inspections. These permanent ballast tanks will, in other words, be sealed off for the life time of the unit, after mooring hookup and installation of initial risers/umbilicals, with no permanent in-place means of emptying or transferring the ballast from the tanks. The ballast system can thus be arranged with two sets of ballast tanks: 1) As a minimum, four large permanent pontoon ballast tanks that will be completely sealed off (both filling and vent pipes) with no permanently installed means for emptying or transferring ballast. Submerged pump filling/emptying & vent arrangements for the permanent ballast tanks (intended for use during mooring and initial riser installation) will be kept onboard, if needed. Vents should be large enough to act as caissons for submerged pumps, once unsealed. It is proposed to fully coat the permanent ballast tanks, but to further minimize corrosion, the permanent ballast tanks may be filled with fresh water and corrosion inhibitor. Figure 1, a schematic tank plan, indicates a suitable position for permanent ballast tanks.

- Trim Ballast System – As a minimum, four trim tanks with permanent active means for adjusting the ballast in the tanks, can be located in the column node. These tanks can be used for corrections of operational changes in weights and center of gravity and to right the unit after damage. Trim tank adjustments may be made using permanently installed submerged deep well pumps in internal caissons in the hull or ejectors. Ejectors are simple and reliable, but deepwell pumps can be withdrawn for maintenance without hull or tank entry. For robustness, it is proposed to install a second internal caisson in the tank as well as a secondary retrievable pump with reel for hose and cable that can be lowered into the second caisson in case of malfunction of the permanent pump. Both caissons will further have the possibility to serve multiple functions, i.e. acting as tank vent and filling pipe in addition as pump caisson. For safety, the vent on top of the caisson shall be fitted with tank vent ball float valves. The tanks shall be fully coated and equipped with sacrificial anodes for corrosion protection. Figure 1, a schematic tank plan, indicates a suitable position for trim ballast tanks.

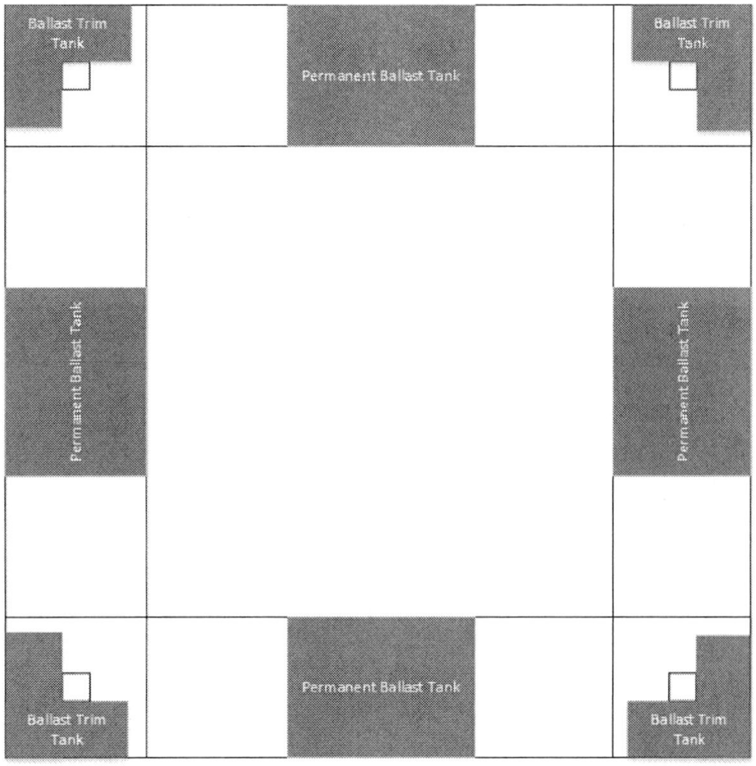

Figure 1 – Schematic Tank Plan

It is recognized that the trim tanks described above will require regular inspection, due to their active use with associated air and water exchange. However, the logistics of inspection for only four tanks will represent a substantial improvement in comparison to a complete inspection protocol. In order to facilitate inspection of the trim tanks transfer of ballast from the trim tank to a nearby void, designated as temporary ballast tank, may be arranged with permanent piping. Filling and vent pipes to the void designated as temporary ballast tank will be sealed off at top of column during normal operation.

From a Class perspective, the recommended simplified ballast system design consists of proven and mature equipment and solutions, all arranged and easily accessible from the column top. Further, the recommended solution provides means for managing changes in weight and center of gravity as well as providing means for righting after damage. Pumping redundancy may be addressed through the spare retrievable pump or pumps described. Suitable capacity for damage recovery must be addressed in detailed design through appropriate sizing of trim tanks and/or compartment subdivision.

Reliability of the recommended system configuration is considered similar or better than a conventional system, since there are far fewer valves in the system and all active equipment and valves are readily accessible from the top of column. Operability is improved as the operating sequence is simplified – i.e. select the proper quadrant, start the pump and open the discharge valve. It is thus apparent that the simplified ballast system will present fewer opportunities for human error and unintentional cross-flooding of tanks. Additionally, piping and cable penetrations across water tight bulkheads are significantly reduced.

With implementation of the ballast system described above, substantial simplification of additional hull systems is enabled, as discussed below:

- Simplified Bilge System – As a result of minimizing ballast equipment and piping in hull, there are no leak sources that require the presence of a permanent bilge system in hull. However, it is proposed to include a void drain system for leak detection from individual voids. Void drains can be led to a bunded area, i.e. a bilge well in the bottom of the access shaft, with level detection. A caisson sized to accommodate the retrievable ballast pump can be arranged to these bilge well areas in the bottom of the access shaft. Draining of the bilge wells can therefore be done by using the retrievable ballast pumps from the column top.

From a Class perspective, it can be noted that similar minimum bilge systems have been approved in the past, however with the possibility to use e.g. de-ballast pumps for emergency bilge. Since there are no internal leakage sources and no equipment in hull that is likely to require emergency bilging, it may be possible to demonstrate that the system meets likely needs.

Operationally, discharging accumulated water from hull requires lowering the retrievable pump into the bilge caisson in the access shaft which is somewhat more cumbersome than for an industry standard permanent system. Fewer valves and equipment results in reduced maintenance requirements and simplifies operability once the pump is in position. Furthermore, the expected infrequent requirement of bilging mitigates the additional effort required for lowering the pump.

Compared with a traditional bilge system the system configuration described above lacks the emergency bilge functionality. However, there is very little equipment in hull, no pump room and as already mentioned no internal leakage sources, piping penetrations across water tight bulkheads are significantly reduced, and the hull is normally unmanned.

Means of detecting accumulation of water in the bottom of the access shaft and means of draining without entering the hull is provided by means of retrievable pump. The safety and robustness of the proposed system configuration may therefore be considered similar to that of a traditional bilge system.

- Simplified Ventilation System - With no permanent equipment in hull and an unmanned hull there is no longer a need for a permanent ventilation system. Thus, it is possible to omit conventional air handling units on the column top and a majority of the ductwork in hull. The access shaft for access to hull should be sealed off in normal operation. The minimum HVAC system configuration will then consist solely of a supply pipe from top of the column to the bottom of the access shaft, and an extract connection on top of the column, both with blind flanges on top of the column for connection to temporary fans for supply and extract prior to any required entry. Air supply to other voids in hull, if needed for entry, will be by temporary fans and ducts.

From Class perspective, the proposed system does not represent a safety risk to the facility in normal operation. Safe personnel entry will be managed and safeguarded by dedicated entry procedures. Operationally, less equipment results in lower required maintenance hours. More effort is required for hull and hull compartment entry to install temporary fans and ventilate the access shaft prior to entry, but regular hull entry is not required. Less permanently installed ducting reduces piping penetrations between compartments. Thus, overall facility robustness and safety are considered to be similar or slightly better compared to an industry standard configuration.

- Tank Vent System – Active vents will only be provided for tanks considered in active use; in effect this means only the ballast trim tanks, as described. The caissons for the ballast pumps will also serve as vents for the ballast trim tanks. The vents for the permanent ballast tanks will be sealed off on the top of the permanent ballast tank caissons. No vents will be installed for the voids in pontoon, column and node. Sealing the compartments not in active use will ensure that the atmosphere inside will become oxygen scavenged and thereby inhibit corrosion and coating breakdown, with a view toward eliminating inspection.

It will be key for Class and regulatory approval of this approach to demonstrate that the structural integrity of the compartments without vents is maintained even in the event of worst-case flooding. Potential improvements in overall safety and robustness are achieved due to the number of down flooding points being reduced, and further, there is less risk of vent being a source of hazardous vapors (H_2S).

- Piping Material - Material selection for piping and equipment would not necessarily change with the unmanned hull configuration compared with an industry standard configuration. However, as the quantity of piping in hull and also the number of watertight penetrations is greatly reduced the cost impact of using more exotic materials is also reduced. The objective of reducing the requirement for maintenance and inspection may also in certain cases drive the application of exotic materials, such as titanium.

ADDITIONAL HULL SYSTEMS

The changes proposed for the primary hull marine systems described above allows simplification of additional hull systems, as discussed further below:

- Hull Electrical Distribution & Lighting - Less equipment in hull results in less electrical distribution equipment needed compared with a typical hull systems design. Elimination of column electrical rooms and co-location of the hull electrical distribution gear with the topsides electrical equipment should be considered. It is proposed to install permanent lights in the access shaft that are positively isolated during normal operation but that can be switched on when required for entry. Only the following electrical equipment, all located on the column tops, will be needed: Motor starters for the ballast pump, the retrievable ballast pump and winch and the valve HPU pump motors, and a feeder for lights in the access shaft and utility stations on the column tops.

- Instrumentation - Hull instrumentation will also be reduced. It is proposed to remove all leakage detectors in voids in hull and rely on void drains and level detection in the bilge wells, as discussed above. The active ballast tanks will have absolute pressure level transmitters fitted. There will be no remotely operated hydraulic watertight doors in hull, and the number of actuated valves is significantly reduced. Further, there is no need for fire, smoke or gas detectors permanently installed in hull. On this basis, the required number of I/Os and the complexity of the control system is also significantly reduced. It is proposed to evaluate co-locating the hull instrumentation and automation equipment with the topside's equipment if possible, thereby avoiding an instrumentation room within columns.

- Hydraulic Systems - As the number of actuated valves are significantly reduced compared to a traditional hull systems arrangement, from (order of magnitude) somewhat above 100 actuated valves to about 30, the valve control HPU and solenoid rack can also be significantly reduced. Consideration should be given to reducing the number of HPUs and solenoid racks.

- Hull Utility Systems – A permanent firewater system likely will not be required in hull, given that there is little or no flammable material or inventory, and the hull is generally unmanned. Temporary fire detection and fire-fighting equipment may be set up if entry is required.

- Hull Storage – Storage of consumables required for production is sometimes implemented within FPU hulls. Typical requirements may include diesel fuel supply, methanol, or other large-volume well chemicals. The hull systems approach described above still leaves the possibility of arranging such storage in stand-alone tanks within voids at the column top. These tanks can be accessed with deepwell pumps from the column top, and local leak and fire detection/suppression means may be provided.

- Topsides Mooring Equipment – An additional opportunity for savings in hull weight, complexity, and maintenance/manpower is the application of in-line mooring tensioners for large FPU moorings. In-line tensioners have been delivered for moderate size moorings up to the range of 130mm chain. Delivery is in progress for a 140mm application. Upsizing is still needed for most FPU applications, where 150-170mm is likely to be required. Although the in-line tensioners are costly, a modest net savings may be realized, given that chain jack or winching systems on the column top are not required. More importantly, the manning, maintenance, and column top arrangement issues associated with chain jacks and chain lockers can be eliminated.

HULL STABILITY CONSIDERATIONS

This section provides an overview of stability and damage response considerations for the "unmanned hull" configuration described. Stability assessment requires comprehensive analysis of a well-defined hull design, with many variations of damage location and wind approach. However, key differences and concerns associated with the proposed configuration are identified. Initial work with a specific configuration has been promising.

- Intact Stability – From a Class perspective, Intact stability of a design that follows the "unmanned" approach outlined should be little changed from the base case of a conventional hull. In fact, there is likely to be some improvement resulting from the reduction in vents, which are typical downflood locations, and the opportunity to cluster vents closer to column centers.

- Operational Stability – Operational stability focuses on the capacity of the unit to adjust trim to keep a suitably level deck for operations (e.g. helicopter landing/takeoff, process equipment operability, crew comfort, etc.). Trim adjustment may be needed due to movement of substantial loads on deck or bulks within tanks. Trimming may also be required to respond to strong/shifting winds and/or currents that create heeling moment on the unit. Initial work with a configuration for a typical semisub FPU of about 80,000 Tonne displacement has shown good operational stability, though further trim tank size optimization is needed. Figure 2 illustrates some of the evaluation performed in consideration of wind moments. Substantial wind strengths up to 70 knots were found to be within trim tank capacities, while retaining damage response capability.

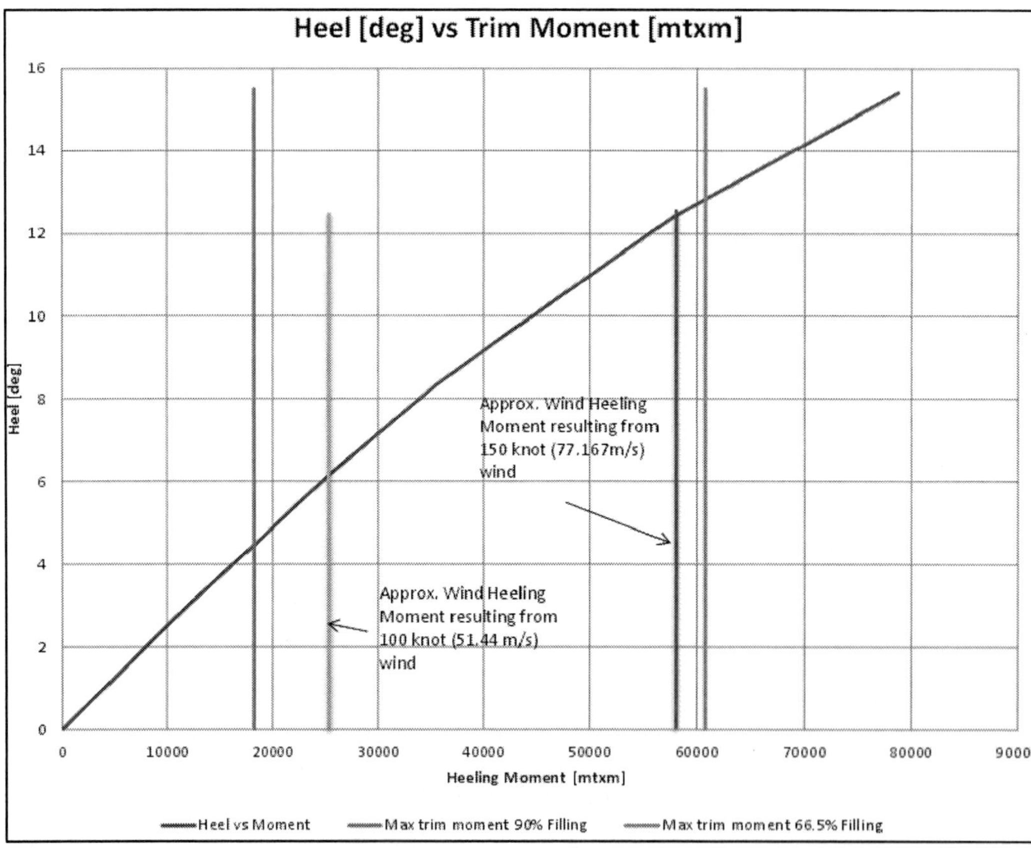

Figure 2 Heel vs. Trim Moment

- Damage Stability – The key difference in the "unmanned" hull design approach, from a viewpoint of damage stability, is the limited capability for active response to damage. This arises from the limited number and size of the trim tanks in each corner. Conventional Class requirements dictate that the unit be able to pump ballast to re-establish level trim within a few hours of damage. Of course, the operational advantages of the "unmanned" hull point toward a smaller number and size of trim tanks to limit inspection requirements, corrosion concerns, etc. Thus, a design optimization is needed to meet the operational and damage stability needs with a suitable, but not excessive, trim tank volume. Compartment subdivision can also be enhanced to produce an acceptable design that meets the agreed requirements.

An element of assurance that may underpin the optimum "unmanned" hull solution for a given set of requirements would be the robustness inherent in the stability of the design. General elements of design that may contribute to typical FPU stability robustness include: generous metacentric height (a stability measure), further compartmentation beyond Class case requirements (particularly in the waterplane area and also vertical access trunks), permanent mooring "righting" moment (not considered in Rules-case evaluations), and limitations on free surface due to typical tank cross-sections (vs. ship or barge shapes).

These elements of robustness may be evaluated in connection with multi-compartment flooding scenarios beyond Class requirements, and/or evaluation of extreme wind speeds at a location (again, beyond typical Class requirements of 50 knots). Additional robustness contributing factors for the "unmanned" hull configuration would include the very limited sources of possible flooding within the hull, and the availability of deepwell pumps onboard that fit the caissons that are integral to the design, for extreme situations that may require removal of water beyond trim/damage tank volumes.

HULL STRUCTURAL CONSIDERATIONS

This section describes a preliminary assessment of the hull structural design approach required to support implementation of the "unmanned" hull concept. The design approach for the structure concentrates the permanent ballast in locations that are not fatigue sensitive, i.e. the middle of the pontoon, while providing void spaces adjacent to fatigue sensitive areas. This arrangement can be seen in the Tank plan shown in Figure 1. This structural concept is intended not only to minimize or eliminate areas requiring inspection, but also to provide easier access to the most fatigue-critical areas, if necessary.

In recent conceptual work for an "unmanned" hull application, an overall scantling-based steel weight savings of several hundred tonnes was indicated, as a result of the simpler compartment layout. A coarse structural global analysis and fatigue screening was performed for the unmanned structural layout, and this showed encouraging results, but also highlighted a few areas of concern for further focus during more detailed evaluations.

Figure 3 is a screen shot from the modeling work done, which shows the location of the permanent ballast tanks in the mid-pontoon region. A design wave approach was applied for both fatigue (FLS)and ultimate conditions (ULS) analysis. Wave period and height was considered based on DNV methodology (Refs. 1,2, & 3) to capture dynamic response based on the metocean data for the region considered.

A SESAM hydrodynamic and structural model was developed and supplemented with an ANSYS model for the top of column frame connections to the column. Figures 4 and 5 illustrate these models. Topsides mass is not shown but is included in a simplified manner on top of each column. Any stiffness contribution from topside structure was ignored to allow for a conservative hull evaluation.

Preliminary results of the structural analysis show some areas where additional design and refined structural details will be required to address the dynamic von Mises and component stress levels. Results of the fatigue analysis identified areas where maximum dynamic stress levels would indicate fatigue challenges when considering the non-inspectable level of a 10X design fatigue factor (DFF). Generally, these areas were in typical expected locations for a semisubmersible, which helps to demonstrate the conceptual validity of the model. In general, it is anticipated that a DFF of 10 is possible to achieve, with attention to design and also fabrication detail, potentially, even without the use of castings. The most challenging local area based on both the structural analysis as well as previous experience is the top of column to topside connections, which would be the most likely to require castings, if detailing turns out to not be an achievable/acceptable solution.

Figure 3 Model showing Permanent Ballast Tanks

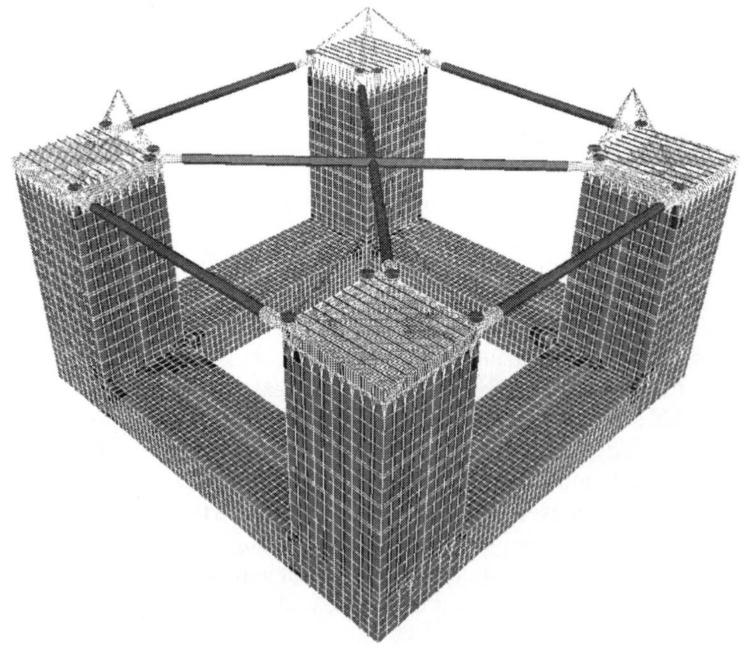

Figure 4 Global Structural Model

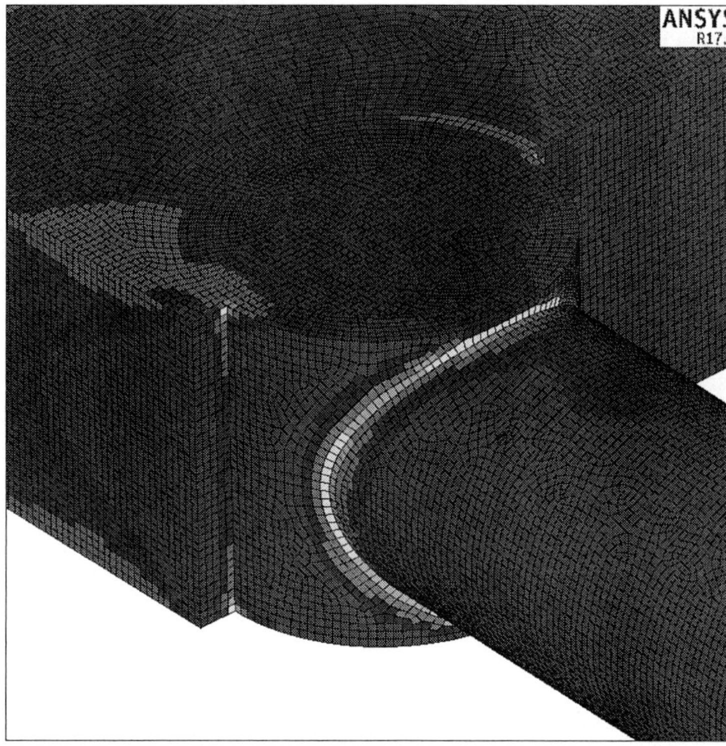

Figure 5 Local Structural Model showing Topsides Support Frame to Hull Connection

HULL/TOPSIDE INTERFACES

In addition to satisfying the marine requirements of the hull itself, the Hull Systems are required to support certain production and utility functions for topside operations. These are unchanged from the base "industry standard" hull approach and include:

- Supply of cooling and fire water to topsides via external hull mounted caissons and deepwell pumps suspended from the deck above the hull. These will be unchanged for the "unmanned" hull.

- Storage of any bulk materials, if required in significant volumes, for topside operations. For the "unmanned" hull, these volumes will have to be arranged in stand-alone tanks in the top voids of the columns, with column top access.

Hull systems will receive power and other utility support from topsides, as follows:

- Normal power
- Emergency power for safety critical consumers, such as ballast pumps, HPU for remote operated valves etc.
- Nitrogen, air, sea & fresh water, and other incidental utilities as required on column top.

The unmanned hull configuration requires only minor changes to the hull and topsides interfaces. There is much less equipment on and in the hull, but the key topsides hull interfaces are little affected. If centralized power distribution and control cabinets are adopted, the interface connection point for these systems will actually increase the complexity of the integration work slightly. Consumption of other utilities typically required in an industry standard hull, as noted above, will be very limited for the unmanned hull configuration and distribution will only be to each of the column tops.

The use of submersible pumps on all four columns will need to be considered in terms of material handling, either by local handling equipment applicable for pulling the pumps or by crane coverage and column top access. However, due to the minimum equipment in hull, there will be less requirement for maintenance operations on or in hull. All hull access will be regarded as confined space entry to be regulated by dedicated procedures, requiring temporary fans, air monitoring, etc. to be set up prior to entry.

CONCLUSIONS AND RECOMMENDATIONS

The vision of an "unmanned" hull that would not require any routine entry for operations, inspection, or regular intervention has been described. Conceptual design work for a specific potential application has shown that simplification of ballast and bilge equipment is key to implement the vision. Work to date has shown that considerable equipment can be removed from the hull through this approach without significantly impacting hull systems performance as well as robustness. Equipment and bulk weight and cost savings of about 50% (compared to an "industry standard" base case) have been estimated. More importantly, the potential to avoid large, disruptive campaign inspection/maintenance with associated peak accommodation demand is potentially very beneficial in terms of safety, uptime, and logistics benefits.

Key elements of equipment simplification include:

- Minimal active ballasting in 4 designated trim tanks with permanent ballast tanks sealed for long-life service without inspection.
- Use of deepwell pumps in caissons from column top for ballast and bilge service to avoid requirement for pump rooms.
- Closed vent system for inactive tanks to minimize corrosion potential through air changes. Temporary ventilation of access shaft via column top fans if entry required.
- Permanent lighting system in place but not normally lit. Minimal hull instruments.

Stability characteristics of the "unmanned" hull concept are discussed, with a particular focus on operational stability (trim control) and damage response capability. The important design balance between compartmentation and trim tank capacity is noted. Elements of stability robustness of semisubmersible FPUs are briefly highlighted.

Preliminary structural configuration and analysis for strength and fatigue is described. Early results from the analysis indicate potential to achieve design fatigue factor (DFF =10) to support the zero-inspection objective in most areas of the hull, without extensive use of castings. Detailed structural analysis needs to be performed to further confirm the preliminary findings.

Hull/topsides interfaces are briefly discussed, and an additional simplification opportunity for the column tops is identified by the elimination of onboard mooring tensioning systems, in favor of in-line tensioners.

ACKNOWLEDGEMENTS

The authors acknowledge the support of management in Woodside Energy Ltd. And Wood PLC for the publication of this paper in order to raise awareness and development interest in the potential advantages and design challenges presented by the "unmanned" hull concept.

REFERENCES

1. **DNV-GL-RP-C205 ENVIRONMENTAL CONDITIONS AND ENVIRONMENTAL LOADS, DNV-GL, AUG. 2017.**

2. **DNV-GL RECOMMENDED PRACTICE, DNV-GL-RP-C103, COLUMN STABILISED UNITS, JULY 2015**

3. **DNV-GL RECOMMENDED PRACTICE, DNV-GL-RP-C203, FATIGUE DESIGN OF OFFSHORE STEEL STRUCTURES, APRIL 2016.**

ROADMAP FOR SELECTING COST EFFICIENT DEEPWATER INVESTMENTS

SHILADITYA BASU
TECHNICAL ADVISOR – MARINE, KBR/GRANHERNE

TIRTHARAJ BHAUMIK
PRINCIPAL TECHNICAL PROFESSIONAL – MARINE, KBR/GVA

Proceedings of the 24th Offshore Symposium, February 20th 2019, Houston, Texas
Texas Section of the Society of Naval Architects and Marine Engineers

ABSTRACT

The precipitous fall of oil prices in the fall of 2014 left most deepwater oil and gas developments at a huge disadvantage. Long lead time to First Oil, constricted supply chain and inflated development objectives put a squeeze on them such that the Final Investment Decisions on deepwater developments came to a screeching halt. Since then deepwater projects have become increasingly cost competitive. While there are multiple deepwater basins under evaluation and development across the globe, the recent discoveries of US shale oil having shorter cycle time, the penetration of electric vehicles in the automobiles market and the emergence of greener alternatives have significantly added competition. Even though crude oil prices have recovered significantly from the low levels observed in 2016, the deepwater industry is on the continuous lookout for strategies to remain cost competitive in the anticipated "lower-for-longer" price environment.

Based on the Industry trends in exploring new unproven horizons in mature basins, multiple development scenarios are developed with a goal to compare CAPEX, OPEX, NPV, schedule effectiveness and cost efficiency. Options are divided in two broad categories of Full Field Development and Phased Field Development. Appropriate economic models are developed for each categories and various components are studied for sensitivity.

Annual cash flows along with revenue and profit margins are estimated and compared for the two options in deepwater US Gulf of Mexico. Risk tornado graphs showing the sensitivity of Net Present Value and payback period to important project parameters e.g. CAPEX, Oil Price and Project Schedule are presented. Conclusions are based on comparing the cash flows of the options through efficient capital planning and schedule management.

In this work, the most promising concepts are identified for their cost-competitiveness in a global and diverse asset portfolio. Key financial parameters are studied with a view to providing a roadmap that can assist in efficient planning of future deepwater field developments. Conventional industry wisdoms for project scoping, financial goals, scheduling, execution philosophy, etc. are challenged to develop a robust screening approach for competitive deepwater investments.

Keywords: Deepwater, Phased Field Development, Economic Models.

INTRODUCTION

The sustained investment in US Gulf of Mexico (USGOM/GOM) offshore deepwater production increase in early 2000s and a surge in U.S shale production since 2009 led to a crude oil supply glut in 2014. The resultant price drop from north of $100/bbl to less than $30/bbl by early 2016 wiped out a significant CAPEX deployment for deepwater projects between 2014 and 2018. Instead of waiting 7-10 years for payback (from Discovery to First Oil) from deepwater discoveries, Operators have started to rationalize their complete portfolio with shorter-cycle developments available in onshore Shale plays. Deepwater projects now need to become more attractive via lower breakeven costs to compete with other projects in the Operator portfolios (Figure 1).

Even though the WTI Crude inched above $70/bbl in 2018, the underlying fear of weakening global demand growth has compelled the Operators to maintain a "lower-for-longer" mindset. Recent analyst presentations by Operators point to a current breakeven level of $50-$55/bbl (Bloomberg, 2018). While deepwater projects have reduced their breakeven cost from $70/bbl down to $42/bbl (Figure 1), further reduction is necessary for future developments (D'Souza, 2018; Bloomberg, 2018). According to recent studies (EIA, 2016), this reduction in the breakeven cost can increase the deepwater project economic feasibility from about 50% to 90%.

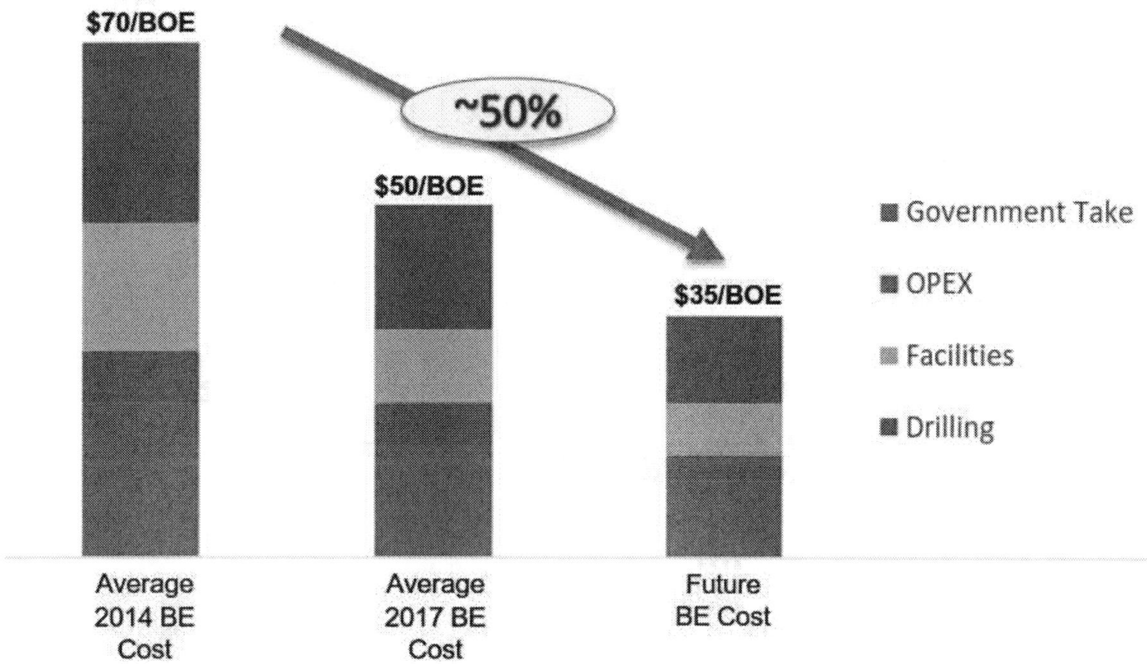

Figure 1 Reduction necessary in deepwater Breakeven costs to kickstart project sanctions

A typical deepwater development with a dedicated floating platform (or a host) often run into billions of dollars with 5-7 year cycle times (preFEED to First Oil). For Supermajors (or Integrated Oil Companies, IOCs), often their internal cash flow is the primary source of financing. For smaller Operators, Independents and others, that luxury doesn't exist. Hence, it becomes essential for deepwater developments to not only remain cost competitive, but also to get to First Oil faster than before. While it will be simplest to develop a deepwater discovery with a suitable converted FPSO and one or two subsea wells tied back to it, most of the value of the reservoir can be lost if proper assessment of full field and phased field developments aren't carried out (Figure 2, D'Souza, 2012).

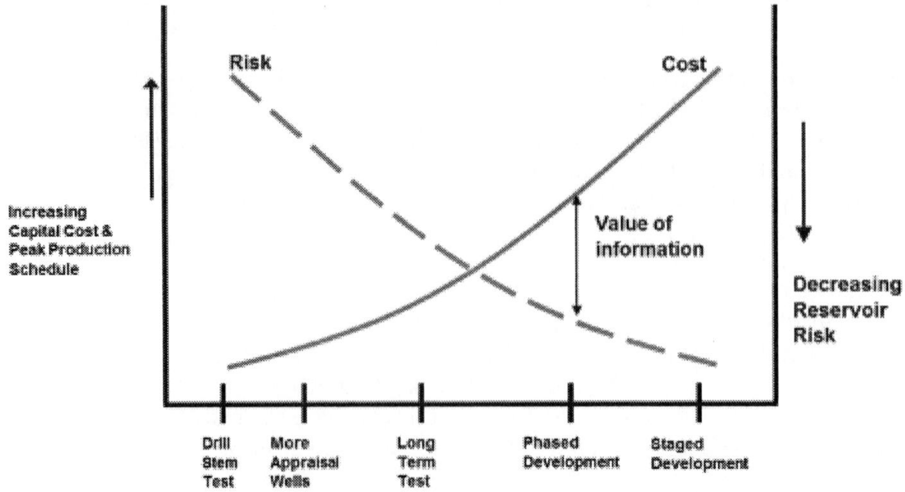

Figure 2 Value of Information, price of decreasing reservoir risk

A recent oil market report published by the International Energy Agency (IEA, 2018) predicts a solid oil demand growth from 2018 to 2023. Though IEA doesn't expect the "Peak Demand" scenario (mentioned in recent presentation by some IOCs) unfolding in this time frame, continuous adaptation of electric and natural gas vehicles will indeed slow down the liquid demand. Mostly driven by demands from India and China, an average growth of 1.2 MMBOPD is expected to lead to a daily demand of 104.7 MMBOPD by 2023. While this is good news for the oil & gas industry as a whole, it is also true that for deepwater, a more competitive business space lies ahead and it remains to be seen how much share of that market the deepwater segment is able to capture.

Offshore industry has used Early Production Systems (EPS) to successfully gather more information from a complex reservoir or from one without an analog. While EPS can gather more reservoir information it also reduces overall NPV by delaying a full field go-now decision. This can be mitigated somewhat by utilizing the EPS as an early producer till the full field platform is ready. In spite of every imaginable simulation, a reservoir can be only known once it starts producing. Thus installing an EPS after initial discovery wells will not only expedite production and generate more information on the reservoir, it can also provide an exit point in case the reservoir does not perform as expected. Petrobras has successfully employed a Pilot Program in most of their prolific basins, where a smaller facility started production at a field while waiting for the much larger full field platform to arrive.

In this paper, a comparative study is made between a Full Field Development (FFD) using a large conventional facility and a Phased Field Development (PFD) using a smaller sized facility, similar to an EPS. The objective is to capture the upside of the reservoir while minimizing the downside of an underperforming reservoir. To that effect, distributing the large CAPEX & DRILLEX required for an FFD is lowered as a PFD case would require much lower initial investment. This approach reduces initial cash flow and allows an exit with lower investment in case of an underperforming reservoir.

METHODOLOGY

The deepwater field in our model is assumed to be situated in US GOM at a water depth between 4000 ft and 7000 ft. While no particular platform type is explicitly modeled or excluded, it is assumed to be a wet tree development where one or multiple drill centers are tied back to the host. Though use of wet trees eliminates the potential of using intervention techniques for higher recovery associated with dry trees, the topsides and host facility designs are simplified and the drilling of initial producers are removed from the critical path.

A single deepwater reservoir has been considered that is expected to have a mid-range or P50 recoverable reserve of approximately 400 MMBO. Based on the history of the US GOM reserve data (BSEE, 2018), this reservoir would be among the top five deepwater reservoirs in the GOM. We also modeled a low-range or P90 recoverable reserve assumed to be 200 MMBO. It is assumed that the reservoir is a uniform "pool" and drained by production wells identical to each other using only natural drive. This allows the model to exclude injection wells, or secondary or tertiary recovery techniques.

Based on EIA data (EIA, March 2016), deepwater well cost in GOM ranged between $120 Million for a Miocene formation well to $225 Million for a Jurassic formation well in 2014-15. Most of the recent prolific discoveries are made in complex sub-salt Miocene, Lower Tertiary or Norphlet formation. Thus it is within reason to assume a "representative" well to cost around $200 Million in 2016. Assuming a 25% reduction in total well cost from 2015 (mainly based on rig rates), individual well cost of $150 Million is used in this analysis. This well cost incorporates both Drilling & Completions costs.

Government Take is modeled as percentage of the total revenue. The Federal Income Tax and Royalty Rates are representative of contemporary US, set at 34% and 16% respectively.

The FFD surface facilities (Hull & Topsides) are designed for 100 MBOPD processing capacity with associated gas and water rates. Corresponding SURF infrastructure is also defined for the full 100 MBOPD. The PFD system is defined assuming a "Design 1 Build 2" approach. The PFD surface facilities and SURF infrastructures are both designed for 50 MBOPD capacity each. The first phase of PFD is installed for the 50 MMBOE production. When the PFD is extended in the second phase, another 50 MBOPD capacity surface facility and SURF infrastructure are added to the first phase facilities. For FFD, the complete SURF infrastructure is installed in one stage. For PFD, the SURF investment is performed in two stages, with each stage consuming 50% CAPEX investment. Similar CAPEX phasing is taken for the surface facilities as well, with 50% investment in the first stage and 50% in the second.

For the purposes of this study, the Hull & Topsides and Surf & Export Lines costs have been estimated to be $4.0/bbl based on previous experience.

It is also assumed that the PFD extension "Go/No-Go" decision is taken 3 years after the first phase First Oil. This allows the initial wells to provide sufficient dynamic reservoir information to determine whether it will be worthwhile to sanction the next phase of PFD. When a "Go-Now" decision is taken, it will take the same construction timeframe as the first phase to achieve the second phase First Oil. This allows us to use the same CAPEX phasing for the facility and SURF infrastructure in both phases.

Both FFD and PFD modeling used the same set of inputs as described here. Particulars of these two approaches are listed in Table 1.

Table 1 Summary of Primary Assumptions

Item	FFD	PFD
Production Capacity	"Full"	"Half", followed by another "Half"
Host	Host_1	Host_2, followed by another Host_2
# of wells	20	
Development wells (#/yr)	2-4-2-2-2-2-2-2-2	2-2-2-2-2-2-2-2-2-2
Reserve	~400 MMBO	
IRR	10%	
NPV Post Tax	Zero	
Break-even Oil Price	Outcome	
Cumulative Cashflow	Outcome	
Payback period	Outcome	

With regard to individual well recovery, a mid-range (P50) cumulative production of ~20 MMBO (area under the curve in Figure 3) over the life of the well with a peak production of 14.5 MBOPD is assumed. Peak production is assumed to reach in Year Two of the well and an immediate fall in production from Year Three at a decline rate of 25%. Production is assumed to cease from a well when the production falls below 2.5 MBOPD. The P90 production profile is nearly half of the P50 production profile, where each well produces an estimated 10MMBO.

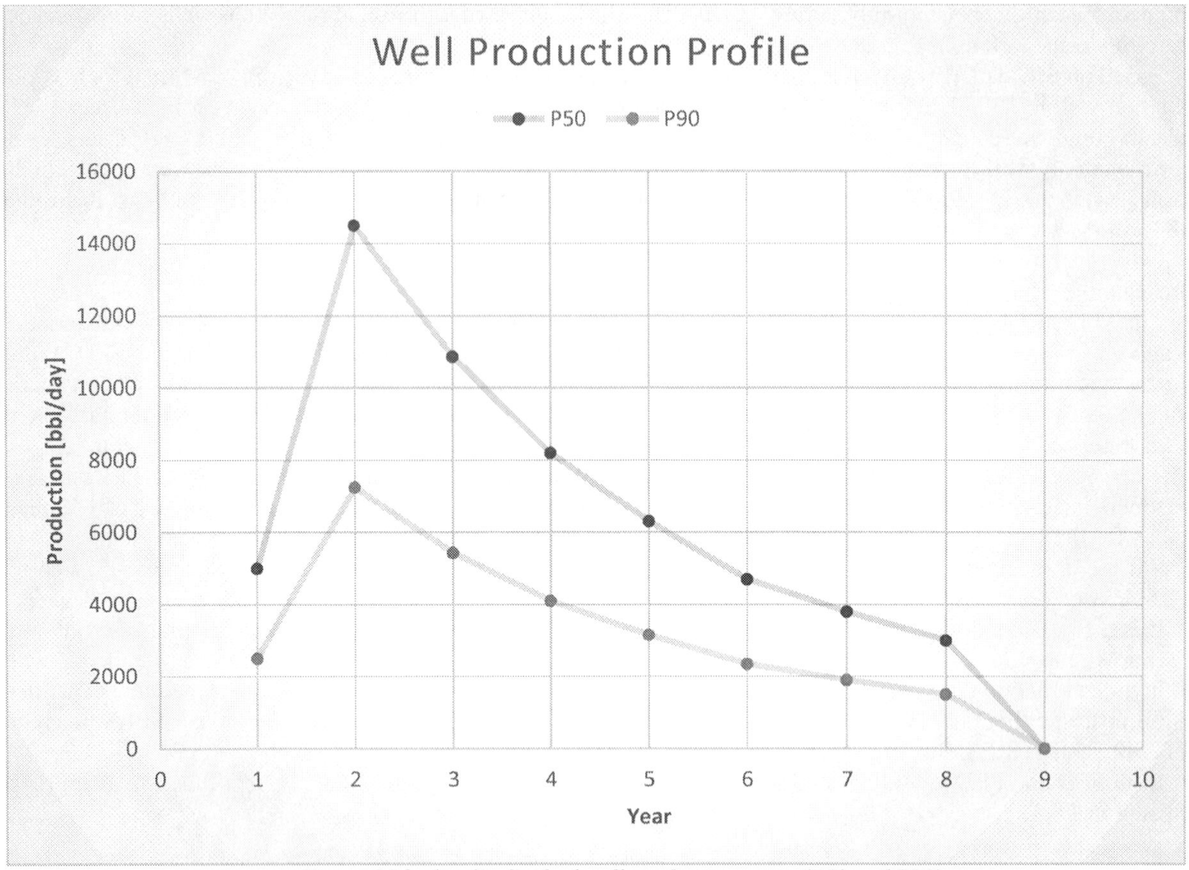

Figure 3 Idealized individual well production curve (P50 and P90)

Based on our assumed well production curve, it will require 20 wells to drain the complete reservoir. Each well is expected to take 6 months from spuding to Ready for Production. All downtimes including weather delay, technical delays, converting from drilling to completion mode, etc. are included in the drilling duration. Thus one drilling rig can be used to drill 2 wells per year. For the FFD First Oil, it is assumed that 8 wells will be available. A drilling campaign is envisioned using two rigs to drill 8 wells over a two-year period including the year of First Oil. Once First Oil is reached, one of the rigs is released. The rig remaining on contract will continue to add two more wells each year until all 20 wells are drilled, completed and available for production.

If we assume that each well is allowed to flow unrestricted then the peak production might exceed the host capacity when many wells are available for production. To avoid that scenario, we assume that the field production is maintained at the host design capacity by means of appropriate measures. These can include completing but not bringing wells online or throttling back on producing from a newer well when the initial wells are still producing at a high rate. Discussion on such methods are not within the scope of this paper.

The main purpose of the drilling of these wells in our model is to spread the DRILLEX appropriately. Detailed considerations of drilling schedule, peak production etc. require nuanced and more thorough treatment of this topic.

Base Case production profile (P50) for the PFD case assumes that 4 wells will be available at First Oil with two more added each year till all 20 wells are available. It is assumed that a single rig is used to drill all 20 wells over a 10 year period. As the production increase is staggered, multiple drilling rigs are not necessary. The FFD and PFD Base Case production profiles are shown in Figure 4.

All technologies are assumed to be industry standard, thus eliminating any qualification programs that might extend project development cycles.

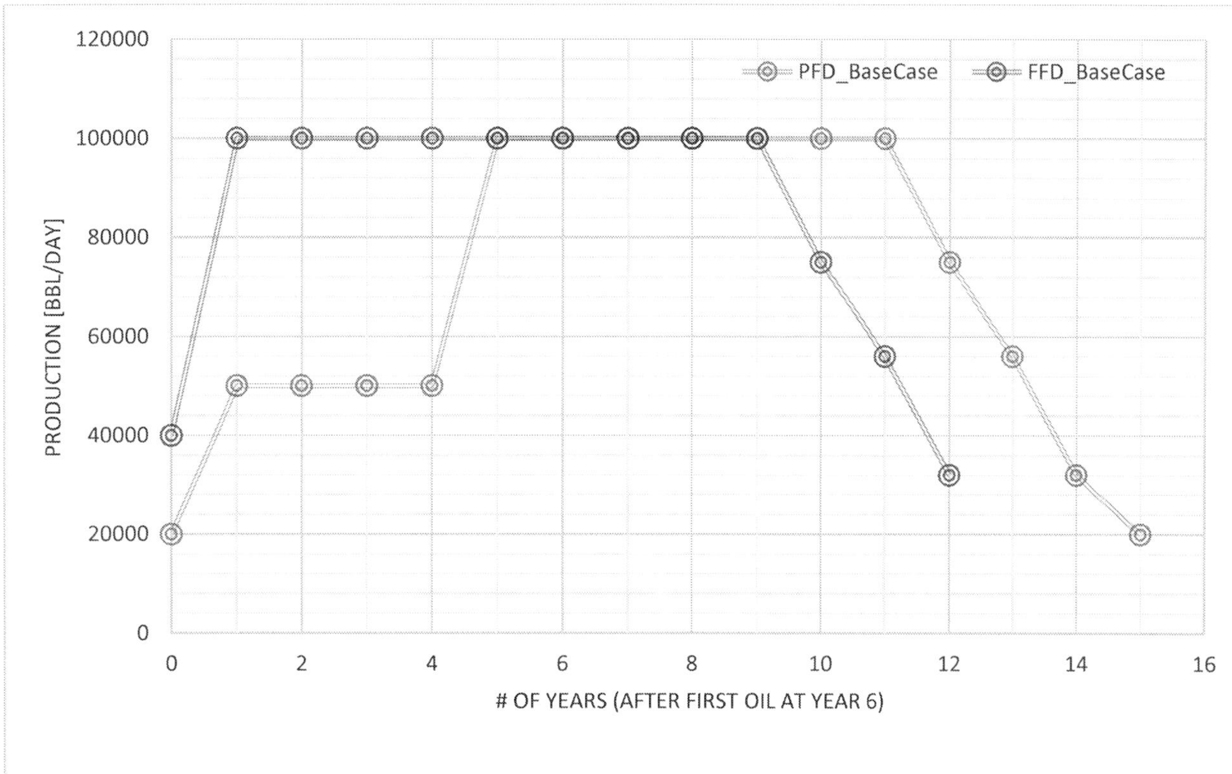

Figure 4 Base Case Daily Production Capacity Comparison Curves

The FFD and PFD Base Case production profiles are shown in Figure 4. The initial FFD production rate is 40,000 BOPD and reaches the plateau production rate of 100,000 BOPD one year after reaching First Oil. This plateau is maintained by appropriately managing the well production from all the available wells. Approx. 12 years after First Oil, the reservoir is "drained" and production ceases.

The PFD Base Case starts off at 20,000 BOPD and reaches the plateau of 50,000 BOPD one year after First Oil. As the initial wells continues to flow at an acceptable rate, the second stage is sanctioned and comes online in Year 5. The field production instantaneously jumps to the "full" capacity of 100,000 BOPD. Again the full plateau is maintained for about 7 year using appropriate measures.

However, wells and reservoirs don't always turn out as expected. Hence, we modeled a low-range or P90 reserve, where each well produces an estimated 10MMBO.

Table 2 Description of Sensitivity Cases

	FFD	**PFD**
Base Case, P50 Recovery	Case 1	Case 2
P90 Recovery	Case 1A	Case 2A
First Oil 1 year early (-1 year), Compressed Schedule	Case 1B	Case 2B (on Phase 2 only)
First Oil 1 year late (+1 year), Delayed Schedule	Case 1C	Case 2C (on Phase 2 only)

Table 2 describes the various sensitivity cases studied. Besides the Base Case, the effect of fabrication and installation schedule were also studied. The First Oil for FFD (at Year 6) and that for Phase 2 of PFD (at Year 10) were varied by +/-1 year to study its effect on the NPV, CAPEX and Break-even oil price. CAPEX phasing due to the surface and SURF facilities remained unchanged from Base Case to these sensitivity cases. The drilling schedule for the FFD Case 1B accounted for early drilling of 2 wells to reach the initial ramp up production of 60 MBOPD. Case 1C did not require such change, as all 20 wells were available to reach plateau after the delayed (+1 year) start of production.

For the PFD cases, the schedule sensitivity is studied around the First Oil of the second phase rather than changing the First Oil of the whole project. The production profiles used for these cases are plotted in Figure 5.

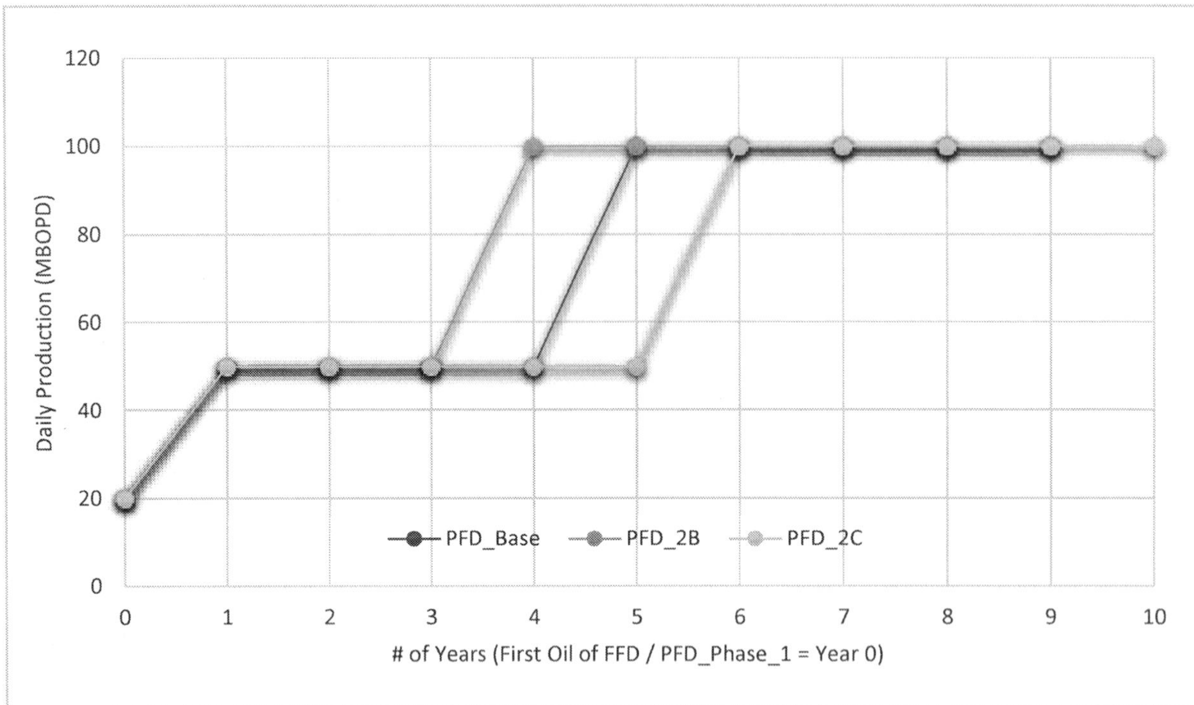

Figure 5 PFD Sensitivity Case Production Profiles

A summary of the pertinent inputs for FFD and PFD cases are shown in Table 3. Water cut, and certain other variables has not been modeled. The assumed GOR is low indicating this is mostly an oil field. The oil and gas price inflation has been assumed to be zero.

Besides studying the effect of the schedule, ancillary cases involving the CAPEX are also studied to generate Tornado Risk diagram. While the CAPEX is typically locked within a +/-10% bound at FID, in our modeling they were varied by +/-20%. The resultant changes in NPV trends are shown.

Table 3 Economic Input Data Summary for the FFD and PFD models

Item	Units	FFD	PFD
Production			
Gas production	Scf/bbl oil	750	
Water cut	% oil	-	
Total field production (P50)	MMBO	400	
Plateau Production	MBOD	100	50 and 50
Prices			
Quality Discount	$/bbl	4.00	
Transport & Tariffs	$/bbl	3.50	
Oil price inflation	% pa	-	
Gas Sales Price	$/mscf	4.00	
Transport & Tariffs	$/mscf	2.00	
Gas price inflation	% pa	-	
Costs			
Development Well Cost	$MM	150	
Hull & Topsides Equipment	$/bbl	4.00	

Hull & Topsides Phasing	%	40-40-20	20-20-10-0-0-20-20-10
Subsea Equipment	$/bbl	4.00	
Subsea Phasing	%	0-50-50-0	25-25-0-0-0-25-25-0
Decommissioning Costs	$MM	50.00	
Fixed Lease Expenses	$MM	60.00	
Fixed O&M Expenses	$MM	50.00	
Variable Operating Costs	$/bbl	3.50	
Taxes & Royalties			
Royalty	%	16%	
Federal Income Tax	%	34%	

RESULTS

Results from the various base cases and sensitivity cases are presented and discussed in this section. While dollar values are reported, they should be considered in a qualitative rather than a quantitative manner. The revenue or cost figures are representative of the trends rather than accurate values.

Summary of the pertinent results are shown in Table 4 for the FFD scenarios for the P50 field development plan. It is seen that with a schedule compression by one year at First Oil (Year 6), the break-even oil price is reduced from the base case value of $52.75 to $48.42. On the other hand, with a schedule delay of one year from the Base Case, the break-even oil price is correspondingly increased to $57.69 as the revenue is pushed farther into the future in the project schedule. This shows that shortening the time to First Oil for an FFD has obvious economic advantages (even though 2 extra wells need to be drilled) as reflected by the average annual break-even oil price reduction. Figure 6 shows the annual cash flow comparison for the three scenarios for the FFD case.

Table 4 Summary Results for FFD Scenarios

		Base Case	**Case 1B**	**Case 1C**
Landed Crude Oil price	$/bbl	52.75	48.42	57.69
NPV 10 Pre Tax	$MM	797	740	891
Cumulative Revenue	$MM	18820	17178	20811
Cumulative Cash Flow	$MM	4373	3401	5431
Peak Negative Cash Flow	$MM	4365	3539	4730

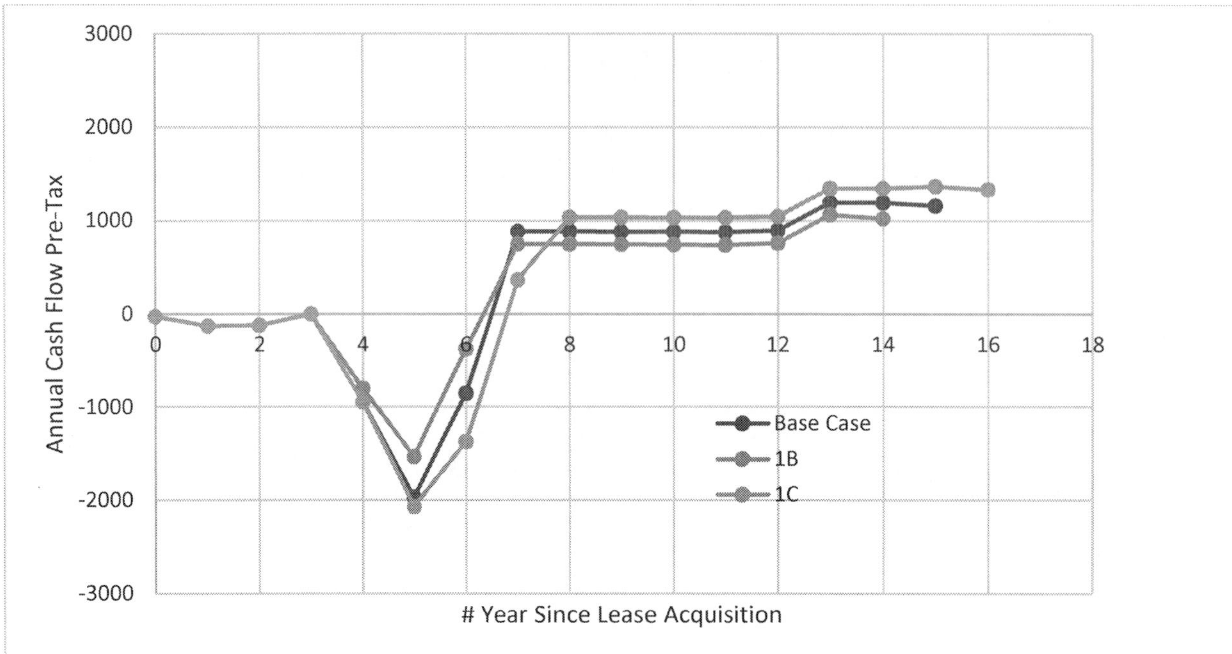

Figure 6 Comparison of Annual Pre-Tax Cash Flow for FFD Cases

Summary of similar information for the PFD cases for the P50 field development plan are presented in Table 5. It is seen that with a schedule compression by one year at Phase 2 First Oil (Year 10), the break-even oil price is reduced from the base case value of $54.21 to $53.35. On the other hand, with a schedule delay of one year from the base case, the break-even oil price is correspondingly increased to $55.85 as the increased cash flows are pushed farther in the future. It also shows that PFD break-even oil price remains within a narrower band compared (~5%) in comparison to FFD case (~18%). This shows that PFD offers a tighter band of return compared to FFD case. It also shows that a later phase "Go" / "No-Go" decision has lesser impact on total project NPV, as the cash flow peak is moved further in the future. Another key takeaway from this table is that the PFD peak negative cumulative cash flow is lower compared to an FFD. All these can be taken together as a substitute for lower investment risk in PFD vis-à-vis FFD case.

Table 5 Summary Results for PFD Scenarios

		Base Case	Case 2B	Case 2C
Landed Crude Oil price	$/bbl	54.21	53.35	55.85
NPV 10 Pre Tax	$MM	708	704	734
Cumulative Revenue	$MM	19409	19099	20014
Cumulative Cash Flow	$MM	4561	4383	4897
Peak Negative Cash Flow	$MM	2906	3070	3205

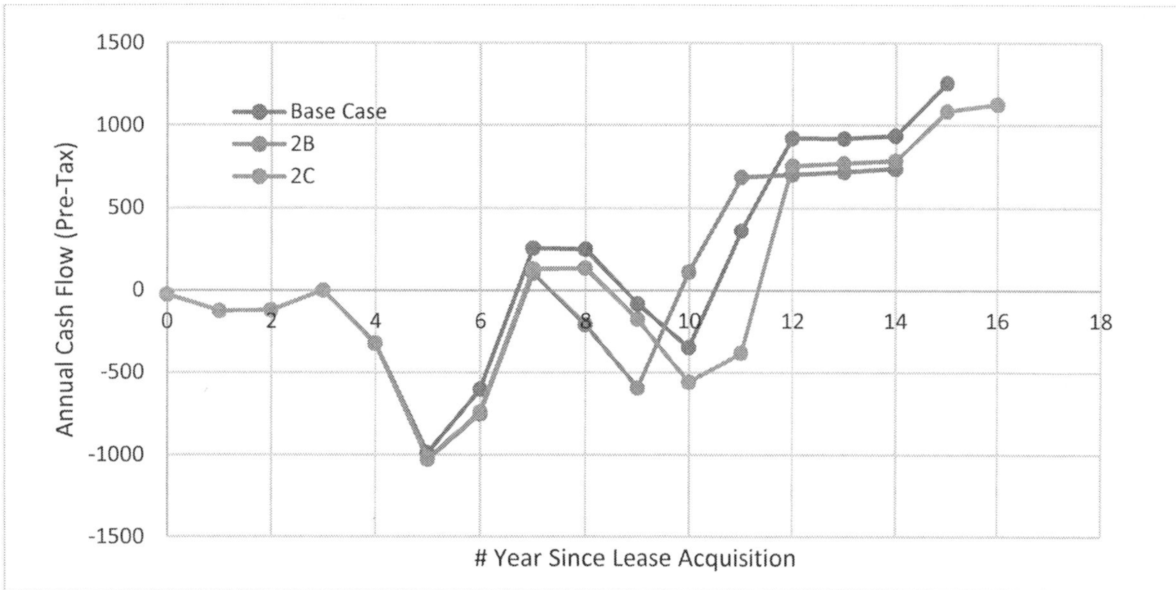

Figure 7 Comparison of Annual Pre-Tax Cash flow for PFD cases

Figure 7 shows the annual cash flow comparison for the three scenarios for the PFD case. The first phase of the PFD is similar in principle to the FFD and we see a little bump for the discovery wells. The large negative cash flow is due to the investment in the facilities and wells. However, the negative cash flow magnitude is much lower as only "half" the investment is made in this phase.

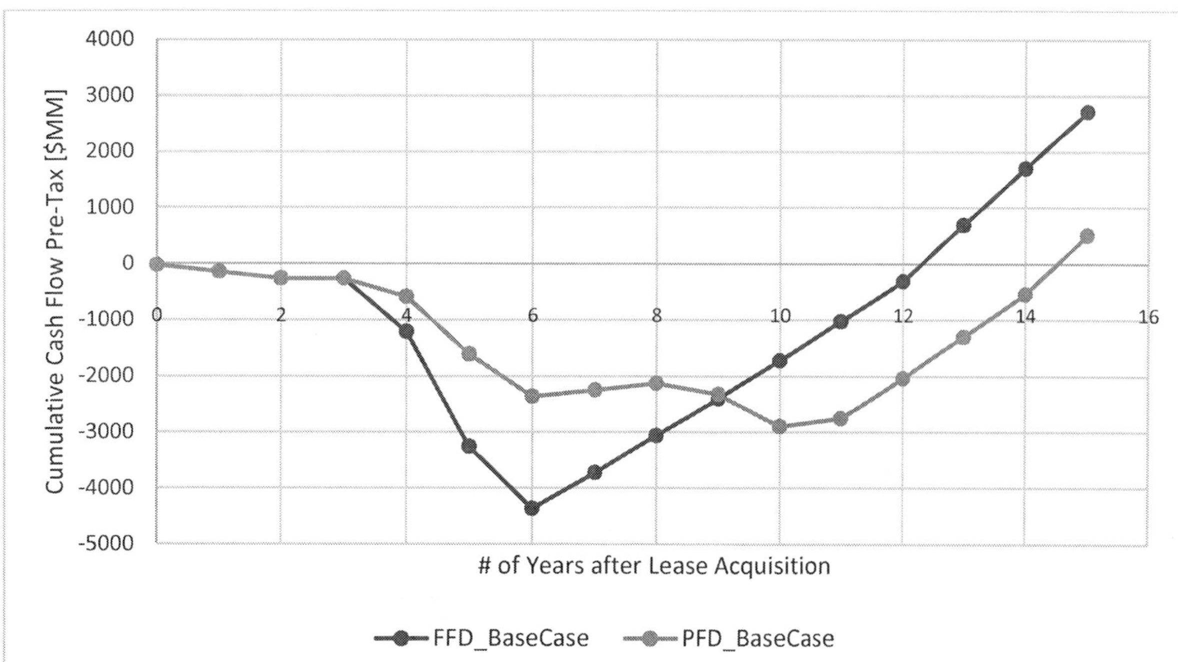

Figure 8 Comparison of Cumulative Cash Flow for FFD and PFD cases

Figure 8 compares the cumulative cash flows among the FFD and PFD base cases. Once production kicks off, the cumulative cash flow turns toward becoming positive from the initial negative values. The payback period corresponds to the time taken for the cumulative cash flow to reach zero from negative. Comparison of the FFD and PFD base cases show that the payback period is increased when moving from an FFD to a PFD plan, however, the peak negative cash flow improves significantly for the PFD case.

The peak negative cumulative cash flow can be used as a proxy for financial risk in the development. The PFD base case has a slightly longer payback period on the downside, however it also has a lower peak negative cumulative cash flow compared to the FFD base case. The biggest contributors for the FFD base case negative cash flow are the facilities expenses and the DRILLEX. The FFD base case requires a bigger facility (both surface and subsea) that needs to be expensed up front. Also to reach the design plateau faster, more wells need to be available at First Oil compared to the PFD base case. These are both significant project risks which appear in the peak negative cumulative cash flow.

Judicious use of the well development plan can be combined along with schedule (time to First Oil) compression to achieve a desired tradeoff between the payback period and development financial risk.

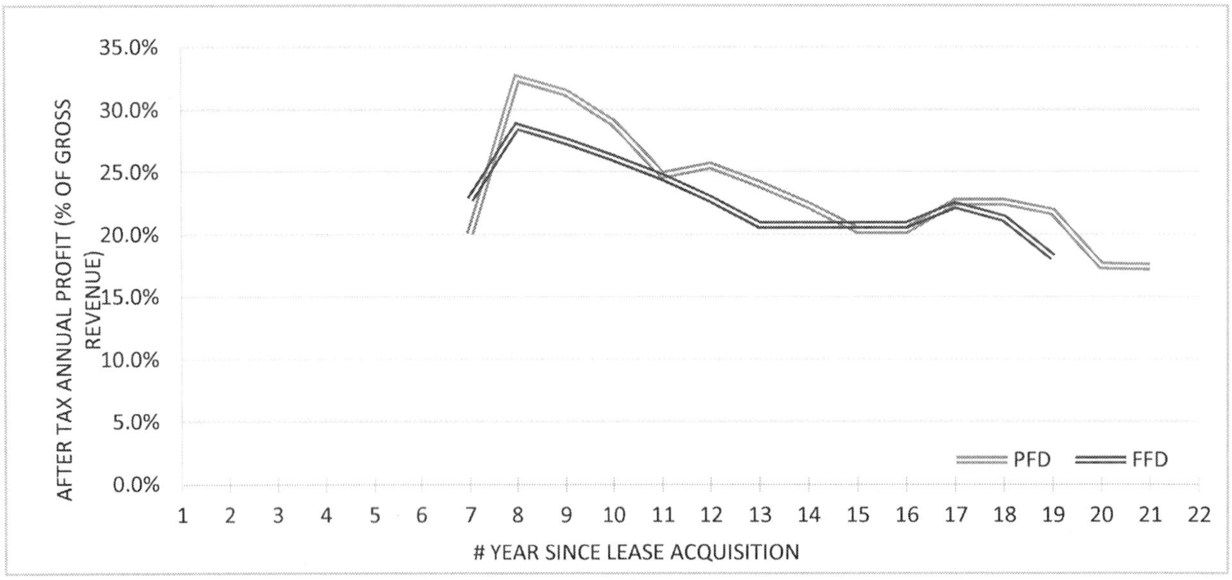

Figure 9 After Tax Profit Comparison – PFD vs FFD

Figure 9 and Figure 10 show the after tax profit margin and CAPEX comparison between PFD and FFD base cases respectively. Since First Oil occurs on Year 6 (Lease acquisition = Year 0) there is zero revenue and hence no profit is calculated. Similarly there is zero revenue after cessation of production on Year 21. Yearly profit appear to track closely with the production. FFD base case reaches plateau early and it is captured in the early peak yearly profit. It is seen that the profit margin is a few percentage points higher in case of a PFD than an FFD.

PFD base case initial plateau is lower which is shown in the lower peak in profit for the PFD curve. However, as the second phase of the PFD comes online, the combined production from the PFD facilities generates a second higher peak.

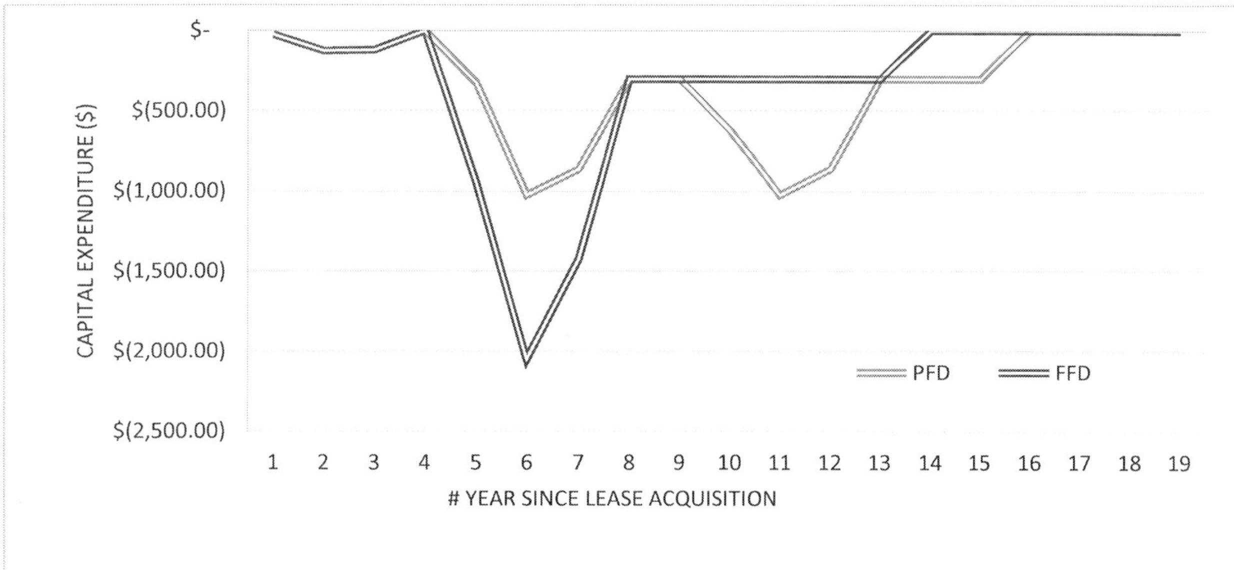

Figure 10 CAPEX Comparison – PFD vs FFD

The risk tornado diagram for Oil Price, CAPEX and Project schedule duration for PFD and FFD are plotted in Figure 11 and Figure 12 respectively for the same total field production of 400 MMBO. The risk tornado graph shows the relative effect of variation of the economic parameters (in terms of ratio of NPV to Total Oil & Gas Production). It is seen that an oil price volatility of +/- $10/bbl has the highest impact on NPV similar to a CAPEX variability of +/- 20%, whereas a schedule variability of +/- 1 year has relatively less impact. The impact is relatively less in case of a PFD than a corresponding FFD production scenario.

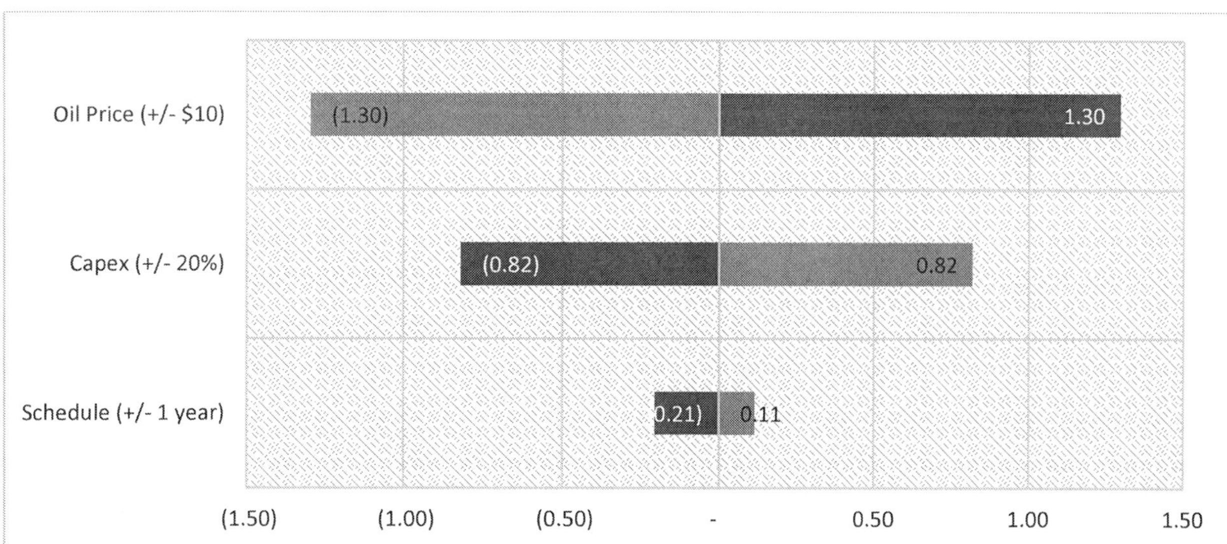

Figure 11 Risk tornado for PFD – Oil Price, CAPEX and Schedule Variability

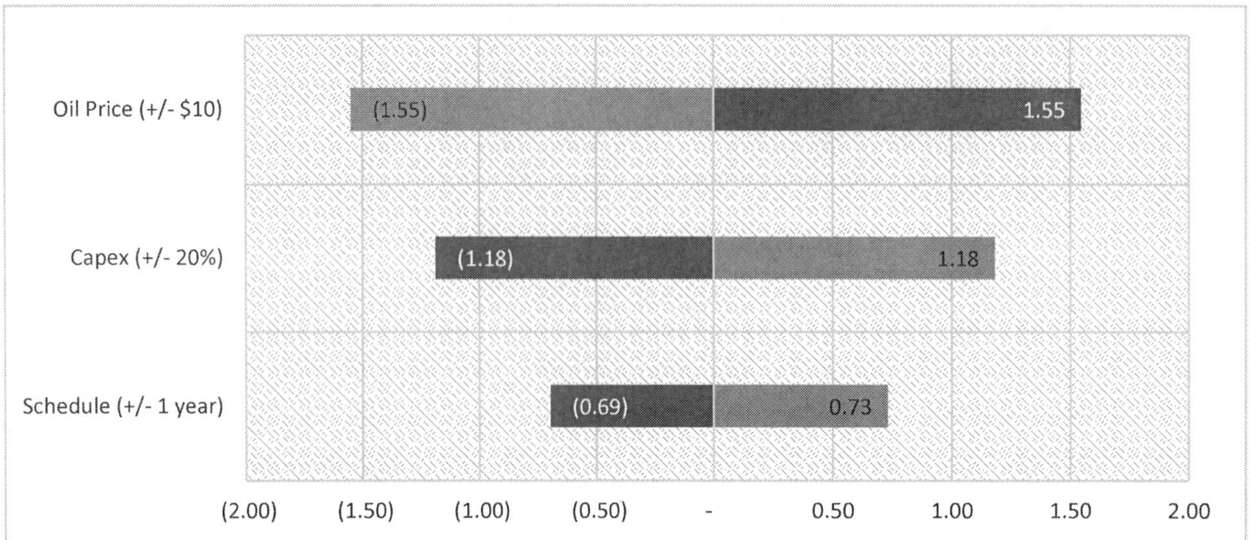

Figure 12 Risk tornado for FFD – Oil Price, CAPEX and Schedule Variability

Following observations can be made from the economic analysis results tabulated in Table 6. The major benefits of using PFD are seen to be as follows:

- Maximum annual CAPEX reduces requiring less start-up investment cost
- Maximum cumulative negative cash flow reduces significantly thus minimizing risk
- Gross Profit margin increases

Table 6 Comparison of FFD and PFD Base Cases

Parameter	FFD-Base	PFD-Base	Outcome (FFD-->PFD)	Effect (FFD-->PFD)
Peak Annual CAPEX ($MM)	($2049)	($1025)	Reduces by 50%	Good
Payback Period (from Lease Acquisition)	12.3 years	14.5 years	Increases by 18%	Worse
Peak Negative Cumulative Cash flow ($MM)	($4365)	($2906)	Reduces by 32%	Good
Gross Profit Margin (%)	28.7%	32.5%	Increases by 13%	Good
Break-even Crude Price ($/bbl)	$52.75	$54.21	Increases by 2.8%	Worse

We also compared the P50 and P90 recoverable cases for the PFD and FFD production scenarios. An illustrative roadmap for each is presented through the flowchart shown in Figure 13. Once a project is sanctioned, it moves forward under the assumptions made at sanction. Contracts are signed and money spent assuming, say the P50 production estimate. However, things don't always pan out the way they are anticipated. To study such a scenario, we kept the cost profile and phasing unchanged from the P50 case but reduced the daily production to match an overall recovery of 200 MMBO or P90 recoverable as described earlier. This drastic drop is manifested in a large swing in FFD pre-tax NPV from a positive value of nearly $800 million to a negative of almost $1550 million, a swing of nearly $2350 million (Figure 14). The PFD on the other hand shows a swing of $2035 million. It is seen that the project NPV increases when the total realized production from the field increases.

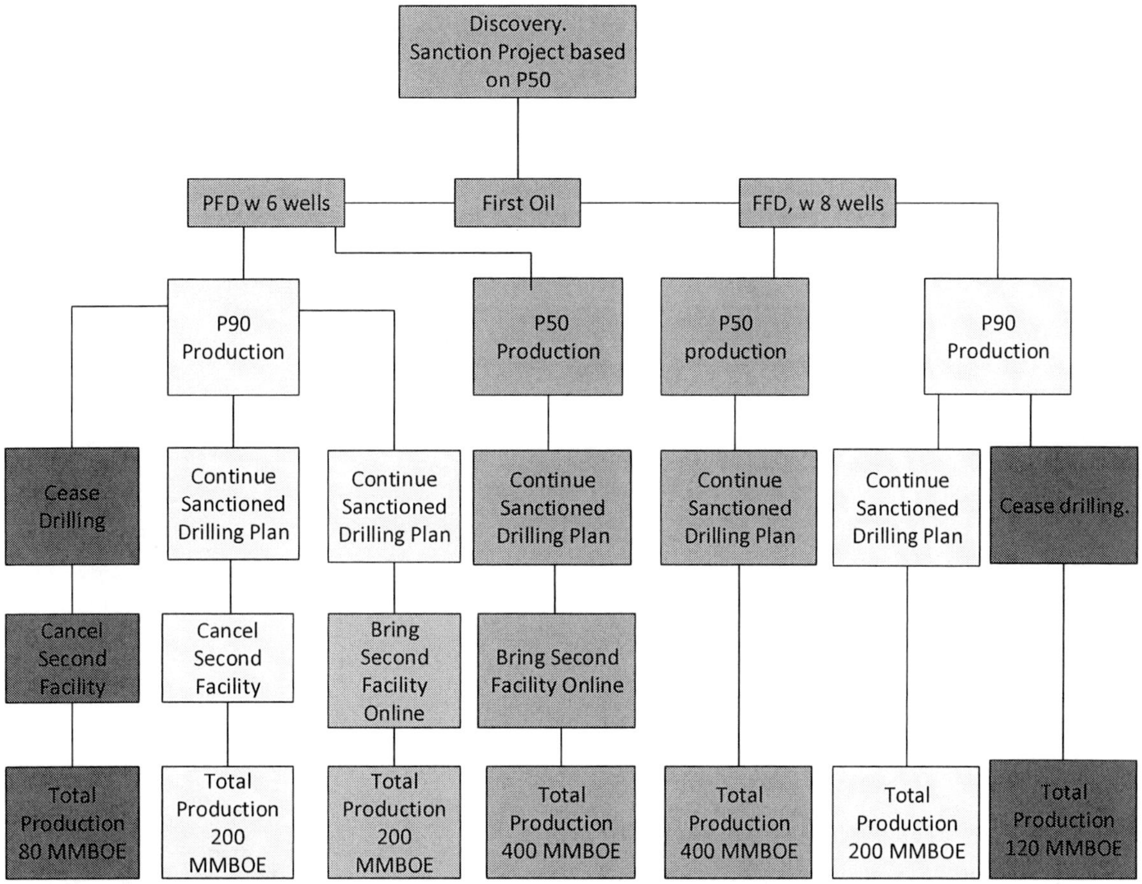

Figure 13 FFD and PFD – Roadmap for P50 and P90 production scenarios

Herein lies the crux of this development dilemma: what to do when faced with unsatisfactory production profile? Should an FFD keep drilling all 20 wells (as per current modeling) even though the first 10-12 wells (drilled by Year Two after First Oil) do not produce at P50 level? The wells are not cheap to drill, but not drilling those remaining wells also mean leaving whatever left of the P90 reserve in the ground. Assuming a suspension of drilling after Year Two, only 60% of the P90 reserves can be recovered, thus painting a bleaker picture for the FFD case.

On the other hand, a PFD has not spent as much money upfront as the FFD case. By Year Two of PFD, when it is apparent that the reserves are P90 rather than P50, only 8 wells have been drilled and 50% of the CAPEX have been spent. This makes the decision to move forward with drilling the 12 remaining wells more palatable. However, the decision of exiting the field with only 40% of the P90 reserves can also be defended.

There are few potential avenues to explore to increase NPV while still working within the P50 and P90 boundaries set here. The biggest risk comes from spending upfront based on P50 while recovering only P90 reserve. As the project brings in two new wells every year, the subsea expenses can be phased with that cycle as well. On one hand, it will handcuff the host from reaching plateau production fast. On the other hand, it also smoothens out the effect of a large expenditure on NPV by distributing the subsea infrastructure cost over a longer period.

Another approach is employing secondary or tertiary recovery techniques to maintain the production plateau. This is usually achieved by pre-investing in Topsides and Hull utility capacities and then installing appropriate equipment as necessary. Two of the most common conventional hull designs, namely FPSOs and Semi-submersibles can reserve space and payload at First Oil with minimal exposure for future. Not committing to a particular recovery method also allows changing course if the reservoir sends a signal different than what was expected. If the original plan was to use water injection, then one can install it on Day 1 spending money upfront. If the reservoir turns out as expected, then the WI module simply takes up valuable deck space and CAPEX/ OPEX. On the other hand, if the reservoir requires gas injection or any other suitable method, then the unusable WI module becomes really expensive by preventing installation of an appropriate set of equipment.

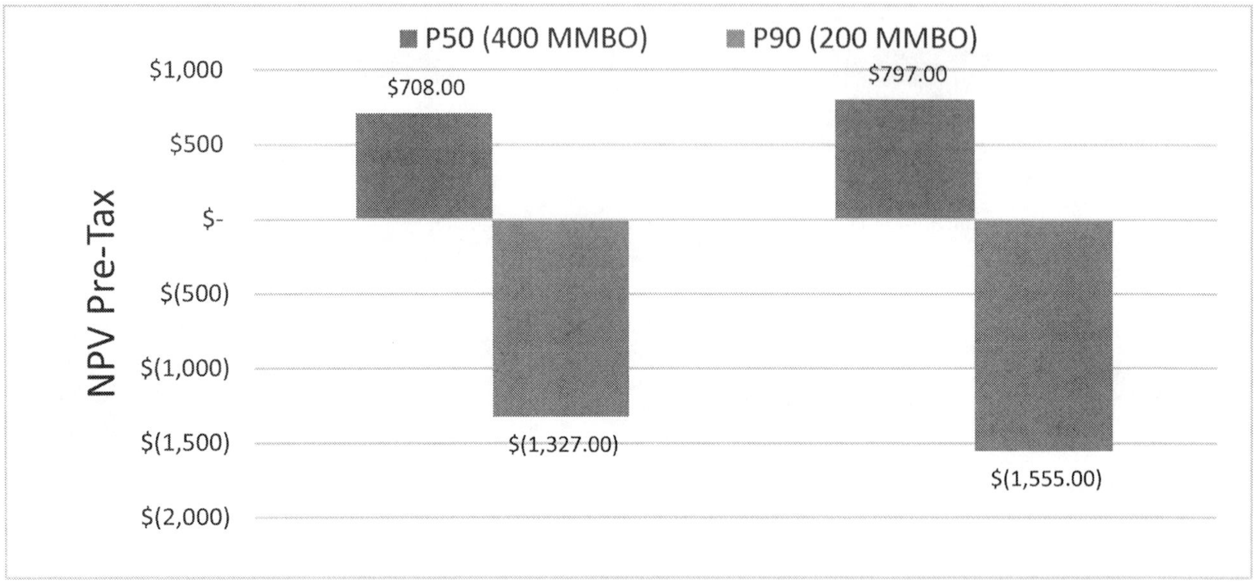

Figure 14 NPV comparisons between P50 and P90 production scenarios, PFD (left) and FFD (right)

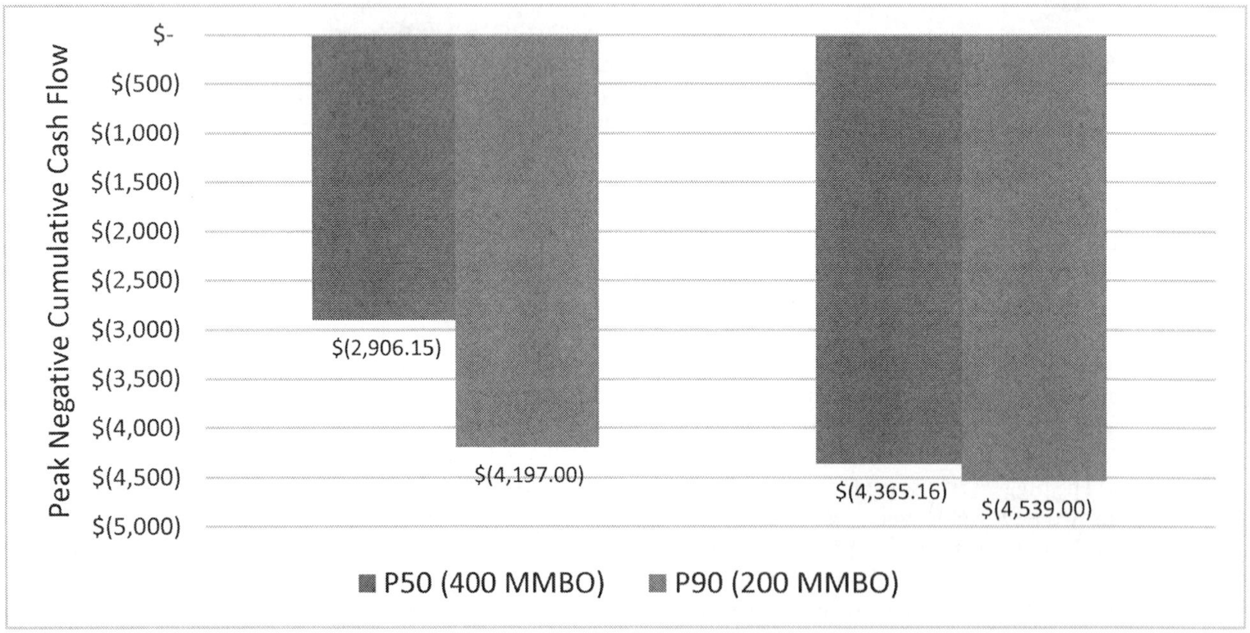

Figure 15 Peak negative Cash flow comparisons between P50 and P90 production scenarios, PFD (left) and FFD (right)

After studying the cumulative cash flows for these two scenarios, it is found that the peak negative cumulative cash flow is higher when the total field production is increased, as depicted in Figure 15. This is expected as infrastructure expenses have been already incurred in both cases which is the sole contributor to the negative cash flow. It starts to turn positive with the start of production or revenue generation. Initial years produce similar level of revenue for both P50 and P90 scenarios, thus showing a small difference in the peak negative cash flow for the FFD case. However, the PFD case shows a much larger negative peak cash flow as the Phase 2 investment is made in spite of lowered revenue.

CONCLUSIONS

Two different approaches to field development have been studied for US GOM using contemporary break-even costs and basing assumptions on recent data of offshore production platforms. The conclusions drawn from them are listed below. Compared to an FFD, using a PFD approach:

- lowers the cumulative cash flow risk profile
- increases the profit margin (percentage of revenue)
- enables using smaller platform processing capacity to lower facility start-up and initial operational costs

The above benefits or advantages are achieved by taking a slight hit on the payback period and at a slightly higher break-even price which are expected from an economic standpoint.

The comparative study evaluated for a phased field development vis-a-vis a conventional full field development provides some perspectives which is believed to be useful when devising a roadmap for future deepwater investments in the near term context.

ACKNOWLEDGEMENTS

The authors acknowledge the support of Granherne and GVA management teams for supporting this work. We also sincerely appreciate the economic tools and course discussions on Energy Investment Analysis at the Bauer College of Business, University of Houston.

LIST OF ABBREVIATIONS

bbL	:	barrels (of oil)
BSSE	:	Bureau of Safety and Environmental Enforcement
CAPEX	:	Capital Expenditure
DRILLEX	:	Drilling Expenditure
EIA	:	Energy Information Agency
EPS	:	Early Production System
FFD	:	Full Field Development
FID	:	Final Investment Decision
FPSO	:	Floating, Production, Storage and Offloading
GOM	:	Gulf of Mexico
GOR	:	Gas to Oil Ratio
IEA	:	International Energy Agency
IOC	:	International Oil Company
OPEX	:	Operating Expenditure
MBOPD	:	Thousand Barrels of Oil per Day
mcf	:	million standard cubic feet (of gas)
MMBO	:	Million Barrels of Oil
NPV	:	Net Present Value
PFD	:	Phased Field Development
SURF	:	Subsea, Umbilicals, Risers & Flowlines
US	:	United States of America
WI	:	Water Injection

REFERENCES

Bloomberg/Rigzone, "*Drop in Crude to Test Big Oil's Devotion to Spending Discipline*", November 14th, 2018 (1)
EIA, "*Trends in U.S. Oil and Natural Gas Upstream Costs*", March 2016 (2)
IEA, "*Oil 2018, Analysis and Forecasts to 2023*", March 2018 (3)
D'Souza, Basu and Khurana, "*A Universal Field Development Approach for Advancing Stalled GOM Deepwater Projects*", OTC-28970, Houston, April 30th, 2018 (4)
D'Souza, Basu and Fales, "*Selecting the Right Field Development Plan for Global Deepwater Developments*", DOT-22, Perth, 2012 (5)

SNAME
THE INTERNATIONAL COMMUNITY
FOR MARITIME AND OCEAN PROFESSIONALS

REVIEW OF OFFSHORE MOORING SYSTEM LIFE EXTENSION – CHALLENGES AND RECOMMENDATIONS

AIFENG YAO
CHEVRON ENERGY TECHNOLOGY CO.

QINGHE ZHANG
AMERICAN BUREAU OF SHIPPING

Proceedings of the 24th Offshore Symposium, February 20th 2019, Houston, Texas
Texas Section of the Society of Naval Architects and Marine Engineers

ABSTRACT

Mooring systems of many long-term/permanent floating offshore installations (FOIs) are approaching the end of their original design life. Some FOI mooring systems are seeking life extension to support continued production/storage or planned tie-back. Worldwide, many offshore mooring systems have been approved for continued service.

This paper presents a state-of-the-art review of the current industry practice on mooring system life extension with an emphasis on fit-for-service technical assessment but excluding the project economic considerations for decision-making. The objective is to initiate an industry-wide approach to improve technical assessment for mooring life extension by identifying gaps and challenges.

A comprehensive survey was carried out on the available policies from the regulatory authority and technical guidance from the industry. The major technical challenges are discussed covering baseline information collection, in-service mooring inspection, damage/degradation assessment of components, mooring design re-analysis, and remaining life prediction. Recommendations on technological development, industry guidance, and life cycle integrity management were made to advance the current industry practice of mooring system life extension.

Keywords: Offshore Mooring, Mooring Life Extension, In-Service Inspection, Mooring Component Assessment, Remaining Life

INTRODUCTION

Life extension studies of offshore facilities initially focused on fixed platforms, such as jackets, and was directed more towards structural integrity evaluation [Stacey et al., 2008]. Recently, there is growing attention to life extension, also known as continued service, of long-term/permanent floating offshore installations (FOIs) as many of them are approaching the end of their original field life, and some of them are seeking continued production from the same or adjacent oil and gas fields. These FOIs include FPSOs, FSOs, SPARs and semi-submersible platforms, as well as offloading buoys.

The life extension of mooring systems deserves more rigorous investigating and assessment. As a safety critical element of a FOI, the mooring system performs the station-keeping function in a corrosive and dynamic marine environment in which inspection is often difficult [Gordon 2015]. It has been well known that mooring failures occurred at an unacceptably high rate [Ma, et al., 2013, Fontaine et al. 2014b] during the original design life, primarily attributed to aging processes such as the ever-present hazards of corrosion, wear, and fatigue, compounded by manufacturing defects, accidental damage, and design shortcomings [Gordon 2014].

Recently, government regulations have been developed to define regulatory requirements for managing aging and life extension of floating offshore facilities, notably in the North Sea and in the Gulf of Mexico. Class societies have published guidance on reassessment and requalification methodology for life extension process. Working with both local regulators and selected classifications societies, some operators have been approved to extend the field life of their offshore assets including the associated mooring systems [Gordon 2015].

This paper presents a state-of-the-art review of the current industry practice on the mooring system life extension. It discusses the limitations in the existing industry guidance and the challenges in technical assessment but excludes the project economic consideration for decision making. The objective is to initiate an industry-wide approach to improve technical assessment for mooring life extension.

REPRESENTATIVE REGULATORY REQUIREMENTS

There are very limited regulatory requirements and industry guidance worldwide with respect to the mooring system life extension/continued service. Back in early 2000s, integrity management and life extension of aging assets started with a focus on fixed offshore installations and then moved on to include FOIs as well. To date, most guidance or requirements are developed for the general life extension procedure of FOIs. Although the mooring system is widely considered as a safety critical element of a FOI, it has not yet been articulated how to perform a mooring life extension study. Below, some representative government regulations on mooring life extension are summarized.

1. BSEE (USA)

To date, BSEE has no well-defined regulations for life extension of permanent installations including mooring systems in Gulf of Mexico (GoM). The regulatory requirement from BSEE usually refers to API standards. Discussion on some API standards relevant to mooring life extension is provided in the next section.

2. USCG (USA)

USCG [USCG policy letter 2016] issued the interim guidance on life extensions for floating OCS facilities, which provides the recommended flow chart for obtaining life extension. This letter requires a baseline survey, engineering evaluation, risk assessment, and mitigation and repair plan mainly on structural, machinery, and stability, and puts emphasis on the stability of tendons and TTMS (Tendon Tension Monitoring System) on TLP projects. Some specific regulatory requirements pertinent to mooring systems are highlighted below:

- The operator shall summarize surveys/inspections performed over the past five years by the classification society/Coast Guard.
- The operator should be as thorough as possible when conducting the baseline survey, as this is critical and will serve as the basis for future additional requirements.
- The baseline survey for the mooring system should include corrosion protection systems and mooring attachments to the hull.
- In additional to the base line survey, the operator should be prepared to perform an engineering evaluation and risk assessment for the facility, including detailed comparison of the facility's original metocean design criteria compared to most recent criteria.
- The operator should document any significant repairs, incidents, weather events and critical equipment changes.

- The operator should report the frequency of future topside, In-Service Inspection Plan (ISIP), and Under Water In Lieu Drydock (UWILD) inspections and changes in Structural Critical Inspection Points (SCIP).
- The operator should submit mitigation and repair plans (if necessary) for problems identified in the baseline survey and risk assessment to the Coast Guard. Any mitigation or repairs should be reviewed and approved by the classification society (if applicable) and the Coast Guard.
- Following completion of the baseline survey, risk assessment, and mitigation and repair plan, the operation shall update all applicable procedures to the facility.

The USCG policy letter outlined the floating facilities life extension process for regulatory compliance. Nevertheless, many technical aspects on such as the baseline survey, engineering evaluation and risk assessment are not discussed.

3. UK HSE

The key guidance from UK HSE relevant to the aging and life extension of offshore installations are summarized below.

UK HSE's Offshore Division (OSD) launched a Key Program 3 (KP3) in 2007 focusing on the Asset Integrity Management (AIM) of all types of offshore installations. An Offshore Information Sheet (OIS) 5/2007 published by UK HSE in 2007 [UK HSE 2007] requires duty holders to periodically reassess aging processes that affect the structure of semi-submersible installations, particularly for installations beyond their intended design life. Reassessment considerations were required for the integrity of aging semi-submersible installations, covering fatigue life, redundancy checks, comparison between the installation and as-built documentation, inspection and maintenance, replacement or repair, etc. In this UK HSE Information Sheet, inspection and maintenance of mooring and positioning systems was explicitly spelled out. Considerations for remaining fatigue capacity, corrosion and thickness measurement were highlighted.

UK HSE Offshore Information Sheet 4/2009 [UK HSE 2009] provides guidance for asset managers, safety managers and safety engineers in the offshore industry on taking account the aging of the installation during a thorough review, which is required by the Offshore Installations (Safety Case) Regulations 2005 (SCR05). This information sheet refers to ISO 19900 series of standards as a good basis for the assessment of life extension.

Following the KP3, UK HSE carried out an Aging and Life Extension (ALE) Key Program 4 (KP4) between 2011 and 2013 [UK HSE 2014]. It identified areas of industry strengths and made recommendations for additional focus, supported by case studies. However, no specific detailed technical guidance is provided for mooring engineers to implement the recommended practices on such as technical assessment, data management, preparation for life extension, and periodic ALE audits, etc.

4. PSA (Norway)

In 2008, the Petroleum Safety Authority (PSA) Norway published the requirements for life extension of aging offshore production installations [PSA, 2008]. This regulation was mainly developed for the aging fixed offshore installations, with defined requirements for information, processes and criteria for the PSA and operator to decide whether the aging installation can continue production in extended life. A generic framework for life extension application was developed, including a wide range of data covering inspection and test report, operating and maintenance records, design and construction data, manufacturers' data, failure and incident report, and technical assessment. This PSA document is not specific to offshore permanent mooring system and therefore, no detailed guidelines for mooring life extension is provided.

The Norwegian Oil and Gas Association revised the guidelines for the management of life extension of a facility and determining the extended life [Norwegian Oil and Gas 2017]. This is one of the most comprehensive guidelines to date which addresses life extension planning, required activities for assessment of system condition and functionality, assessment of technical lifetime of systems, total risk assessment, and approval and verification of life extension. Given the regulatory nature of this Norwegian guidelines document, no specific discussion was made on how to address the acknowledged technical challenges and uncertainties in assessing system conditions and functionality, and determining the extended life.

5. ANP (Brazil)

Motivated by the applications to extend the oil and gas concessions, Brazilian National Agency of Petroleum, Natural Gas and Biofuels (ANP) has begun work on the development of life extension regulations since 2015 [Pires et al. 2016]. Up to date, no specific life extension regulation has been established except the enforcement of the existing ANP Resolution 43/2007. The safety management practices in ANP Resolution 43/2007 are stated to be valid throughout the life cycle of the installations. For an aging production facility, continued operations should comply with ANP Resolution 43/2007 to maintain the valid concession agreement. The regulatory compliance encompasses mainly the continuous asset integrity management from design to decommissioning. Given a broader and more complex work scope in life extension than that regulated in Resolution 43/2007, ANP is expected to develop a formal framework to regulate the life extension of aging offshore facilities.

6. NOPSEMA (Australia)

Similar to the incipient nature of ANP regulation, the Australian National Offshore Petroleum Safety and Environmental Management Authority (NOPSEMA) does not have specific rules on life extension. Bell (2016) presented NOPSEMA's position on the regulation development for asset integrity management and life extension. The philosophy of ALARP (As Low As Reasonably Practicable) is required for the risk reduction. A performance-based regime is acceptable to enable the facility operator to choose the methodology for risk management. The primary regulatory aim is to prevent loss of containment and the prevention of structural failure. Some pitfalls of poor practices were identified. These included poor maintenance of Safety Critical Equipment, omission of 3rd-party verification, lack of appropriate performance standards or lack of compliance with them, and inadequate evaluation and mitigation of inspection anomalies, etc.

REPRESENATATIVE INDUSTRY GUIDANCE

Although there is a wealth of industry standards and codes on offshore moorings [API 2005, ISO 2011, et al.] and Class rules and standards, the development of accurate and current guidance on mooring assessment life extension is very limited and extremely overdue. This is particularly true in the areas of in-situ inspection, components degradation assessment, and prediction of the remain life. While not exhaustive, a list of industry standards and codes usually referenced during a mooring life extension study is provided in Table 1.

Table 1 A List of Representative Industry Guidance on FOI Life Extension

Mooring Design
API RP 2SK (2005), "Design and Analysis of Stationkeeping Systems for Floating Structures,", Third Edition, October 2005.
ISO 19901-7 (2013) "Stationkeeping systems for floating offshore structures and mobile offshore units," Part 7 of Petroleum and natural gas industries - Specific requirements for offshore structures, Second Edition, May 2013.
API RP 2SM (2014), "Design, Manufacture, Installation, and Maintenance of Synthetic Fiber Ropes for Offshore Mooring," Second Edition, July 2014.
API RP 2FPS (2011), "Planning, Designing, and Construction Floating Production Systems," Second Edition, October 2011.
DNVGL-OS-E301 (2013), "Position Mooring," July 2015.
ABS (2018), "Guide for Position Mooring Systems," May 2018.
Mooring Inspection and Assessment
API RP 2I (2008), "In-service Inspection of Mooring Hardware for Floating Structures," Third Edition, April 2008.
API RP 2MIM (under balloting), "Mooring Integrity Management," First Edition, 2017.
DNVGL-RP-E304 (2017), "Damage Assessment of Fibre Ropes for Offshore Mooring," July 2017.

Mooring Manufacturing/Class Certification
API Spec 2F (1997), "Specification for Mooring Chain," Sixth Edition, June 1997, Reaffirmed June 2015.
API Spec 9A (2011), "Specification for Wire Rope," 26th Edition, May 2011.
API RP 2SM (2014), "Design, Manufacture, Installation, and Maintenance of Synthetic Fiber Ropes for Offshore Mooring," Second Edition, July 2014.
ABS (2017), "Guide for the Certification of Offshore Mooring Chain," May 2017.
ABS (2011), "Guidance Notes on the Application of Fiber Rope for Offshore Mooring," August 2011, updated February 2014.
DNVGL-OS-E302 (2015), "Offshore Mooring Chain," July 2015.
DNVGL-OS-E303 (2015), "Offshore Fibre Ropes," July 2015.
DNVGL-OS-E304 (2015), "Offshore Mooring Steel Wire Ropes," July 2015.
Bureau Veritas (2012), "Classification of Mooring Systems for Permanent Offshore Units," April 2012, Rule Note NR 493 DT R02 E.
Lloyds Register (2013), Rule and Regulations for the Classification of a Floating Offshore Installation at a Fixed Location. Part 2 Rules for the Manufacture, Testing and Certification of Materials, July 2013.
Mooring System Reassessment/LE
API BULL 2INT-EX (2007), "Interim Guidance for Assessment of Existing Offshore Structures for Hurricane Conditions," May 2007.
API RP 2FPS (2011), "Planning, Designing, and Constructing Floating Production Systems," Second Edition, October 2011.
ABS Guidance Notes on Life Extension Methodology for Floating Production Installations, May 2017

Notes:

1) Many of the governing standards on mooring design analysis (API and ISO) were published 5-10 years ago.

2) Codes and standards developed by API, ISO and Classes provide a good basis for the mooring system design, components manufacturing, and in-services inspections, but they mostly do not provide specific guidance on mooring life extension.

3) The most frequently referenced industry standard on mooring inspection is API RP 2I Recommended Practice for In-service Inspection of Mooring Hardware for Floating Structures [API 2008]. However, API RP 2I needs significant revision to keep up with the latest regulatory and industry requirements on the mooring inspection and assessment. More discussion and recommendations are made in the subsequent sections of this paper.

4) API RP 2MIM (currently under balloting phase) provides guidance on the inspection, monitoring, evaluation of damage, fitness-for-service assessment, risk reduction, mitigation planning, and the process of decommissioning. This recommended practice incorporates and expands on the Integrity Management recommendations found in API RP 2I (2008) and API RP 2SK (2005). With various assessment initiators including life extension request, API RP 2MIM also specify the below targeted criteria for fit-for-service together with the risk assessment and risk reduction recommendation.

    a. Analysis results satisfy the current code or recommended practice references (e.g., API RP 2SK 2005 and API RP 2SM 2014), or

    b. Analysis results indicate that some individual components do not meet current code or recommended practice requirements, but the risk of mooring failure is assessed as being acceptable.

    However, API RP 2MIM is still a high-level recommendation practice on permanent mooring system life extension and does not present the detailed guidelines on many challenging aspects, which are discused in the subsequent sections " in this paper.

5) API BULL 2INT-EX was published in 2007 to address the reassessment of the existing floating installations within their original design life due to the significant changes of hurricane parameters especially in Central GoM, which requires three-level checks including a survival check under a 100-year latest environmental condition with an FOS of 1.0. The criteria specified in this interim guideline was lower than the newly designed installation including mooring system. The guidance in this document is neither intended for use in designing new platforms nor for life extension purpose.

6) API RP 2FPS (2011), though not intended for life extension, provides the guidelines on the assessment of existing floating structures design. For floating structures subject to hurricane conditions, this API code provides specific guidance on the update of metocean conditions and three levels of engineering checks - design, survival, and robustness. In addition, the acceptance criteria are also specified for the fit-for-service assessment of existing floating structures. Specific acceptance criteria for the mooring system assessment are developed for all three levels of engineering checks.

7) Very recently, ABS published the *Guidance Notes on Life Extension Methodology for Floating Production Installations* [ABS 2017]. This is the first industry guidance that outlines the life extension process for the offshore structure, mooring system, stability, and machinery and systems, etc. In the following section, current mooring life extension practice is described in line with this ABS Guidance Notes. Examples of recent mooring life extension projects approved by ABS were presented.

CURRENT MOORING LIFE EXTENSION PRACTICE

Based on the recent mooring system life extension projects approved by ABS, the current mooring system life extension practices followed by industry are demonstrated in the below flowchart (Figure 1).

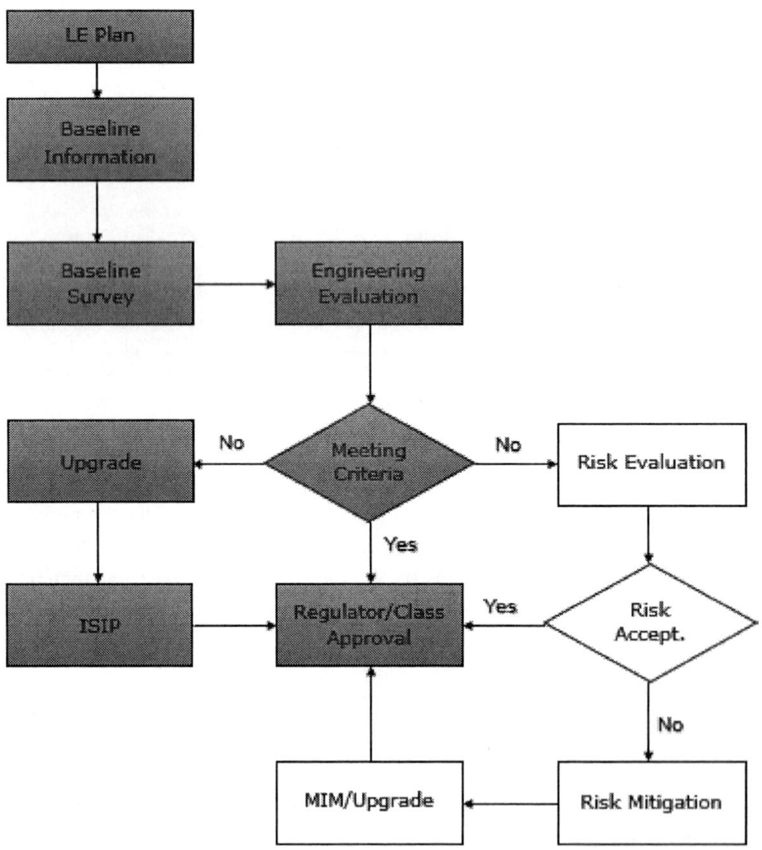

Figure 1 Sample flowchart of the mooring life extension process [ABS Guidance Note, 2017]

1. Life Extension Plan

In order to get approval of life extension projects from local authority and classification societies, a life extension plan denoting how to proceed should be developed that includes inspection, engineering work, risk assessment, risk mitigation, and response plan, etc. as applicable.

2. Baseline Information Gathering

Generally, floating production installation design and operational information is to be collected to allow an engineering assessment of mooring system integrity. It is essential to have the original design reports, documents, original and as-is plans, specifications, survey records during fabrication, installation and past service. The operator should ensure that any assumptions made are reasonable and information gathered is both accurate and representative of actual conditions at the time of the assessment. For mooring system life extension, the following information is important and to be collected:

- Updated metocean data, metocean history;
- Original and/or intermediate drawings & analysis reports for vessel and mooring system including at least the following:
    - Vessel
    - Mooring system
    - Off-vessel mooring components
    - On-vessel equipment
    - Foundation
- Mooring system previous upgrade history including repair, replacement, and operation history
- Records of installation such as pretension, piles orientation, inclination, etc.
- Measured load history of mooring, risers, tendons, etc.
- Survey and gauging reports

3. Baseline Survey

The baseline survey is the critical step in evaluating the mooring system, based on which the baseline condition of the mooring system is built. The following general steps may be followed for the mooring baseline survey:

- Review of reports of previous surveys and maintenance
- Development of an inspection procedure
- Complete inspection (including underwater inspection) to verify that an accurate assessment of the installation's condition is obtained
- The scope of baseline survey is to be not less than the next upcoming Special Survey specified by Class society
- The baseline survey information can be collected from the surveys leading up to the expiration of the design life. All survey items are to be up to date and examined within the previous 5 years. All anomalies are to be resolved or accounted for in the life extension plan

The detailed inspection requirement and expected information for mooring system evaluation are to be achieved during the mooring baseline survey:

- Carry out general visual inspection (GVI) of mooring components and connectors, and top of piles
- Carry out GVI of fairleads and mooring equipment (i.e., windlasses, chain jacks, chain stoppers, etc.). Additional inspections such as close up visual inspection (CVI) and/or NDE (wet magnetic particle inspection (MPI) or similar) may be applicable for suspect areas identified by the GVI. Additional/alternative inspection methods can be approved on a case-by-case basis by the regulators or class societies

- Inspect for cracks, corrosion, dimensional checks, and wear, as accessible. A reasonable length of each mooring chain is to be cleaned to ensure the overall condition of each chain can be satisfactorily verified.
- Inspect wire ropes and synthetic ropes for mechanical damage, twist and sheathing conditions, including anodes, on wire sockets (if installed).
- Non-inspectable areas are to be discussed and agreed upon on a case-by-case basis upon between Class and Owner/Operator.
- The vessel weight survey to verify vessel drafts, center of gravity and mass inertia information.
- Mooring system pretension, configuration such as polyester rope length changes, pile location, etc.
- Any other anomalies that may affect the mooring system performance such as trenching, etc.

4. Engineering Reassessment

Based on the baseline information and baseline survey, the re-evaluation on the strength and fatigue of the mooring system including mooring components, equipment, and foundation are to be conducted to meet the criteria specified by the regulators and class society. During the re-evaluation process, the following items require special attention:

Metocean Criteria:

Under the condition when the environmental condition is changed since the original design, the latest environmental conditions are to be used for the reassessment.

Mooring System Reassessment:

Mooring system models are to be developed and updated incorporating wastage in the model while assessing mooring strength and remaining fatigue life. The vessel, mooring, and riser information to be used should be on as-built condition and calibrated by the baseline inspection.

The fatigue damage assessment occurred in the past service life is to be based on the available historic information such as measured metocean, tension history, etc. The corresponding safety factors can be reduced due to the uncertainty associated with the past fatigues are reduced. If such information is not available, the past fatigue damages should be treated same as those for the future years, and the safety factors applied to the past fatigues should not be reduced any more.

Applying the same corrosion verified for chains to components, detailed calculations for individual components using as-is conditions and actual loads are to be carried out. The degraded mooring components such as corroded/worn chain and wire rope beyond its original design life require special attention. Like-for-like replacement for such components is recommended. However, if the retrieved components are available for strength and/or fatigue testing to demonstrate their performance still meet the rule requirement, the life extension still can be granted for such components with mooring integrity management such as inspection and monitoring implemented.

If possible, the segment of the degraded mooring components is to be retrieved and tested for the remaining strength and fatigue performance to support the numerical calculation.

For foundation system, the reassessment of the anchor holding capacity is required if the maximum mooring load responses are increased.

Acceptance Criteria

Mooring system models are to be developed and updated incorporating wastage in the model while assessing mooring strength and remaining fatigue life.

The mooring strength reassessment is to be performed for the latest environmental conditions For the extended life, the Safety Factors for mooring strength are to meet the ABS rule requirement, i.e., 1.67/1.25 for intact and damage condition, respectively.

The Safety Factors for Fatigue Life are to be as required by the Class rules. In calculating the accumulated damages or used-up fatigue life, the original Safety Factors for Fatigue Life may be reduced providing that the technical justifications submitted by the Owner/Operator reflect a reduction of the uncertainty in the original design and subject to the following conditions:

- No past findings
- Corrosion rate to be within the original design assumption
- Reliable load/metocean history

However, the Safety Factors for Fatigue Life for the accumulated damage are to be not less than the lowest safety factors such as three (3) specified by ABS *Guidance Notes on Life Extension Methodology for Floating Production Installations* [ABS 2017]. For the extended life, the Safety Factors for Fatigue Life are to be in accordance with the current Rules.

5. Outcomes of Mooring Life Extension Study

Regulatory/Class Approval

Once the reassessment of the mooring system including mooring components, equipment and foundation system is completed and the results show the mooring system can still meet the rule-required criteria, then the life extension may be granted. Otherwise, as demonstrated in the procedure flowchart, the upgrading of the mooring system either partially or completely may be warranted since those degraded components may compromise the mooring system integrity.

In the case that the mooring system design criteria cannot be met due to reasons such as environmental condition changes while the mooring system itself still in good condition, or the degradation of the mooring components are insignificant, the risk associated with those reduced performance or degraded components is to be assessed. The risk reduction/mitigation plan may be developed to reduce the associated risk, which can include the load reduction, limiting the operating condition, evacuation plan in the event of emergency, to avoid any potential impact to the HSE issue. However, risk assessment for mooring system still did not reach industry consensus for the methodology and acceptance criteria.

In addition to numerical assessment, the In-Service Inspection Plan (ISIP) is also important to consistently monitor the mooring system condition, especially the degraded components if not replaced during the life extension process. This ISIP is to be submitted for regulator or Class society approval. For mooring system monitoring, it could be project-specific and so far the regulatory and Class rules do not have mandatory requirements on the means to monitor the mooring system. Similar to inspection, it relies on the available technology and the reliability of such technology.

GAPS AND CHALLENGES

The offshore industry has witnessed the evolution of practices in extending the field life of offshore assets including the associated mooring systems through the engagement and collaboration of all stakeholders including operators, regulators, classes and engineering services providers. Recognizing the not well-established practice and the lack of industry consensus, this section presents some gaps and challenges.

1. Availability and Quality of Baseline Information

The lack of historical data (part of the baseline information) is the first challenge to overcome when a mooring system life extension is initiated. Following the ABS *Guidance Notes* [ABS 2017], it is often difficult to gather the following documentation, incorporating all the changes over the field operations.

- ***Detailed vessel and topsides design drawings and up-to-date hydrostatic data of the FOI***: These include, but are not limited to, original vessel drawings, changes in dimensions, structure and equipment weights and distribution, ballast tank levels, updated trim and stability parameters as well as wind and current coefficients.
- ***Manufacturing data books of all mooring components, both on-vessel and offshore vessel***: A mooring system often includes various components manufactured by a number of vendors. All manufacturing data books contain

the traceable certified records material properties, dimensional and weight data, key manufacturing processes, quality control examinations and proof/breaking load tests, and dimensional control measurements. It is critical for a mooring life extension study to gather all the components data books to establish the as-manufactured baseline information. For example, chain link dimensional measurements can be used to determine the corrosion and wear over time.

- ***Original design analysis reports and drawings of the mooring system and design reports of individual mooring components*:** These reports provide the design premises on the original metocean criteria, field life, load cases, factors of safety, corrosion and wear allowance or protection measures, and the applicable design codes and standards. Mooring design drawings define the system layout, fairlead and anchor locations, line payout and pre-tension, etc. All these documents are important references for the mooring life extension re-analysis.
- ***As-built survey of the mooring system*:** This is usually carried out right after the completion of the mooring installation or before the facility production. A quality as-built report often yields valuable benchmark data for both mooring design model calibration and future mooring inspections. Typical survey data include the mooring line tracks, touchdown point locations, pile inclination and locations (if applicable), line profiles defined by connector depth, lengths of chain and polyester segments, and measured line tensions etc. Right after the as-built survey, the mooring design model can be updated and calibrated with as-built information. For a mooring system encompassing polyester ropes, the model calibration often involves the check of selected stiffness and measured rope lengths.
- ***Operational history:*** this includes the records of mooring system modification, equipment of component repair and replacement. Like "health checks", findings from periodic surveys as part of the integrity management or after a major storm should be well documented. Consistent inspection methods should be used as in the as-built survey. Mitigation activities for the discovered anomalies should be tracked. It has to be recognized that not every FOI has well maintained and properly functioning mooring monitoring equipment. However, any measured and validated load history of the mooring lines could provide vital information during the life extension study.

2. Acquisition of Current Conditions of Mooring Components

A baseline survey as per the regulatory requirements [e.g. USCG 2016] serves as a means for acquiring the most recent conditions of the mooring components. However, to fulfill those requirements, the following gaps are identified.

- *Inadequacy of industry guidance*: As indicated in API RP 2I, deterioration of permanent mooring components is usually characterized by corrosion, wear, and damage, which are often attributed to the mooring failures as a result of reduced strength and fatigue resistance. Very limited guidance is provided in API RP 2I on the measurement/inspection for corrosion and wear for steel components. For chain diameter measurements, two readings in two perpendicular directions are required after removal of corrosion scale and marine growth. In the interlink grip areas, only the double diameter measurement in the longitudinal direction can be practically made. Wire ropes of permanent mooring are usually of the sheathed spiral strand construction. The inspection for wire ropes typically only includes the visual observations by ROV (or limited diver access in shallow water) to check for broken wires or internal corrosion where the sheath is damaged. For other steel mooring connectors protected by anodes, only the depletion of the anodes is used as an indication or corrosion protection. Note that the above-mentioned inspection and measurement methods are far from sufficient to provide the corrosion and wear data for mooring life extension, particularly:
  - Pitting corrosion of the mooring chain cannot be quantitatively measured by simple diameter measurements.
  - The cross-section of the grip zone of a chain link often assumes an oval shape due to the bar bending during manufacturing. As allowed by Classes, the diameter in the longitudinal direction at the grip zone could be less than the nominal diameter (so called negative dimensional tolerance). Measuring two diameters at the grip zone is not able to unbiasedly capture the corrosion and wear by only measuring the diameter in one direction.

- The prescribed inspection method for sheathed wire ropes is not going to detect any internal corrosion or breaks unless a damaged sheath exposes those deterioration.
- For mooring connectors, only qualitative guidance was provided for visual inspection.
- There is more ambiguity in the inspection of polyester rope. Destructive inspection and test on recovered inserts is denoted in APIR RP 2I. Nevertheless, in the absence of testing of the inserts, there is little information that can be obtained from in-situ inspection other than the visual observation of the jacket integrity.

- ***Limitations of In-Situ Inspection Technology*****:** Unlike the MODU moorings, the long-term or permanent mooring system of an FOI can only be inspected in-situ during its service. In most of the cases, general visual inspection (GVI) and close-up visual inspection (CVI) are carried out on both on-vessel and off-vessel mooring components to gather qualitative description, supplemented by non-destruction examination (NDE) or wear and corrosion measurements for steel components [API RP 2MIM, currently under balloting]. For example, ROV-operated GVI and CVI were used to inspect wire rope and polyester rope twist, mechanical damage and sheathing integrity, monitor local scour/trenching around anchor piles. GVI and CVI are more readily used for the condition evaluation of on-vessel equipment, such as the fairleads, chain jacks and chain stoppers, or mechanical parts of a turret mooring system. Magnetic particle inspection (MPI) is usually employed to for surface crack detection. Recently, underwater stereoscopic photographic and video imaging techniques are being developed and deployed [Hall and Trower 2011, Allan et al 2013] for detailed measurements of rope cross sectional area and 3D representation of chain links and connectors. Despite the constant progress, many technological challenges and limitations have been recognized for the in-situ mooring inspections. Some of these challenges include:
  - Removal of hard marine growth
  - Inspection mooring components covered by hard marine growth
  - Dimensional measurements of cross sections of chain links
  - Three-dimensional reconstruction of mooring components after underwater scanning or imaging with desirable speed and accuracy (as follow-up inspection after severe deterioration such as pitting, wear or deformation has been confirmed)
  - Quantifiable cross-sectional examination of sheathed /jacketed wire or polyester ropes when the sheath/jacket is compromised
    - Detection of cracks on metal surface or on the wire rope sheath
    - Inspection of the worn chain link at the touchdown area
  - Inspection of buried mooring components (e.g. ground chain and anchor piles)

3. Fit-for-Service Assessment of Aging Mooring Components

There are very limited guidelines available on how to evaluate the strength and fatigue performance of the degraded mooring components/equipment such as chain, wire rope, polyester rope, socket, subsea connectors, mooring equipment, etc.

- ***No rationalized technical basis for discard/acceptance criteria:*** API RP 2I remains as the main industry code providing the discard criteria for various aging mooring components. However, no technical justification is given. For permanent mooring chain, the allowable strength reduction is 10% based on end of life breaking strength after corrosion and wear. When applied to the field chain diameter measurements, this discard criterion is often interpreted as either 1) an average of two measured diameters less than 95% of the nominal diameter or 2) a diameter in any direction less than 90% of the nominal diameter. The same allowable 10% strength reduction is specified for wire and polyester ropes. Without considering different safety margins used for different mooring systems, this discard criteria of 10% allowance strength reduction for chain, wire and polyester ropes, though specific and simple, is not rational. No quantitative discard criteria is currently available for mooring connectors. The chain discard criteria specified in APR RP 2I could be more problematic when applied to chain links subject to pitting corrosion or abrasion. It was found in the SCORCH JIP that pit depth growth rate could be up to 10 times the uniform corrosion rate [Fontaine et al. 2014a]. The correlation of loss of strength with loss of metallic area due to uniform corrosion and pitting corrosion are saliently different. Based on the FEA (FEARS JIP Finite Element Analysis of Residual Strength) and breaking test data, Rosen et al. (2015) reported that the relationship between capacity loss and cross section area reduction could be non-linear for some regions of the chain link that typically limiting the strength of the link. Though chain FEARS JIP aimed to address discard criteria for mooring chains, no analytical model or set of guidelines were provided for application to generic cases. In addition, industry consensus on the technologies used in those JIPs are yet to be reached.

In the mooring reassessment for life extension, the industry still did not yet reach the consensus on acceptance criteria such as the governing design standards in general and factors of safety specifically. As per ABS Life Extension Guidance Notes [ABS, 2017], the original standards and codes are targeted to be met for the life extension studies. However, as per the drafted API RP 2MIM, the latest mooring standards and codes are to be met. The mooring system strength and fatigue safety factors in the past decades did not change significantly, variation in the values of factor of safety does exist among such as API standards and class rules. In addition, some topics raised by the mooring industry in recent years are not well covered in the original design codes such as chain Out-of-Plane Bending (OPB) fatigue.

- ***How to estimate the remaining strength and fatigue life?*** There is no adequate knowledge in the industry to readily estimate the remaining strength of an aging mooring component. For mooring chain subject to uniform corrosion, empirical formulae is available to estimate the remaining strength [DNVGL 2015], which has been known to be conservative. When other chain degradation mechanisms are present, such as abrasion, wear, or pitting, a more sophisticated method has been proposed combining Finite Element Analysis (FEA) and/or laboratory break tests [Rosen et al. 2015, Crapps et al. 2017]. However, all the FEA methods require the three-dimensional geometry of degraded chain links, which is most favorably acquired from in-situ measurement without unnecessarily recovering the chain from the field.

There are more unknowns on how to estimate the remaining strength of aging wire ropes. Only qualitative industry knowledge in estimating the field life has been gathered from limited number of tests of aging wire ropes retrieved from the field (e.g. SCORCH JIP). It has been found once the cathodic protection coating is depleted and wires are exposed to seawater, the corrosion process begins, resulting in rapid reduction in rope strength. To apply knowledge from the SCORCH JIP to the assessment of the aging wire ropes, it seems the key information needed is the corrosion conditions of individual wires. Also pointed out in this JIP is the development of more rigorous numerical models, instead of the current empirical approaches, to estimate the service life by taking into account field-specific environmental and loading conditions, rope design and properties, and progression of wire corrosion.

DNVGL published a recommended practice [DNVGL-RP-E304 2017] on damage assessment of fiber ropes including the strength and fatigue life assessment. This is the most comprehensive guidance available in the industry. However, all the specified assessment methods require the access to and inspection of the subropes, which are most likely not available from the ROV-assisted underwater inspection.

Remaining strength and fatigue performance of mooring connectors components are not being studied in detail. Mochet et al. (2014) reported a forensic joint study program on limited quantities of decommissioned mooring connectors after at least 10 years in service (some after 20+ years). The test programs focused on the dimension check against as-built data, mechanical and break load tests, and chemical composition analysis. One alarming finding from the tests is the measured Charpy toughness of various aged connectors including shackles, tri-plates and open sockets are significantly lower than the current standards (e.g., 50J at -20°C),. In some cases, mechanical tests of tri-plates over 20 years in field service revealed the decrease in the toughness by approximately 40% when compared against the original values manufacturing certificate. Based on a review of the existing data available then, Sauvage and Aubert (2008) suggested that this non-conformance problem of Charpy toughness appeared to be more wide-spread than originally thought. The Mooring Components Assessment (MCA) JIP has been formed since then to address this problem and more detailed findings and recommendations are expected to be published. In the context of mooring life extension assessment, uncertainties and consequences of reported lower Charpy toughness of aged mooring components remain to be rigorously investigated and mitigated. It's unknown how to fully understand its impact on mooring connector structural integrity and to avoid brittle failure under extreme or cyclic loading.

SUMMARIES AND RECOMMENDATIONS

Extending the life of a mooring system is one of the key contributors to achieve the economics of the continued service of a long-term FOI. Considered as a safety critical element, offshore mooring systems are well known for their complexity in their design, manufacturing, installation, and importantly their monitoring and inspection to ensure the life-long integrity in the harsh marine environment.

Amid the continuous development of regulatory requirements and industry guidance, quite a few FOIs have extended the life of their permanent mooring systems. This paper presents an up-to-date review of the industry practice on mooring life extension and identified the following recommendations.

1. Further development of industry standards and codes

Many industry standards and codes for offshore moorings, such as the API RP 2SK and ISO 19900 series, are overdue for a major update to provide guidance for mooring life extension. Pioneered by ABS, other Class Societies are also developing respective rules to re-certify the long-term offshore facilities including the mooring system. However, no specific reassessment guidance for various mooring components have been established. The critical knowledge gaps mainly exist in field inspection, data acquisition, understanding of the component deterioration mechanism, and the evaluation and estimation of the residual strength and fatigue life. Consistency and clarity in the industry guidance is essential for the rationalized fit-for-service reassessment of the mooring system and components. Industry guidance should capture leading practice and technology development and be based on rigorous research and field proof.

2. Life-Cycle mooring data management

Life extension reassessment of an aging mooring system requires the history data retention and availability over all phases of the system development including design, manufacturing, installation, and operation. The historical data could include the original design criteria based on the earlier version of the code, material properties of individual components, manufacturing quality control records, as-installed survey reports, as well as monitored in-service mooring performance. In addition, findings and measurements from periodic mooring surveys provide the vital progressive "health" checks from the aging process or accidental damage. The rapidly evolving digital technology (e.g. digital twin) is promising to provide a centralized data management for easy access and efficiency in maintaining and updating.

3. Development of new in-service inspection technology

Development of new technology for in-service inspection of offshore moorings deserves more industry efforts. Effective evaluation and decision-making for mooring life extension requires detailed knowledge of the current state of the components. To supplement the existing ROV-based GVI/CVI, development of new in-service inspection technology for permanent mooring systems is needed to provide more quantifiable data for life extension assessment. The focus of the new technology development will be placed on the quantification of the well-recognized mooring degradation mechanisms and failure modes [Ma et al. 2013, Majhi and D'Souza 2012, Gordon et al. 2014]. Solutions to those specific technological challenges mentioned in the previous sections should be rigorously pursued. The key performance indicators of a new technology will rely on the data acquisition speed, measurement accuracy, spatial coverage, and flexibility for different inspection objects and locations. It is expected that various in-service inspection technologies can be deployed for different levels to suit the survey needs ranging from global screening for anomalies, localized additional quantification, and detailed in-depth examination and measurement.

4. Research on aging mooring component assessment

More fundamental research on rigorous assessment of aging mooring components is warranted, via industry collaboration and information sharing. Beside the scarcity of field data, another source of uncertainty in the mooring life extension assessment is the lack of knowledge to estimate the remaining strength and fatigue life of the degraded mooring components. Motivated by the life extension needs, corroded mooring chain samples were retrieved and tested [Fredheim et al. 2013, Gabrielsen et al. 2017]. The research emphasis was placed on the chain fatigue capacity as a result of corrosion, wear, and other anomalies. All the retrieved chain samples underwent thorough laboratory cleaning, inspection, and fatigue tests. Post-mortem mechanical test and metallurgical analyses were also performed. Collectively, all the available test data is still very limited to develop generalized engineering methodology that can be readily used for quantifying the effect of aging on mooring chain. It is noted that study on the remaining fatigue life of corroded chain is much more complicated than new chain given the plethora of forms from time-dependent degradation and site-specific loading. These forms of degradation could be manifested by the degree of corrosion, types of corrosion, mechanical wear and damage, as well as manufacturing imperfection. Supported by the laboratory tests, FEA in conjuncture with fatigue design curves of corroded chain is also under development to help predict residual fatigue life [Farrow et al. 2017 OMAE]. More industry collaboration on corroded chain assessment is expected to establish the industry-wide guidance on the discard criteria.

Other critical mooring components such as wire ropes and connectors recognized by mooring failure statistics [Fontaine et al. 2014b] have received much less attention for research on degradation assessment. It is important that more extensive research on mooring degradation mechanisms and tests on decommissioned mooring components be performed to help better understand the impact of time-dependent and site-specific deterioration processes to enhance the reliability of mooring life extension.

5. Risk assessment and risk mitigation

As per API RP 2MIM, risk assessment techniques should always be used to demonstrate the fitness-for service of a mooring system. A robust risk assessment depends on the accuracy and precision of the available data that could well capture the as-is conditions and the degradation processes. Risk reduction measures could be implemented to achieve the likelihood or consequence of a mooring failure to an acceptable level.

Besides the four risk assessment cases identified in API RP 2MIM (combinations of original design, original criteria, as-is conditions, and current criteria), it is important to include the case of projected mooring conditions towards the end of the extended field life. In line with the framework developed by, the following recommendations were made for the risk assessment of a mooring system intended for life extension:

- The original design standards and codes should be gathered with gaps identified after comparison with the latest requirements and regulations
- The original design premises including the metocean conditions and functional requirements should be compared with the intended purposes during the extended life.
- The as-is mooring conditions should be well documented based on factual data collected from inspection and monitoring.
- The as-is mooring conditions should be assessed with the current metocean criteria and current design criteria.
- The projected mooring conditions towards the end of extended life should be assessed with the current metocean criteria and current design criteria.
- The non-conformance of the mooring system to current codes and standards in the as-is and projected conditions should be evaluated in the risk matrix of failure probability and consequence. Failure mode, effects and criticality analysis (FMECA) could be carried out at both the component and system level to understand the likelihood of failure and the consequence.
- To mitigate the non-conformance, risk reduction measures could include the enhanced mooring integrity management plan including monitoring and inspection, or the replaced or strengthened mooring components, or modified operational procedures, etc. The key is to establish the agreement among all the stakeholders including regulators and classification societies to the conclude a mooring system is fit-for-service.
- Implementation of risk reduction measures should be periodically audited and reviewed for continuous improvement.

Given the gaps and uncertainties discussed in the previous section, it is obvious that uncertainties associated with risk assessment are always present. It should also be noted that risk reduction may not always work under any condition. For example, if the numerical assessment shows that the code-required safety factors had been significantly compromised, this risk assessment and risk reduction may not able to limit the likelihood or consequence of the mooring failure to the acceptable level. Although some industry guidelines such as API RP 2MIM attempt to address this topic, the procedure and acceptance criteria for mooring risk assessment/mitigation should be investigated in depth and offshore mooring industry consensus should be reached.

ACKNOWLEDGEMENTS

The authors thank the management of Chevron ETC, ABS and CNOOC for permission to prepare this paper. The views expressed in the paper are those of the authors and do not represent those of the affiliated companies or organization.

REFERENCES

1. ABS (2017), "Guidance Notes on Life Extension Methodology for Floating Production Installations," May, 2017.
2. Allan S., T.d. Gier and A. Hall (2013), "Mooring System Life Extension Using Subsea Inspection Technologies," OTC 24184, Houston, Texas, 6-9 May, 2013.
3. API BULL 2INT-EX (2007), "Interim Guidance for Assessment of Existing Offshore Structures for Hurricane Conditions", American Petroleum Institute, May 2007.
4. API RP 2FPS (2011), "Planning, Designing, and Construction Floating Production Systems," Second Edition, American Petroleum Institute, October 2011.
5. API RP 2I (2008), "In-service Inspection of Mooring Hardware for Floating Structures," 3rd Edition, American Petroleum Institute, April 2008.
6. API RP 2MIM (under balloting), "Mooring Integrity Management," First Edition, American Petroleum Institute, 2017.
7. API RP 2SK (2005), "Recommended Practice for Design and Analysis of Stationkeeping Systems for Floating Structures", 3rd Edition, American Petroleum Institute, October 2005.
8. API RP 2SM (2014), "Design, Manufacture, Installation, and Maintenance of Synthetic Fiber Ropes for Offshore Mooring," Second Edition, American Petroleum Institute, July 2014.
9. Mochet, C., F. Parquereau and N. Decultot (2014), "Mooring Integrity: Forensics programs on Used Mooring Connectors – Lessons Learned, Present Practices & Futire Development," DOT-2014-79, Deep Offshore Technology International Conference, Aberdeen, Scotland, 14-16 October 2014.
10. Crapps, J., H. He, and D. Baker (2017), "Strength Assessment of Degraded Mooring Chains," OTC-27549-MS, Offshore Technology Conference, Houston, Texas, USA, 1-4 May 2017.
11. DNVGL-RP-E304 (2017), "Damage Assessment of Fibre Ropes for Offshore Mooring," July 2017.
12. Farrow, Gary, Andrew Potts and Daniel Washington (2017), "Investigations into Fatigue Performance of Offshore Mooring Chains," Proceedings of the ASME 2017 36th International Conference on Ocean, Offshore and Arctic Engineering, OMAE2017-62218, 25-30 June 2017, Trondheim, Norway.
13. Fontaine E., J. Rosen, A. Potts, KT Ma and R. Melchers (2014a), "SCORCH JIP – Feedback on MIC and Pitting Corrosion from Field Recovered Mooring Chain Links," OTC 25234, Offshore Technology Conference, Houston, Texas, USA, 5-8 May 2014.
14. Fontaine E., A. Kliner, C. Carra, and D. Washington (2014b), "Industry Survey of Past Failures, Pre-emptive Replacements and Reported Degradations for Mooring Systems of Floating Production Units," OTC-25273-MS, Offshore Technology Conference, Houston, Texas, USA, 5-8 May 2014.
15. Fredhelm, S, SA Reiholdtsen, L. Haskoll, H.B. Lie (2013), " Corrosion Fatigue Testing of Used, Studless, Offshore Mooring Chain," Proceedings of the 32nd International Conference on Ocean, Offshore and Arctic Engineering, Nantes, France, 9-14 June, 2013
16. Gabrielsen, Øystein, Kjell Larsen and Svein-Arne Reinholdtsen (2017), "Fatigue Testing of Used Mooring Chain," Proceedings of the ASME 2017 36th International Conference on Ocean, Offshore and Arctic Engineering, OMAE2017-61382, 25-30 June 2017, Trondheim, Norway.
17. Gordon, Robert, Martin Brown and Eric Allen (2014), "Mooring Integrity Management: A State-of-the Art

Review," OTC 25134.

18. Gordon, Robert (2015), "Considerations for Mooring Life Extension," Proceedings of the 20th Offshore Symposium, February 2015, Houston, Texas, Texas Section of the Society of Naval Architects and Marine Engineers.
19. Hall, Anthony and Alastair Trower (2011), "Mooring System Integrity: Deteriorative Mechanisms on Mooring Systems and Appropriate Inspection Techniques," Brasil Offshore Technology Conference, OTC 22615, Rio de Janerio, Brazil, 4-6 October 2011.
20. ISO (2011), "Stationkeeping systems for floating offshore structures and mobile offshore units," Part 7 of Petroleum and natural gas industries - Specific requirements for offshore structures, ISO 19901-7.
21. Ma, K.T., A. Duggal, P. Smedley, D. L'Hostis, H. Shu (2013), "A Historical Review on Integrity Issues of Permanent Mooring Systems," OTC 24025.
22. Majhi, Sai and Richard D'Souza (2013, " Application of Lessons Learned from Field Experience to Design, Installation and Maintenance of FPS Moorings," OTC 24181.
23. Norwegian Oil and Gas (2017), "122 – Norwegian Oil and Gas Recommended Guidelines for the Management of Life Extension," Norsk Olje & Gass, No: 122, May 2017.
24. Pires T.S., C.P.M. Morais, R.D.A. Freitas, A.L.T.O. Monteira (2016), "Discussion on the Life Extension of Offshore Production Facilities," Rio Oil & Gas 2016 Expo and Conference, 24-27 October 2016, Rio de Janeiro, Brazil.
25. PSA (2008), "Requirements for Life Extension of Ageing Offshore Production Installations," Petroleum Safety Authority, 17554/1/08, January 2008.
26. Rosen, J., A. Farrow, A. Potts, C. Galtry, W. Swedosh, D. Washington, and A. Tovar (2015), " Chain FEARS JIP: Finite Element Analysis of Residue Strength of Degraded Chains," OTC-26264-MS, Offshore Technology Conference Brasil, Rio de Janeiro, Brazil, 27-29 October 2015.
27. Sauvage, Jean-Pierre and Jean-Michel Aubert, Mooring Components Assessment JIP Scope of Work, FPSO Research Forum, Bilbao, Spain, 12 November 2008.
28. Stacey, A., M. Birkinshaw and J. V. Sharp (2008), "Life Extension Issues for Ageing Offshore Installations," Proceedings of OMAE2008 27th International Conference on Offshore Mechanics and Arctic Engineering, OMAE2008-57411, 15-20 June 2008, Estoril, Portugal.
29. UK HSE (2007), "Ageing semi-submersible installations," Offshore Information Sheet No. 5/2007.
30. UK HSE (2009), "Guidance on Management of Ageing and Thorough Reviews of Ageing Installations," Offshore Information Sheet No. 4/2009.
31. UK HSE (2014), "Key Programme 4 (KP4) - Ageing and Life Extension Programme," A report by the Energy Division of HSE's Hazardous Installations Directorate, May 2014.
32. USCG (2016), "Interim Guidance on Life Extension (Continued Service) Requirements for Floating OCS Facilities,", Unite States Coast Guard, January 2016.

ASSESSMENT OF NEARSHORE FSRU/FLNG MOORING

AMITAVA GUHA
AMERICAN BUREAU OF SHIPPING

SUE WANG
AMERICAN BUREAU OF SHIPPING

Proceedings of the 24th Offshore Symposium, February 20th, 2019, Houston, Texas
Texas Section of the Society of Naval Architects and Marine Engineers

ABSTRACT

Nearshore solutions for storage and regasification of liquefied natural gas using large floating structures such as FLNGs (Floating Liquefied Natural Gas) and FSRUs (Floating Storage and Regasification Units) are becoming popular due to their economic and safety advantages. Permanent mooring systems are designed specifically to keep these large floating structures within an acceptable excursion range for the site-specific environmental conditions and design life of the structure. The typical jetty mooring arrangement commonly used for temporary berthing is upgraded to nearshore position mooring of FLNG/FSRUs. The higher mooring loads and longer design life require consideration of several factors, which are discussed here. A case study has been carried out based on the ABS Guidance Note on Nearshore Position Mooring to provide a detailed analysis procedure. The effect of bathymetry, water depth, tide, directional wind, wave and current, interaction between the mooring lines, fenders and the vessel, interaction between the LNGC (Liquefied Natural Gas Carrier) and FLNG are studied. Sensitivity of mooring load prediction to various modeling parameters is also presented. Finally, an overview of the latest industry effort of developing guidelines for the assessment of environmental loads and nearshore mooring system is given.

Keywords: FSRU, FLNG, LNGC, Jetty Mooring, Nearshore Mooring, Shallow Water

1 INTRODUCTION

Floating terminals are large infrastructures that provide a convenient means for the storage and regasification of liquefied natural gas. These vessels, known as FLNG (Floating Liquefied Natural Gas) and FSRU (Floating Storage Regasification Units), receive LNG from LNG carriers, store it, perform pressurization and regasification, and finally deliver the gas to on-shore gas pipeline grid. There are now 29 FSRUs in service and 12 on order (Corkhill, 2018). The flexibility of location, low environmental impact and related economic benefits are the main drivers of the global rising interest in these facilities.

The transfer of LNG from carrier to terminal (FSRU/FLNG) is carried out using cryogenic hoses and unloading arms affixed to the terminal. As the operations are dependent on the global motion of the terminal as well as relative motion between the LNGC and the terminal, the mooring system must be designed accordingly. The project site location and metocean conditions are important factors that dictate the operability and the cost of mooring systems. The FSRU/FLNG vessel can be moored either within a port against a berth or at a purpose-built mooring system to maintain position. Nearshore locations are preferred for easy connectivity to land, however, there has also been increased interest in moving LNG activities from inside harbors to less sheltered locations. In this case, the vessel may encounter increased wave, wind and current loadings.

The technology required for nearshore position mooring is mostly borrowed from that of the temporary berthing of ocean-going vessels; however, there are also a number of similarities with offshore mooring due to the long duration and operational requirements. For example, mooring equipment onboard the facility is controlled on the facility, and the production is continuous for a relatively longer period of time. The strength assessment is recommended to use a 100-year return period environment similar to offshore mooring as well. The factors which are similar to temporary berthing are direct access to the shore, site-specific environmental conditions, short in air mooring lines with short natural period mooring systems, and easy to adjust and easy to inspect mooring arrangements. In comparison with offshore deep water conditions, the nearshore shallow water environment imposes different challenges (Guha and Falzarano, 2016). The FLNG/FSRU is often subjected to loads from passing vessels (Vasudevan and Nallayarasu, 2017) and surrounding structures. The bathymetry could also influence the response characteristics of the FLNG/FSRU (Hauteclocque, 2007) and the local wave conditions that act on the FLNG/FSRU. The interaction between the mooring system and the first order vessel motions is more pronounced than that in deep water conditions due to its relatively shorter and stiffer mooring lines.

The industry has for many years used the jetty mooring configuration for temporary berthing and has guidelines for the design and operations of these systems. Oil Companies International Marine Forum (OCIMF) and Permanent International Association of Navigation Congresses (PIANC) published a number of guidelines to support the safe operation of temporary jetty moored facilities. However, guidelines for the use of jetty mooring for FSRU/FLNG are limited. This paper provides an overview on the challenges of the use of jetty mooring systems for FSRUs/FLNGs and a class perspective on this evolving methodology. The reader may refer to the ABS Guidance Notes on Nearshore Position Mooring (American Bureau of Shipping, 2018) for further details.

2 Jetty Mooring Guidelines

A Jetty is a bottom-fixed structure typically projecting from land out into water to which a floating vessel can be fastened using the jetty mooring system. Typically, the jetty mooring system consists of:

- Mooring lines made of fiber ropes, chain, wire ropes or a combination thereof
- Fenders
- Fitting and accessories
- Quick release hooks
- Mooring equipment on board the floating structure
- Jetty structure with mooring bollards and other mooring equipment

Different mooring arrangements are applied based on size of the vessels, jetty configuration and cargo handling ability. The most common configuration used during loading and unloading are shown in Fig. 1 and listed below:

- Side-by-side Mooring
- Tandem Mooring
- Cross Dock Jetty Structure Mooring

The mooring system serves to keep the large floating structure in place and supports operations such as berthing and un-berthing of the LNGC or other carriers, loading and offloading of gas and other products, disconnection and departure of floating structures, and other equipment on board that could be affected by the mooring system and the floating structure.

The design life of the mooring system for FLNGs/FSRUs is much longer than usual temporary berthing, generally in the order of 20 years. It could be different than the design life of the floating structure to which it is connected. The actual life of the mooring line components, fenders and equipment can be extended depending on ease of repair or replacement. Such components should be periodically inspected during their service life. The complete mooring system should be designed considering survival condition, operation condition, one component damage condition and disconnecting condition, if applicable. Where spare components are available, and replacements can happen quickly and can be assumed that there is no effect on the position holding capability due to damage of the component, the damage condition of such component may be excluded. Fenders may be considered as a no-fail component.

The jetty mooring system is designed based on site-specific environmental conditions such as wind, waves, swell, seiche, current, tide, water depth, storm surge, bathymetry, ice and snow, marine growth, seismicity, and soil conditions. 100-year return period conditions for wind, wave, current and maximum high water and minimum low water should be considered for survival or disconnecting condition. Operating condition assessment may be performed for one-year return period environmental conditions. The mooring system should be assessed under the most unfavorable combination of wind, wave, and current directions that can be reasonably assumed to occur.

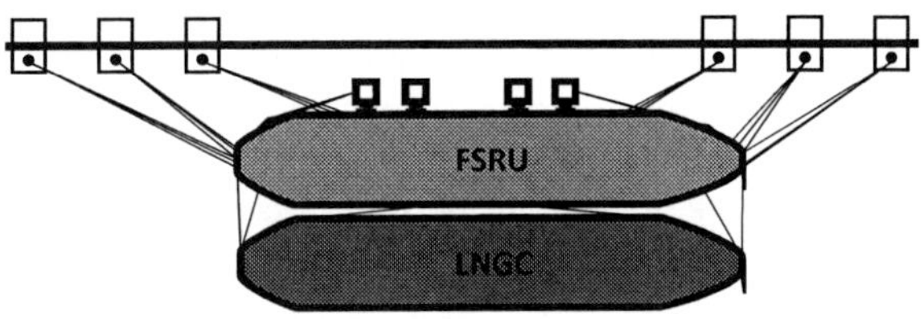

a) Side-by-side Mooring

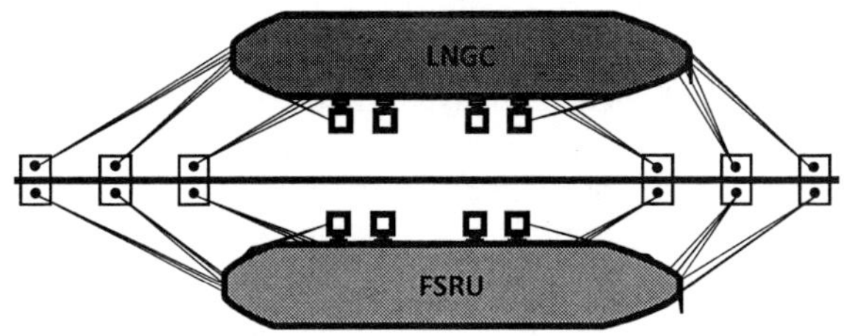

b) Cross Dock Jetty Mooring

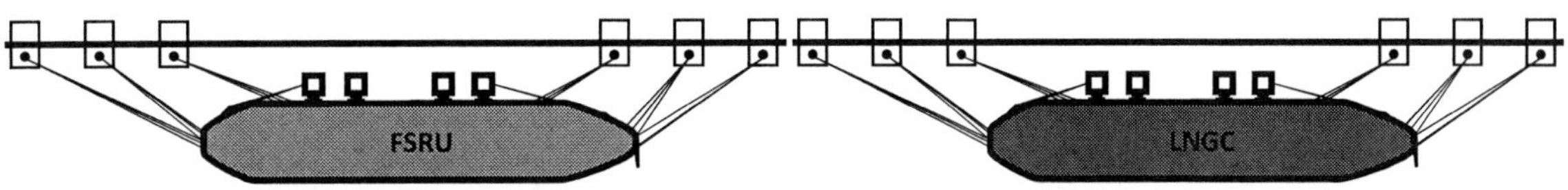

c) Tandem Mooring

Fig. 1. Mooring arrangement for loading and unloading operations

2.1 Environmental Loads

Estimation of environmental loads on the structure is the basis of mooring system design and should be given due consideration. The wind load should be calculated for all design conditions. There are two methods for estimating wind loads: constant in direction and speed measured as the 1-minute average, or fluctuating wind with a steady 1-hour average velocity combined with time-varying wind gust. The wind load is calculated based on wind profiles and windage areas and then summing up the forces on each windage area. For ship-type floating structures with relatively small superstructures typical of LNGCs, wind forces can be calculated using the coefficients presented in OCIMF MEG4 (Oil Companies International Marine Forum (OCIMF), 2018).

Currents could be the dominant force in jetty mooring system due to low under-keel clearance. Current force should be calculated using current direction and current profiles. Where appropriate, flutter and dynamic amplification due to vortex shedding should be taken into account. Coefficients presented in OCIMF MEG4 can be used for ship-type floating structures.

For the jetty mooring system, the natural period of the coupled vessel/mooring system for all degrees of freedom (surge, sway, heave, roll, pitch and yaw) could fall between 10 to 30 second range, which is where the typical wave spectrum lies. This makes the first order wave-induced loads very important for the dynamic behavior of the mooring system. The wave loads are typically obtained using radiation-diffraction analysis which provides the first and second order forces and moments, hydrostatics, added mass and radiation damping, and first order motion RAOs. Care should be taken in shallow water, as the radiation-diffraction theory has limitations on the ratio of vessel draft to water depth. Hydrodynamic interaction between multiple floating structures in close proximity may also be solved using radiation-diffraction analysis. This capability may be used to obtain side-by-side loading/offloading condition assessment. The drift forces require full quadratic transfer function to be evaluated for shallow water case. If the bathymetry has a significant change in character, its effect on wave loads needs to be assessed accordingly.

Other types of loadings that may need to be considered are the effects of piers or passing vessels. Loads due to passing by vessels depend on vessel size, distance from the terminal and speed. Surge and sway forces and yaw moment due to passing by vessel needs to be calculated and applied in mooring design assessment. Piers constructed on piles are effectively transparent to waves, but solid piers can have significant influence on the structure due to wave reflection and hydrodynamic interaction with the floater.

2.2 Design Assessment

A strength analysis should be performed to determine the maximum load and maximum offsets for the design conditions. Transient analysis should be performed to determine the maximum offsets for one component damage condition. Mooring lines and fenders should be designed to withstand these loads with additional safety factors. Allowance for corrosion and abrasion of mooring lines should also be taken into consideration where applicable. The recommended factor of safety for mooring lines is given in the ABS FPI Rules (Section 6-1-1)(American Bureau of Shipping, 2019) and shown below in Table 1. Mooring systems may require adjustment for different loading conditions (e.g. full load and ballast) as tidal influence combined with loading condition could result in significant mooring loads. However, active mooring adjustment should not be considered in the mooring analysis of maximum design conditions. Fenders should be selected in accordance with the industry standards and should be able to withstand the maximum environmental loading conditions.

Table 1. Factor of Safety for Anchoring Lines

		Factor of Safety
All Intact		
Dynamic analysis	(DEC)	1.67
Quasi-static	(DEC)	2.00
One Broken Line (at New Equilibrium Position)		
Dynamic Analysis	(DEC)	1.25
Quasi-Static	(DEC)	1.43
One Broken Line (Transient)		
Dynamic Analysis	(DEC)	1.05
Quasi-Static	(DEC)	1.18
Mooring Component Fatigue Life w.r.t Design Service Life		
Inspectable areas		3.00
Non-inspectable and Critical Areas		10.00

Another aspect of near shore mooring system design is the fatigue load analysis for the mooring line and its components. The components which can be repaired and replaced easily, fatigue analysis for such components may not be necessary and a replacement schedule should be prepared. The fatigue life estimates are made by comparing the long-term cyclic loading in a mooring line component with the fatigue capacity of that component. For the jetty mooring, fatigue analysis methodology and guideline as per the position mooring guideline can be referenced to the ABS FPI Rules (Section 6-1-1/3.5). In brief, the following methodologies are typically applied for fatigue life estimation:

Loads due to steady wind, current and wave forces (first-order and slow drift) are considered for fatigue load assessment. Due to the nonlinear nature of the jetty mooring system, the use of a time-domain approach for fatigue life assessment is recommended. However, a frequency domain dynamic analysis approach may be used in the fatigue analysis, provided nonlinear effects are appropriately linearized. Details and guidance on this can be found in API RP 2SK (American Petroleum Institute (API), 2015a). The two methodologies mentioned above have following steps:

a) Static and frequency domain analysis approach

- Static analysis under steady wind, current and mean wave drift forces

- Frequency domain analysis for first order wave loads
- Frequency domain analysis for dynamic wind load and second order wave loads

b) Time-domain analysis approach

- For each fatigue sea state, perform time domain analysis
- Obtain tension cycles based on rain-flow counting
- Calculate fatigue life based on given mooring line fatigue capacity

Some of the key distinction between temporary berthing of vessels to permanently moored structures are:

- Environmental criteria
- Direct first principal-based motion analysis of vessels
- Coupling effects between vessel's motions and mooring line/fender load
- Mooring line fatigue considerations

2.3 Other Design Considerations

There are several other factors requiring careful attention while designing a jetty mooring system. Guidance on mooring equipment is available to assist in the selection of appropriate mooring components. For mooring chains, the ABS Guide for the Certification of Offshore Mooring Chain(American Bureau of Shipping, 2017) can be referred to. Fiber rope selection can be made using the ABS Guidance Notes on the application of Fiber Rope for Offshore Mooring(American Bureau of Shipping, 2014). Wire rope-related details can be found in API Spec 9A(American Petroleum Institute (API), 2016) and API RP 9B(American Petroleum Institute (API), 2015b). For other components, the reader is referred to OCIMF MEG 4(OCIMF, 2018). The mooring supporting structures, such as chain stoppers, fairleads, chocks, bitts, rollers, etc. should be designed based on the principle of mooring line failure before fitting failure and fitting failure before hull structure or foundation failure. The ABS FPI Rules(American Bureau of Shipping, 2019) and the ABS FLGT Guide(American Bureau of Shipping, 2016) provide further guidance on structural strength, buckling and fatigue assessment methodologies for the mooring supporting structures.

Corrosion, wear and rope creep also affect mooring system design. The degradation of chain and wire due to corrosion and wear requires consideration during the design phase. Special attention should be given to components that go through chocks, sheaves and bend shoes. Ropes are susceptible to wear and abrasion over time. Fiber ropes, such as HMPE (High Modulus Poly Ethylene) are susceptible to creep and has potential for creep rupture. These ropes may also require retensioning after a certain period of service due to high creep rate.

The safety of personnel working near mooring systems must be given the highest priority. All snap-back zones require appropriate demarcation. Synthetic tails added to a mooring line provide additional elasticity to the system enabling it to withstand high dynamic loading. However this increases the snap-back hazard which must be considered appropriately during design. A detailed description of human factors in mooring system design has been given in OCIMF MEG4.

3 FSRU/FLNG Jetty Mooring Modeling

In this section, a simplified jetty mooring system is modeled numerically using commercially available software. The jetty is assumed to be made of piles which do not significantly affect the hydrodynamics of the FSRU/FLNG. Hydrodynamic interaction between the FSRU/FLNG and the LNG carrier is studied using a comparatively smaller sized LNG carrier, which shows that the motion of FSRU/FLNG is not significantly affected. However, if the two side by side vessels are of similar size, the interaction effect becomes more prominent. The mooring system is composed of typical jetty mooring arrangement with 24 mooring lines and 4 fenders as shown in Fig. 2.

The mooring system design is expected to provide 20 years of service life and give due consideration to operational requirements for permanent nearshore mooring such as berthing and unberthing, loading and unloading, and regasification functions. A 100 year return period for the metocean parameters are thus recommended for survival condition design. The design criteria are based on the minimum braking load of the mooring lines and the rated reaction force of fenders with appropriate safety factors for intact and damaged conditions.

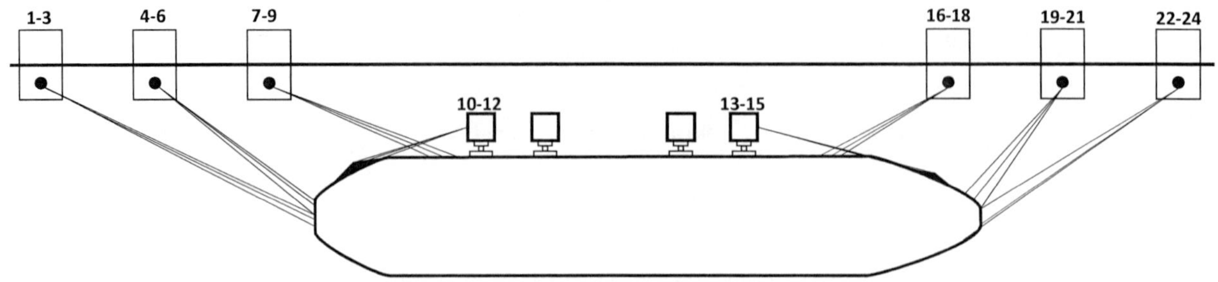

Fig. 2. Jetty mooring arrangement

The finite element method-based mooring and riser analysis software OrcaFlex (Orcina Ltd., 2018) is used for the numerical modeling of the jetty mooring system. As shown in Fig. 2, the mooring lines act to pull the vessel towards the jetty while the fenders provide an opposing force. This allows the vessel to maintain its position as various environmental forces act upon the vessel. In terms of modeling the jetty mooring arrangement, the two key challenges are:

- Defining mooring line lengths and number of segments per line
- Modeling the fender and vessel contact

First, pretensioning the 24 mooring lines to a reasonable value requires both manual trial and error and automated optimization techniques. The "*Line Setup Wizard*" functionality available in OrcaFlex is useful in this respect. However, its capabilities are still quite limited as it highly depends on the initial line length setup.

Secondly, modeling the fenders in a realistic way so that they only act when they come in contact with the vessel and then become inactive when separated requires due consideration. To model the fender numerically, a single segment line element with infinite bending stiffness and zero axial stiffness is used as a guide to hold a spring damper link element. The spring damper link is then given the properties of the actual fender. A typical nonlinear stiffness profile of a cone fender is shown in Fig. 3.

The stiffness of the vessel hull which meets the fender also needs to be modeled carefully. Modeling the impact load on the vessel contact surface (see Fig. 4(a)) may require a very small timestep or otherwise cause the numerical simulation to become unstable. As a three-hour simulation is required to capture the maximum mooring and fender load, having a too small timestep could require an impractical number of hours of simulation time. To avoid this, a variable stiffness model for the contact surface is used as shown in Fig. 4(b). A sensitivity study is performed to quantify the effect of modeling the vessel hull stiffness as a constant value versus a variable stiffness profile. The results showed that the variable stiffness profile does not significantly alter the mooring and fender loads or the vessel motions.

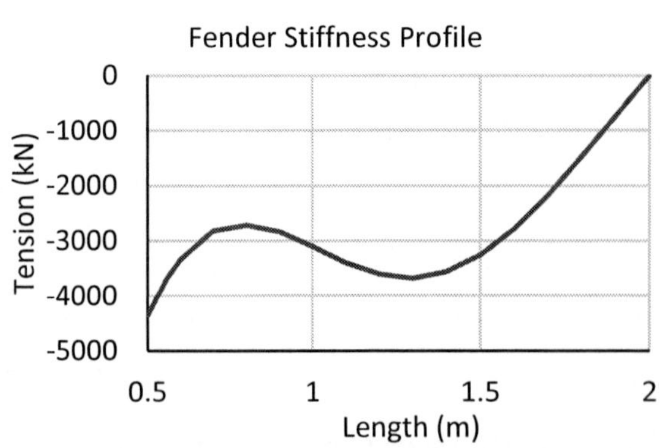

Fig. 3. Nonlinear fender stiffness modeled using OrcaFlex spring element

(a)

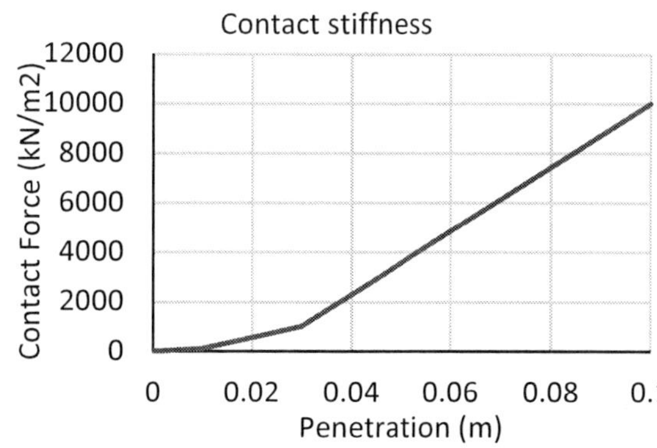

(b)

Fig. 4. (a) Fender and vessel contact model (b) Variable contact stiffness used to model ship fender contact to avoid numerical instability during impact

3.1 Determining Environmental and Vessel Loading Conditions

The nearshore permanent mooring system should be designed to endure the most unfavorable combined effect of wind, waves, tides and currents expected at the specific site. The water depth variation is decided based on bathymetry, seafloor subsidence and tidal conditions. The rate of seafloor subsidence of the site could be of importance, as the mooring system is designed for long term and changes of water depth could affect loads on the vessel (e.g. current loads). A 10-year storm surge level is typically added to the highest astronomical tide (HAT) to obtain the maximum water depth for the operational environment and a 100-year storm surge is added for extreme environment condition. The vessel draft corresponding to ballast and fully loaded condition is considered as shown in Fig. 5. The water depth-to-vessel draft ratio is one of the important factors for numerical analysis, as the potential flow method applied in obtaining the hydrodynamic coefficients and wave induced loads has limitations at shallow water depths. This will be further discussed here. Also, the mooring line vertical inclination angle should be given due consideration while designing the mooring system, as repetitive line contact with the jetty structure could damage the lines.

Wind and current conditions are obtained from site specific metocean data as shown in Fig. 6. For wind, a constant wind speed for a given direction is used here with an NPD wind spectrum. The current direction is typically along the shoreline near the coast line. This reduces the number of current dominated analysis cases. It is essential to ensure that both collinear and cross direction of wind, waves and current is considered to capture the worst-case scenario.

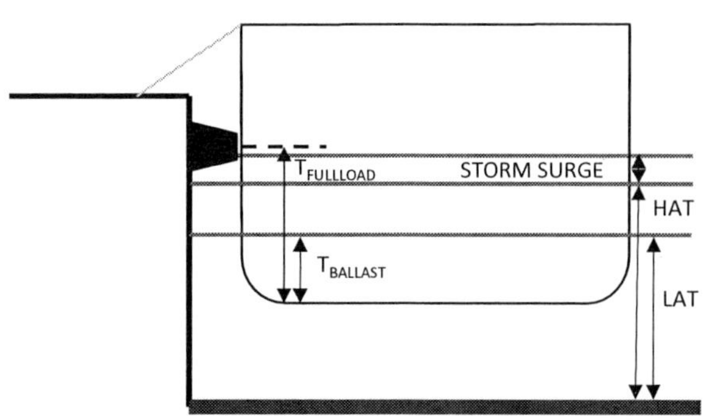

Fig. 5. Vessel draft and water depth considerations

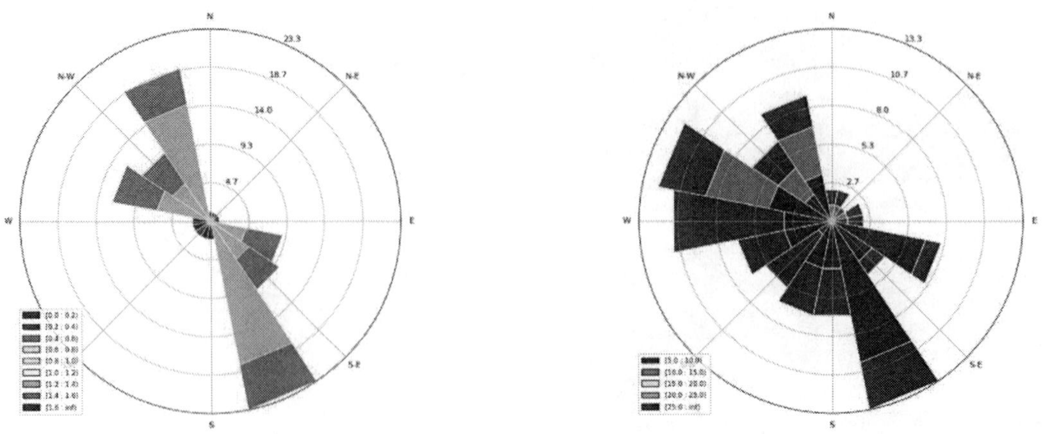

(a) (b)

Fig. 6. (a) Current speed and direction (b) Wind speed and direction

3.2 Effect of Water Depth on Hydrodynamic Coefficients

Hydrodynamic coefficients such as added mass, radiation damping, and wave induced forces are typically obtained using a potential theory-based program. These programs use a source distribution over the surface of the vessel and apply a finite water-depth Green Function to solve for the velocity potential around the vessel. Bernoulli's equation is used to obtain the pressure on the vessel surface which is then integrated over the vessel surface to calculate the wave induced forces and hydrodynamic coefficients. The capability of such programs is limited by both numerical and theoretical constraints. As the under-keel clearance becomes smaller, viscous forces can no longer be ignored. Also, as the distance between the seabed and the vessel keel becomes smaller, the numerical evaluation of the influence matrix due to source distribution over the vessel surface becomes affected. As shown in Fig. 7 and Fig. 8, the heave added mass and roll added inertia for very shallow water ($D/T = 1.5$ or 1.75) are calculated to be significantly higher than the deep-water case. This needs to be validated against model test or in-situ data, as this has a significant effect on the natural period and vessel motion predictions as shown in Fig. 9 and Fig. 10. The limitation of potential theory-based programs with respect to water depth is well known. As a rule of thumb, water depth may be distinguished as articulated in (Hooft, 1974) as:

- Deep water or infinite depth: $D/T > 4$
- Restricted depth or finite depth: $2<D/T<4$
- Shallow water depth: $D/T<2$

Where D is the water depth and T is the vessel draft. The use the potential flow codes is recommended for deep and restricted water depths, and caution is advised at shallow water depths.

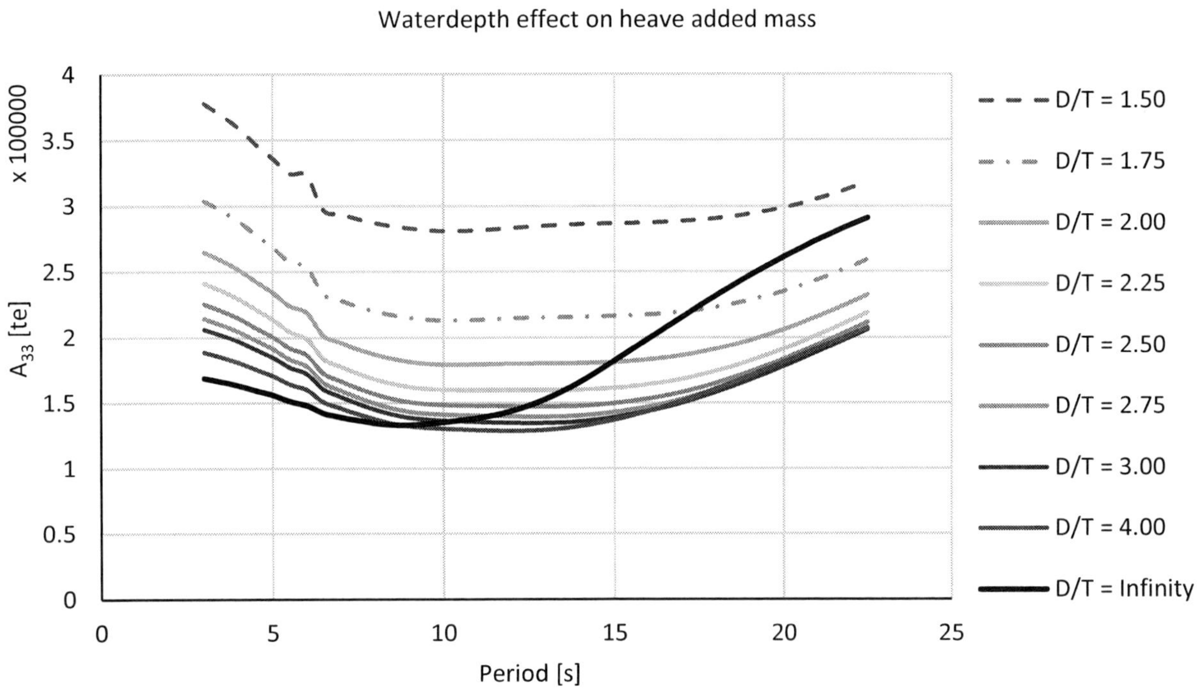

Fig. 7. Heave added mass calculated using potential flow theory increases significantly with decreasing water depth.

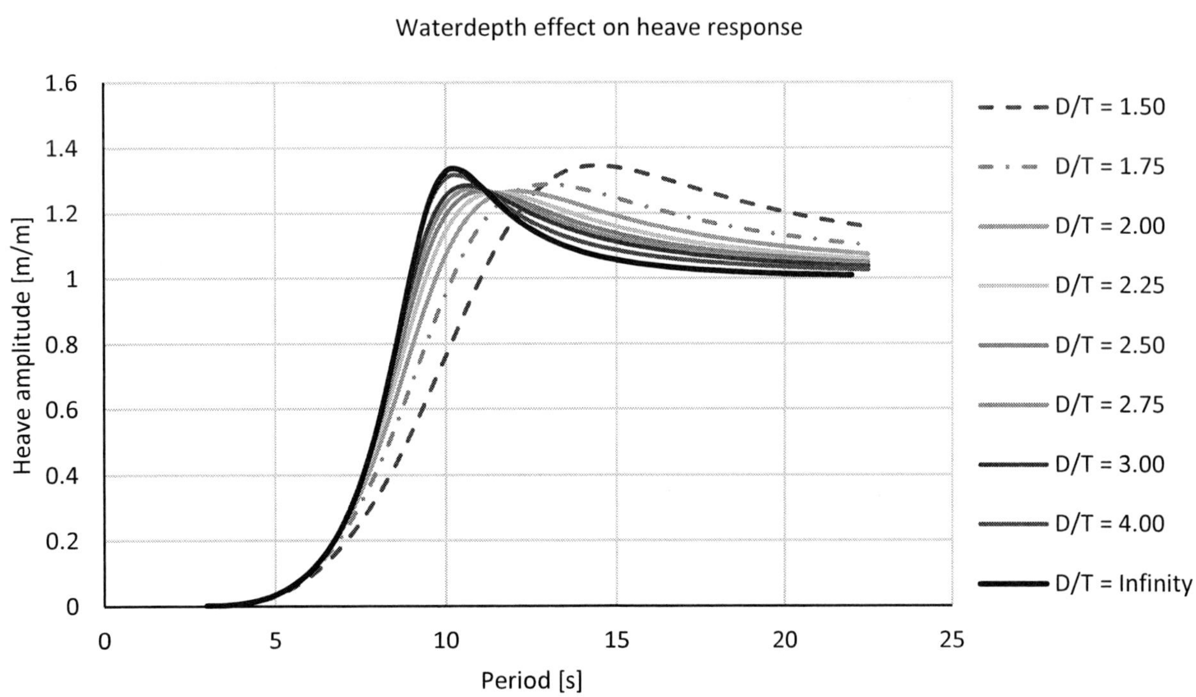

Fig. 8. The effect of increased added mass in shallow water is shown in the heave response amplitude operator. Both amplitude and natural period is significantly affected.

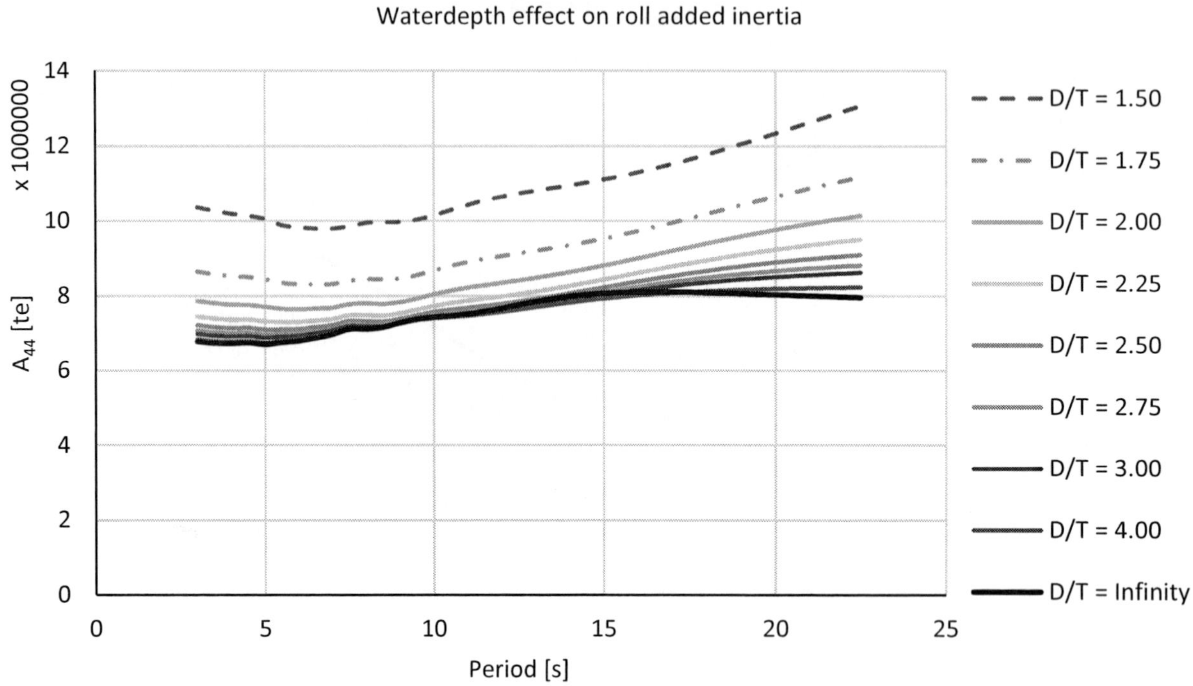

Fig. 9. Roll added inertia calculated using potential flow theory increases significantly with decreasing water depth.

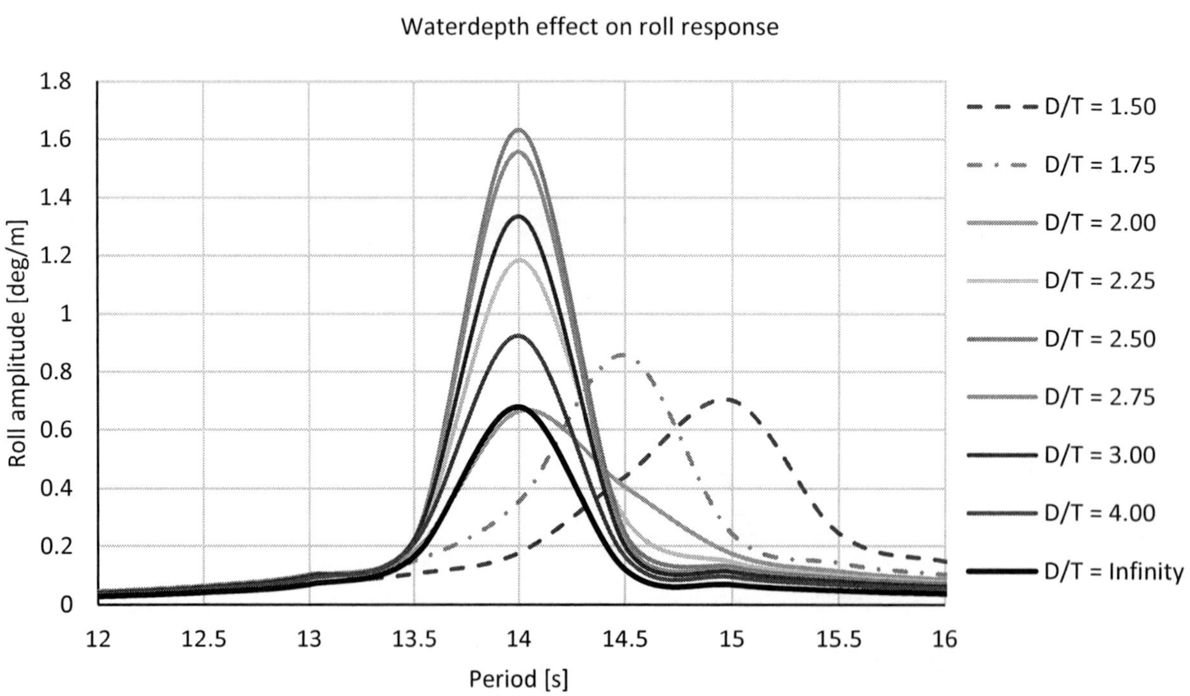

Fig. 10. The effect of increased added inertia in shallow water is shown in the roll response amplitude operator. Both amplitude and natural period is significantly affected.

3.3 Hydrodynamic Interaction Effect between LNGC and FLNG

The LNG produced by the floating terminal is typically stored temporarily in its tanks and then offloaded to an LNG carrier for transportation. The side-by-side offloading configuration is considered to be the preferred mode of operation as the existing midship coupling manifolds on the present fleet and standard mooring arrangements can be used. For the nearshore jetty mooring perspective, it is important to understand whether there are any adverse effects on the FLNG motions which may introduce additional loading on the mooring systems.

Here, a side-by-side configuration of an FLNG and an LNG carrier is studied where the vessels were kept 4m apart. The vessel arrangement is shown in Fig. 11. The wave body interaction is analyzed using the frequency domain tool WAMIT(Lee, 2013) . The head sea condition showed little effect on the vessel motion due to multibody interaction. However, the beam sea case, where the waves approach from the LNGC side showed significant effects on the LNGC motion as shown in Fig. 12. While FLNG motion is not significantly affected, the mooring load between the LNGC and FLNG will be affected. If the size of the LNG carrier is closer to the size of the FLNG, then the jetty mooring system should consider the effect of multibody interactions between the two bodies.

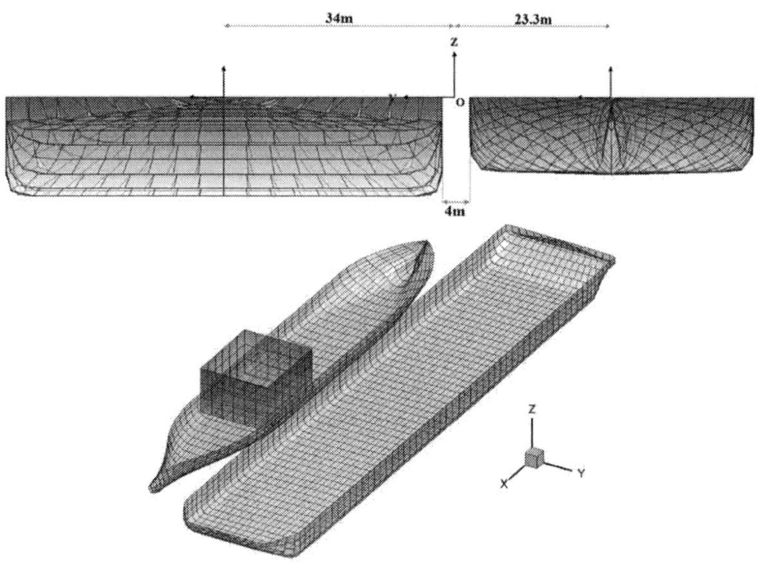

Fig. 11. LNGC with Internal Tank and FLNG Side-by-Side Layout

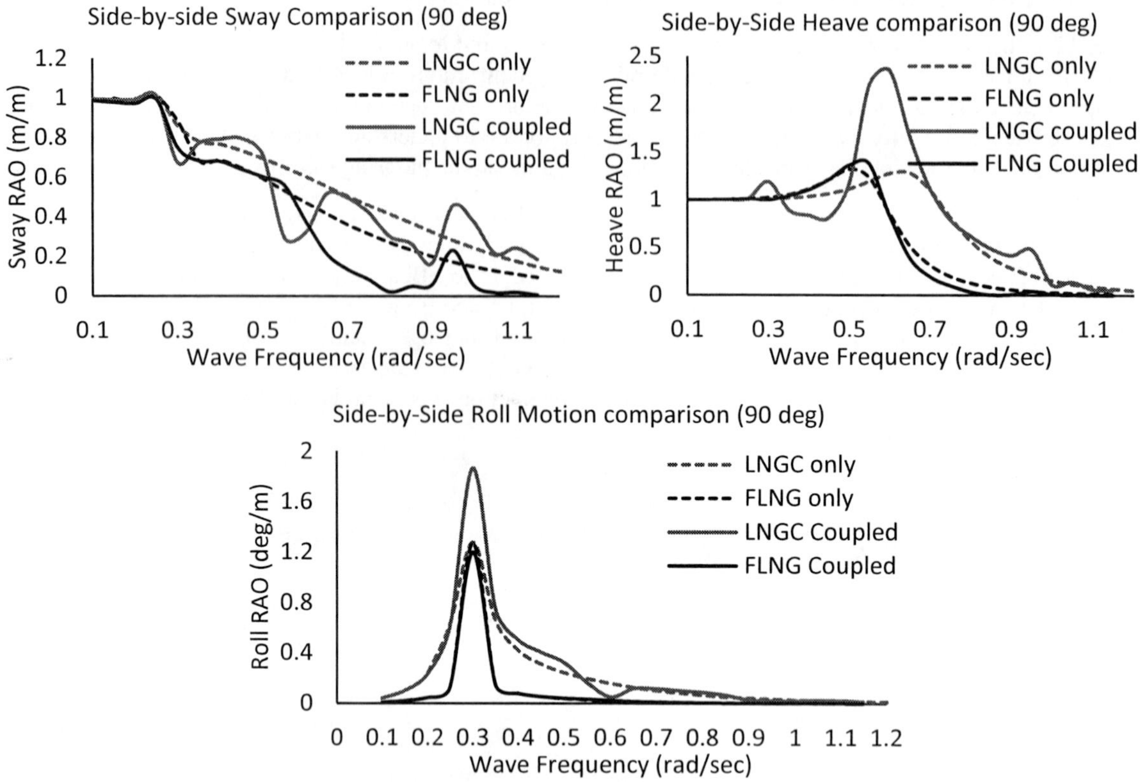

Fig. 12. Comparison of Sway, heave and roll motions for LNGC-FLNG and LNGC-only Interactions for 138K LNGC and FLNG

3.4 Bathymetry Effects on Wave Field

Metocean data is typically obtained using weather buoys deployed in the ocean away from the shore. Thus, it is generally difficult to perform nearshore jetty mooring analysis for the actual wave conditions at the location. The method for converting and applying buoy data to the nearshore location for the application of mooring load prediction is still under development. See (Jaouën et al., 2016) for more details on this topic. Currently, the wave height and period are selected based on nearest available data with engineering judgment.

3.5 Sensitivity Analysis

The numerical model requires conditioning to perform a 3-hour simulation for over 360 load cases. Several measures are taken to reduce the simulation time, keeping the integrity of the results and capturing the critical parameters. The three key parameters that have significant effects on the analysis are:

- Segment length of the mooring line model
- Result logging timestep
- Contact stiffness of the elastic shape used for vessel hull modeling

3.5.1 Line Segment Length

A large number of segments produce a more accurate representation of the mooring line. However, the number of segments modeled per mooring line greatly affects the simulation time. It is essential to model a minimum number of segments so as to capture the behavior of the mooring line without over burdening the computational requirements. A comparison study between a fine segmentation of 50 segments per line and a coarse segmentation of 10 segments per line is performed. The 10-segment case is shown to produce results with similar accuracy to the 50-segment case as shown in Fig. 13 (a) and used for the load case analysis.

3.5.2 Result Logging Timestep

Storing the resultant data points at selected timesteps takes up significant computational resources, both in terms of storage space and simulation time. It is possible to keep a smaller calculation timestep to maintain simulation integrity, while having a larger timestep for recording the results. At the same time, it is crucial to capture the peaks of the results accurately. Fig. 13 (b) shows a comparison between result logging at every 0.1s versus every 1s, where the coarse timestep result logging was successful in capturing the peak data points. Using a 1s timestep reduced the result file size by a factor of ten and significantly reduced the simulation time.

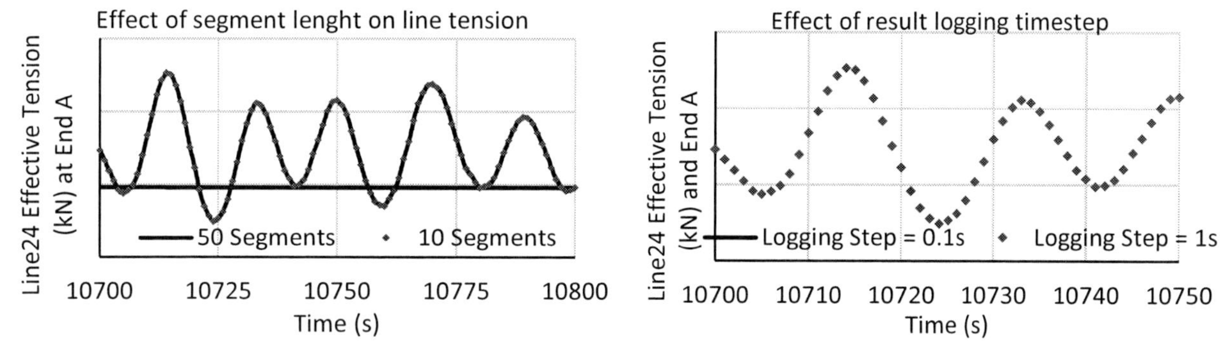

Fig. 13. Result of sensitivity study of line segment length (a) and result logging time step (b)

3.5.3 Contact Stiffness of the Elastic Shape

An elastic shape is used to model a portion of the ship hull that comes in contact with the fenders. The oscillation of the vessel due to wave action causes the vessel to repeatedly impact against the fenders. As these impacts are a short duration phenomenon, having a stiff vessel surface model may require a very short calculation timestep. It was found that the simulation may become unstable after a certain period of time if the stiffness is not modeled appropriately. While using a variable stiffness as shown in Fig. 4(b), it is important to ensure the results are not severely affected. Fig. 14(a) shows that the vessel motions are within an acceptable range and Fig. 14 (b) shows that the variable stiffness model produces slightly higher reaction forces in the fenders compared to the constant stiffness model. The comparison of maximum mooring line tensions between the variable stiffness model and constant stiffness model is shown in Fig. 15. The constant stiffness case is shown to produce slightly higher mooring line tension due to high impact loads. The difference in the maximum line tension calculation compared to minimum braking load is negligible. Similarly, the difference in fender reaction force compared to the rated reaction force is negligible.

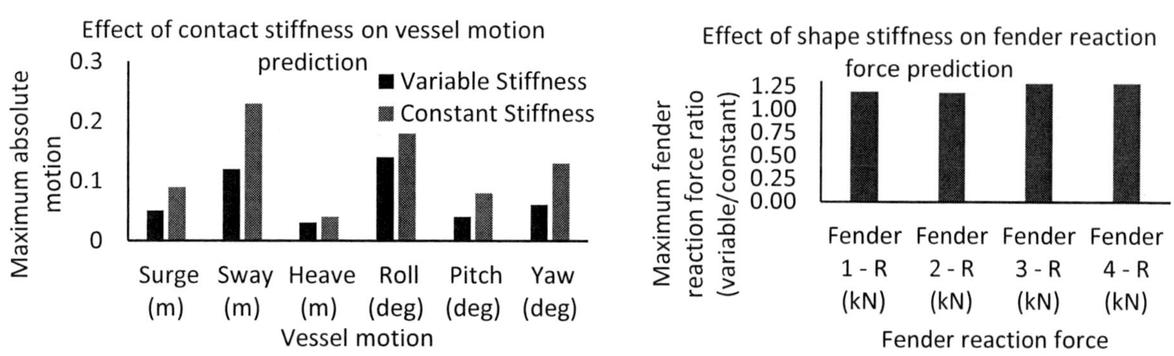

Fig. 14. Effect on vessel motion (a) and fender reaction force (b) for using a variable stiffness instead of a constant stiffness in modeling fender and vessel contact

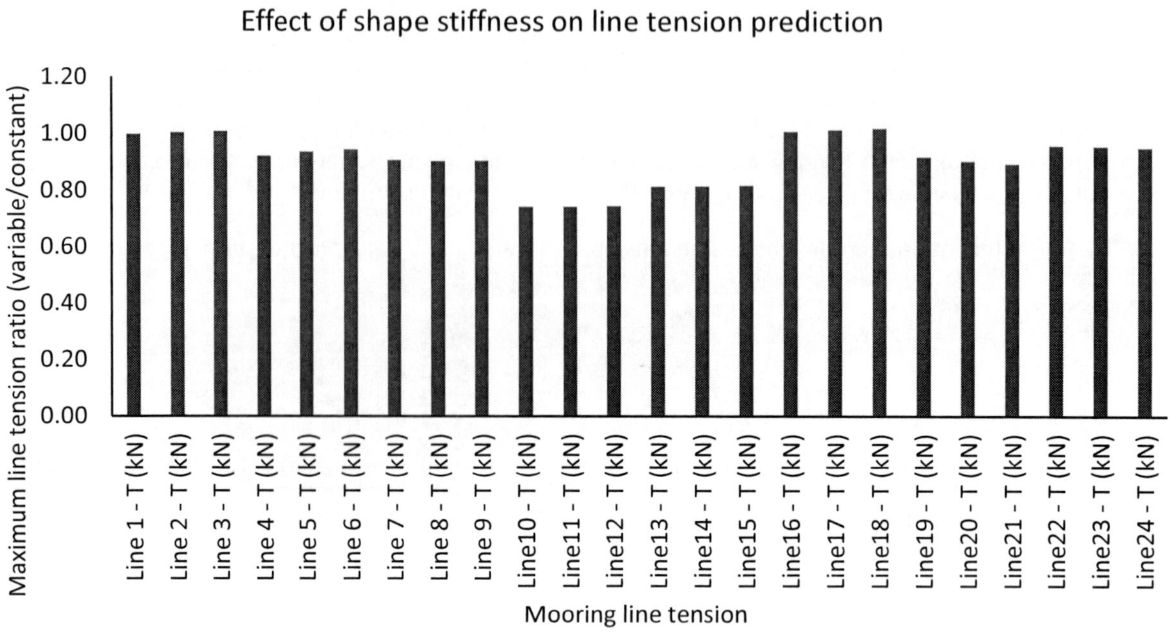

Fig. 15. Effect on line tension for using variable stiffness instead of a constant stiffness in modeling the fender and vessel contact

4 CONCLUSIONS

This paper focuses on the permanent mooring of large floating terminals (FLNGs/ FSRUs) in nearshore locations and the challenges associated with it. Critical factors requiring consideration while designing mooring systems for such structures were reviewed. Reference is made to relevant industry guidance provided for particular component designs.

An example numerical analysis of jetty mooring system is presented with discussion of important factors affecting the load calculations. Sensitivity studies showing numerical challenges of performing coupled dynamic time domain analysis are also presented.

REFERENCES

American Bureau of Shipping, 2019. Rules for Building and Classing Floating Production Installations. Houston,TX.

American Bureau of Shipping, 2018. Guidance Notes on Nearshore Position Mooring. Houston,TX.

American Bureau of Shipping, 2017. Guide for the Certification of Offshore Mooring Chain. Houston,TX.

American Bureau of Shipping, 2016. Guide for Building and Classing Floating Offshore Liquefied Gas Terminals. Houston,TX.

American Bureau of Shipping, 2014. Guidance Notes on the Application of Synthetic Ropes for Offshore Mooring. Houston,TX.

American Petroleum Institute (API), 2016. Specification for Wire Rope (API SPEC 9A). Washington, D.C.

American Petroleum Institute (API), 2015a. API RP 2SK, 3rd ed.

American Petroleum Institute (API), 2015b. Application, Care, and Use of Wire Rope for Oil Field Service (API RP 9B). Washington, D.C.

Corkhill, M., 2018. Bringing floating LNG regasification up to date [WWW Document]. LNG world Shipp. URL https://www.lngworldshipping.com/news/view,bringing-floating-lng-regasification-up-to-date_53908.htm (accessed 12.12.18).

Guha, A., Falzarano, J., 2016. Optimization of a Parametric FLNG in Finite Water Depth, in: Proceedings of the ASME 2016 35th International Conference on Ocean, Offshore and Arctic Engineering. ASME, Busan, South Korea, p. V007T06A071. https://doi.org/10.1115/OMAE2016-55087

Hauteclocque, G. de, 2007. Local bathymetry effects on sea-keeping, HAWII JIP (No. ATA: 1094B).

Hooft, J.P., 1974. The behaviour of a ship in head waves at restricted water depths. Int. Shipbuild. Prog. 21, 367–390.

Jaouën, F., Waals, O., Jon, M. de, Hout, A. van der, Christou, M., 2016. Methodology for the design of LNG terminals, in: Proceedings of the ASME 2016 35th International Conference on Ocean, Offshore and Arctic Engineering. Busan, South Korea, pp. 1–15.

Lee, C.H., 2013. WAMIT User Manual. Chestnut Hill, MA, USA.

OCIMF, 2018. Mooring Equipment Guidelines-4th Edition. Livingston.

Oil Companies International Marine Forum (OCIMF), 2018. Mooring Equipment Guidelines (MEG4), 4th 2018 E. ed.

Orcina Ltd., 2018. OrcaFlex.

Vasudevan, N., Nallayarasu, S., 2017. Simulation of Passing Vessel Effects on Moored Vessel Mooring Response due to Environmental Loads, in: Proceedings of the ASME 2017 36th International Conference on Ocean, Offshore and Arctic Engineering. Trondheim, Norway. https://doi.org/10.1115/OMAE2017-61593

SNAME
THE INTERNATIONAL COMMUNITY
FOR MARITIME AND OCEAN PROFESSIONALS

THE EVOLUTION OF DEEPWATER DISCONNECTABLE MOORINGS

JASON D. PASTERNAK, PE
PROJECT MANAGER - R&D LEAD, DELMAR SYSTEMS, INC.

DANIEL J. ALONSO
SOFTWARE ENGINEER, DELMAR SYSTEMS, INC.

JOHN T. SHELTON III, PE
VP OF ENGINEERING, DELMAR SYSTEMS, INC.

GINNY R. WHISENHUNT
STAFF ENGINEER, DELMAR SYSTEMS, INC.

Proceedings of the 24th Offshore Symposium, February 20th 2019, Houston, Texas
Texas Section of the Society of Naval Architects and Marine Engineers

ABSTRACT

Deepwater drilling and production has been in existence for decades, and with it, stationkeeping philosophies and technologies have evolved with time and experience. In years past, moorings were designed purely with robustness and simplicity in mind, dropping anchors as rigs arrived on location, with the expectation to weather the storms, but now, with increased strengths seen in tropical rotating storms (TRS), ice floes, and deeper operating water depths, more sophisticated mooring components have been developed. Chain has yielded ground to wire and synthetic ropes. Stockless anchors have been replaced with piles, gravity installed anchors, and sophisticated high holding capacity (HHC) drag anchors. The connecting hardware has become "smart," evolving from kenters and c-links to sensor and sonar-equipped remotely releasable systems.

The main drivers for this evolution have been environmental – extreme metocean events, corrosion, wear and fatigue – from the floater to the seabed and below.

This paper will present how "dumb steel" mooring systems have evolved into sophisticated and detailed engineered foundations that span miles of ocean real-estate, while allowing for a new level of vessel mobility that reduces risks to people and assets.

Keywords: Acoustic, Quick Release, Turret Mooring, Releasable, Stationkeeping, Risk Mitigation

INTRODUCTION

"Deepwater" is a fairly subjective term that is used widely in the offshore industry. In the late 1950's and early 1960's, the term "deepwater" was commonly taken to mean a water depth in excess of 300 ft. This depth was outside of the reaches of jack-ups and submersible Mobile Offshore Drilling Units (MODUs) of the time, and drillships performed poorly in the larger waves present at such depths. In 1961, the 300-ft barrier was shattered as seven years of highly secretive development by Shell culminated in the birth of the world's first semi-submersible drilling unit, the *Blue Water 1*. The *Blue Water 1*, as seen in Figure 1, was a converted submersible MODU that was capable of drilling in water depths of 600-ft (National Commission on the BP Deepwater Horizon Oil Spill, 2010). An 8-line mooring system composed of wire rope and the largest ship anchors ever produced were used to moor the rig in-place (Priest, T., 2007). One year later, in 1962, the first turret-moored vessel was launched, The Offshore Company's *Discoverer 1*, a drillship (Schempf F. J., 2007).

Figure 1: Shell's Blue Water 1 (Offshore at 60)

In the 1970's, deepwater began to mean 1,000-ft of water depth, a commonly used definition of the term today. As MODUs ventured into these deeper waters, new innovations and mooring standardizations geared towards the offshore industry were required to survive the more intense metocean conditions. Oil Rig Quality (ORQ) chain, a higher grade than commercial shipping chains, was introduced for these harsher environments (Noble Denton Europe Limited [NDEL], 2006). More specialized and higher capacity anchor types were introduced as well, replacing ship anchors with the likes of the Bruce "Claw" style anchor, Vryhof's Stevin, and Anker Advies Bureau's Flipper-Delta drag anchors (Sincock, P. & Sobdhi, N., 1993). Additionally, in 1975, one of the first commercially available mooring quick release devices, the Rig Anchor Release (RAR), was introduced (Robb, S., 1975). A year later, the RAR 6500 was used in a multi-year drilling campaign (1976-1979) by Canadian Marine Drilling, Ltd. off both the East and West Coasts of Canada – in the North Sea off Labrador and in the Beaufort Sea. The original intent of the RAR during this campaign was to avoid dangerous ice floes, but as the RAR gained credibility and exposure, it began to be used to extend the drilling season until ice packs were too dense to allow for proper stationkeeping (Lasch, J. E., Pearlman, M. D., & Riotte, E. C., 1980).

The 1980's brought more improvements in strength and holding capacities of mooring equipment and a few innovations that were utilized more in future decades. R3 chain was standardized, with an increase in break load of 10% over that of ORQ (NDEL, 2006). The Bruce Twin Shank and Vryhof Stevpris anchors were developed, increasing drag anchor holding capacities by ~50% over previous generations (API RP2SK, 2005). Due to the successful implementation of RARs in the "cut and run" mooring system concept used in the 1970's, when compared to the more robust but harder to release systems examined, the RAR-releasable systems were again chosen for a drilling campaign off East Canada. A new type of anchor, the suction caisson, was deployed in 1981 in Shell's North Sea Gorm Field on a Catenary Anchor Leg Mooring (CALM) Buoy, allowing for significant loads and pin-point installation accuracies (McCarron, W. O., 2011). Though suction caissons allow for significant upward loads at the anchor end of the mooring line, greatly reducing the required lengths of moorings, and are able to be placed with precision while maintaining high holding capacities, they are large, costly to build and install, and require special installation equipment, such as divers with pumping skids or ROV's. The suction caisson, or pile, rapidly became the anchor of choice for Long Term Moorings (LTMs) in the 80's and 90's (See Figure 2). In 1987, building on a concept that was first introduced in 1962, SBM Offshore supplied the first offshore production unit with a disconnectable external turret, the FPSO *Jabiru Venture*. The disconnectable turret isolates the mooring and risers from the production vessel, and acts as a buoy to support the system and facilitate reconnection. The reason for selecting and implementing this design was simple, the facility was to be installed in Tropical Cyclone-prone Australian waters (SBM Offshore, 2018); this is one of the first examples of "cut and run" mooring systems being used with LTMs.

Figure 2: LTM Suction Piles for LLOG's Who Dat Facility - May 2011 (Courtesy of Delmar Systems)

The last decade of the 1900's brought similar advances to the previous decades in mooring component strengths and anchor capacities, but it also brought several advances that allowed for faster and safer mooring operations for MODUs and LTMs. New anchoring concepts, mooring line components and the introduction of subsea connection / disconnection equipment are among these milestones.

The first successful offshore synthetic mooring component, polyester rope, was tested in 1995 by Petrobras in the Campos Basin, Brazil. Polyester ropes have been proven to increase station-keeping performance while increasing mooring survivability, increase variable deck loading allowances on vessels due to mooring line weight reductions, and they are immune to corrosion issues that plague steel systems (Flory, J. F., Banfield, S. J., & Berryman, C., 2007). Polyester and other synthetic fibers are now widely in use in both the MODU and LTM markets.

Multiple new and drastically different anchor concepts came about in the 1990's. The Bruce DENNLA and Vryhof Stevmanta anchors were both tested offshore in 1996 (Aubeny, C. 2018). These drag-installed Vertically Loaded Anchors (VLAs) allowed for some of the vertical loading characteristics of suction piles, but without the complex and costly considerations that come with piles, making them good solutions for MODUs. In an attempt to lower the costs of anchoring solutions, the Suction Embedded Plate Anchor (SEPLA) concept was tested in the USGOM in 1998. This concept uses a large suction pile "follower" to push an inexpensive and simple plate anchor to a suitable depth. The SEPLA allows for several less expensive anchors to be produced, making it the single cheapest anchor available in the market; however, the installation time of a set of anchors is closer to that of conventional suction caissons (Aker Marine Contractors, 2000). The Torpedo anchor was tested by Petrobras for mooring applications in December of 1999 (Wilde, B. 2009). This new type of gravity installed pile has been extensively used by Petrobras due to its low fabrication costs and relatively fast installation times. It should be noted that the mooring and lowering attachment point for torpedoes is universal, at the top of the pile, which requires them to be oversized versus some other anchors, and they are often unrecoverable.

In 1998, the first subsea connectable and disconnectable mooring line connector, the Delmar Subsea Connector (DSC), was developed. The DSC allowed for mooring systems in full or in part to be installed ahead of an arriving MODU or LTM facility, or "preset," to reduce critical path down time of expensive production and drilling floaters while the mooring system and anchors were installed. This preset concept also became more and more important as mooring systems have grown in complexity due to increasing metocean and growing subsea infrastructure that congests many drilling and production locations.

In the 2000's, the preset mooring concept is common-place. With the introduction of several new subsea connector types available on the market, foam and steel subsea buoyancy modules, and MODUs capable of both Dynamically Positioned (DP) and moored stationkeeping, speed onto and off of the well location are the main focus. The OMNI-Max™ was introduced and used as a MODU anchor in 2007. This concept incorporates the speed of installation of a torpedo anchor with the capacity of a suction caisson at ~⅓ of the size. Quick release devices have been installed in several preset systems to allow for rapid storm or ice floe evasion without compromising on mooring reconnection times. The RAR 6500 was used in Shell's Alaska campaign (2012-2015) and ExxonMobil's arctic drilling (2014). The Delmar Quick Release (DQR) was used in the USGOM to increase rig move efficiencies in 2016. Both of these quick release devices led to the development of the RAR Plus™ in 2018, which is now in use globally, as the idea of rapid deployment and rapid release mooring systems gains ground.

Releasable Mooring System Considerations

Releasable mooring systems are now being used in both permanent and temporary mooring applications. The primary benefits of releasable mooring systems include:

- Tropical Revolving Storm Avoidance
- Ice Floe Avoidance
- MODU Rig Move Efficiency
- Risk Mitigation for Well Control Events

For storm and ice avoidance, releasable mooring systems offer an alternative to the traditional design approach that requires a mooring system to survive extreme loading events. The purpose of a releasable mooring system is for the offshore structure to evade the extreme event completely. For permanent systems, this allows for cost savings and efficiencies when designing and selecting mooring component types, mooring component sizes, number of mooring lines, and installation vessel capabilities. For MODUs that have fixed mooring design capacities to work with, releasable mooring systems can significantly reduce the risk of mooring failure due to extreme loading events by employing the release-and-run methodology.

Releasable mooring systems provide many MODUs with the ability to work in certain regions of the world and during seasons that otherwise would not have been marketable due to having mooring systems not capable of meeting the minimum design condition. In other markets where ice and tropical revolving storms are non-existent, releasable mooring systems offer rig move efficiencies by saving time during the disconnection for each location. The cost savings with reduced rig time and Anchor Handling Tow Supply (AHTS) vessel-time can be significant, especially for short duration locations where weather can otherwise delay AHTS operations during disconnection. With the right release tool and procedures, rig disconnection can be completely independent of AHTS operations. For DP capable rigs, this can save days (sometimes weeks) in getting off location.

There are numerous factors to consider when choosing a releasable mooring system. Beyond the typical mooring accessory considerations, the following factors should be well understood:

- Track Record of Deployments and Releases
- Redundancy in Systems (FMEA)
- Release Load Rating
- Backup Release Capability and Methods
- Procedures for Efficient Disconnection
- Manufacturing, Testing, and Certification
- Ancillary Equipment
- Training of Rig Personnel

Mooring Release Technology

The various mooring line release technologies available today have evolved around the aforementioned factors for the last ~45 years. Starting with the first deployment of the Rig Anchor Release (RAR) 6500 in the 1970's for drillship ice avoidance offshore Canada, mooring releases have been available to the offshore drilling rigs. However, it was not until after more experience with the device during exploration offshore Alaska in 2012 that mooring releases started being seriously considered for other purposes. Instead of just the one-off "what-if" emergency release scenario in remote regions, it was envisioned that mooring releases could be used repeatably for efficiency purposes and for storm avoidance.

RAR 6500

The RAR 6500 in Figure 3 was first developed in the early 1970's by oceanographic equipment designer and manufacturer Interocean Systems based in San Diego, California.

Figure 3: Rig Anchor Release (RAR) 6500 on Deck Ready for Deployment

The operating design principle of the RAR is an electro-mechanical system in which coded acoustic signals are transmitted by the command deck unit to the RAR to cause disconnection of the mooring line. The acoustic signals in the water are received by any of the four acoustic transducers and transferred to the RAR electronics unit by underwater rated cables and connectors. Upon receipt of the proper coded acoustic signal, the electronics send power to the two redundant electro-mechanical hydraulic valves, which permit the flow of hydraulic pressure from the accumulators to the activation cylinder that retracts the locking shoes, allowing the anchor end of the RAR to separate from the main body.

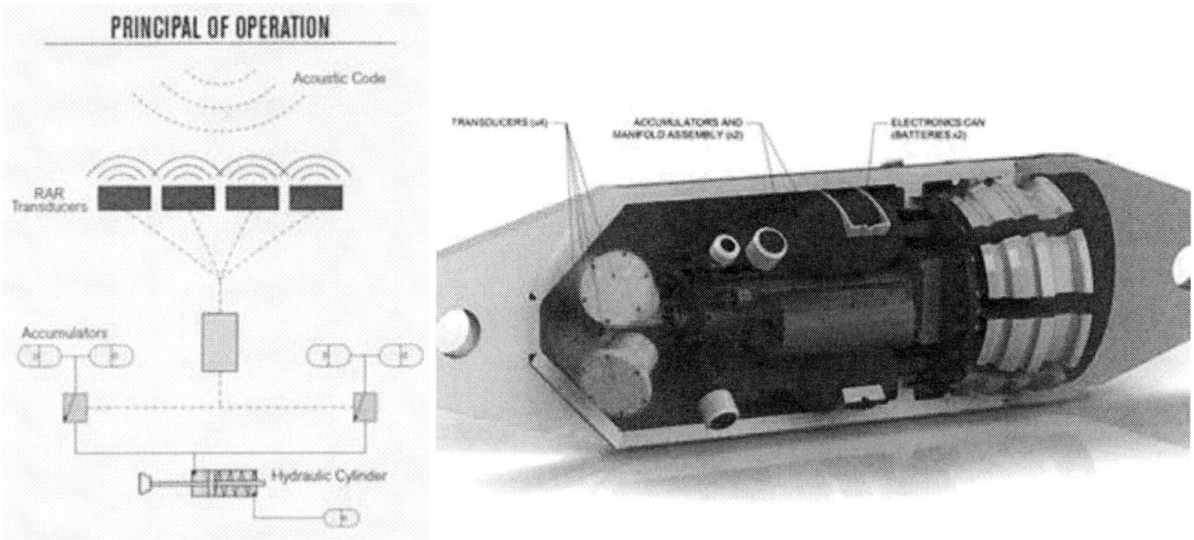

Figure 4: RAR 6500 Operation Principle and Redundant Systems Overview

DQR (DELMAR QUICK RELEASE)

The DQR (Figure 5) underwent two rounds of full-scale release tests- one in January 2014 at Holloway Houston, and one in September 2015 while hooked up to the Transocean *Deepwater Nautilus* at Keppel AmFELS in Brownsville, TX.

Figure 5: Delmar Quick Release (DQR) on Deck Ready for Deployment

The operating design principle of the DQR is a mechanical system in which a trigger sleeve is push/pulled against the DQR to cause disconnection of the mooring line. The DQR comprises of two halves that interlock and hinge about a shackle pin. A shear pin is used to prevent premature release. The DQR is triggered by a trigger sleeve pressing down on the DQR, overcoming friction to break a shear pin, applying a force to the lever arms, opening the release. Unlike traditional disconnections, the DQR does not require the lines to be recovered by a secondary vessel, or the use of an ROV, allowing the rig to depart location faster.

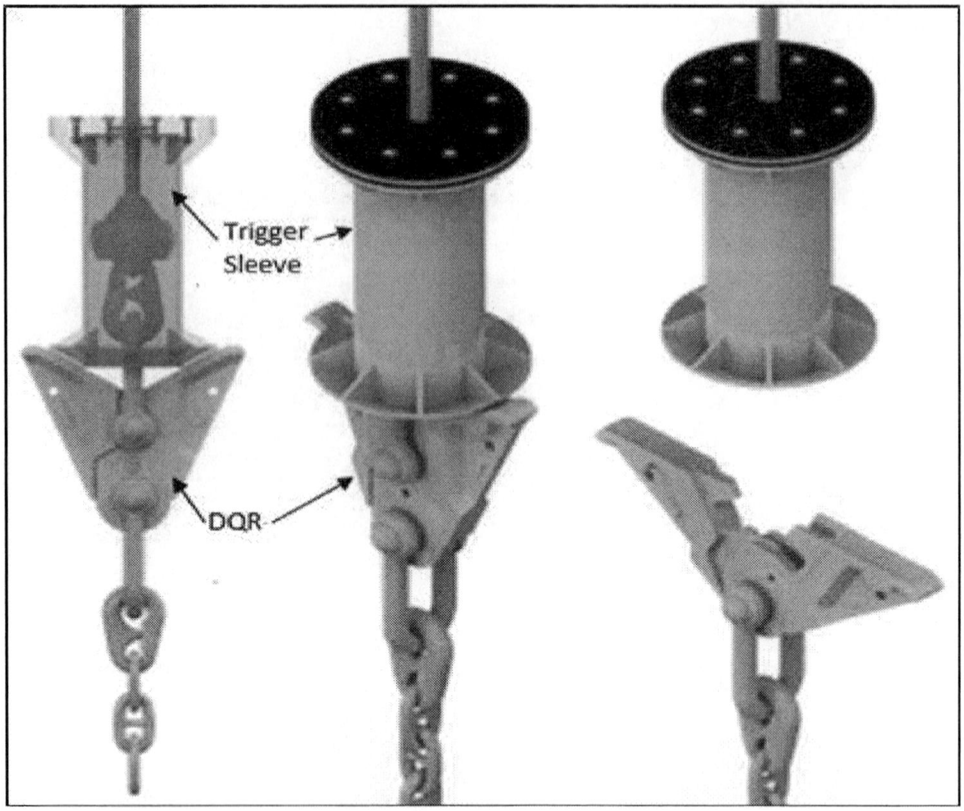

Figure 6: DQR Disconnection

In 2016, the DQR was used on a taught-line preset-moored semi-submersible MODU in the USGOM at a location with a ~7,700-ft water depth. Seven lines were successfully released using the fairlead or AHTS and modified chaser methods. Several of the lines were released during metocean conditions that would have prevented the use of AHTSs, ROVs and/or the rig crane.

It is important to note that Table 1 presents linearly extrapolated time and cost savings results of the moored rig disconnection that occurred in the Gulf of Mexico using the DQR system, as only 7 of the 12 lines were disconnected using the DQR. The DQR was used in a trial application to make a standard rig move more efficient and cut costs for future drilling and exploration programs (i.e. reduce T-Times); the respective data is based upon a post-job analysis and timeline conducted by Delmar Systems. The quick release saved an average of 12-hrs per line during the disconnection, not including other factors, such as eliminating down time due to weather. This rig did not have DP capability, so AHTS vessels were still required to be on location for heading control and tow operations. Approximate savings of ~$1.75 million could have been realized through using the DQR on all 12 lines by reducing overall disconnection time, which saved rig time and AHTS time. It should be noted that only one line was released at a time; further time savings could be realized by conducting multiple line group releases. Furthermore, if this were a rig with DP capability, then AHTS vessels would not be required for disconnection operations, which yields further savings.

Table 1: Rig Disconnection Case Study for 12-line Moored Rig Without DP Capability in the Gulf of Mexico

12-line Moored Rig Disconnection (no DP Capability) – Gulf of Mexico			
Without Line Release		*With DQR Release*	
	w/o delays	w/o delays	Time Savings
Avg. Disconnect Time Per Leg	15.3 hrs	3.3 hrs	78% time reduction
Avg. Total Rig Time for Disconnect	7.6 days	1.7 days	Saved 5.9 rig days or 78%
Avg. Total AHTS Time for Disconnect	7.6 days	1.7 days	Saved 5.9 AHTS days or 78%
	Rate ($/day)	# Days Saved	Est. $ Saved
Rig Spread	$200,000	5.9	$1,188,000
AHTS Vessel	$77,500	5.9	$460,000
Tow Vessel	$32,700	3.0	$97,000
		Est. Total $ Saved	**$1,745,000**

RAR PLUS

The next generation of the RAR is the RAR Plus™, and it comes with some important improvements to the earlier model. Perhaps the most important development is the addition of a mechanical backup release feature, which ensures the ability to release the moorings if the remote acoustic transmission fails to actuate the RAR. The backup mechanical release bypasses the acoustic, electronic, and hydraulic systems, and it can be actuated by either the rig itself or with a nearby support vessel. An ROV is not required to activate the backup release. Using the onboard MODU winches, the RAR is pulled into the trigger sleeve and forced against the fairlead structure. The trigger rod and pivot arm action manually force the hydraulic cylinder end cap to retract and release the system. Alternatively, the trigger sleeve can be used as a chaser and pulled into the RAR using an anchor handling vessel.

Figure 7: RAR Plus on Deck During Installation and Recovery after a Release

After acoustic or manual activation of the release, the RAR body can be pulled taut with the rig's winch and nested with the trigger sleeve against the rig's fairlead (Figure 8). With the ability to self-nest, there is no need for AHTS operations to disconnect the rig. With the RAR securely nested, the rig can sail away without needing to rely on AHTS vessel operations that can delay a rig move if weather downtime or vessel breakdown issues occur.

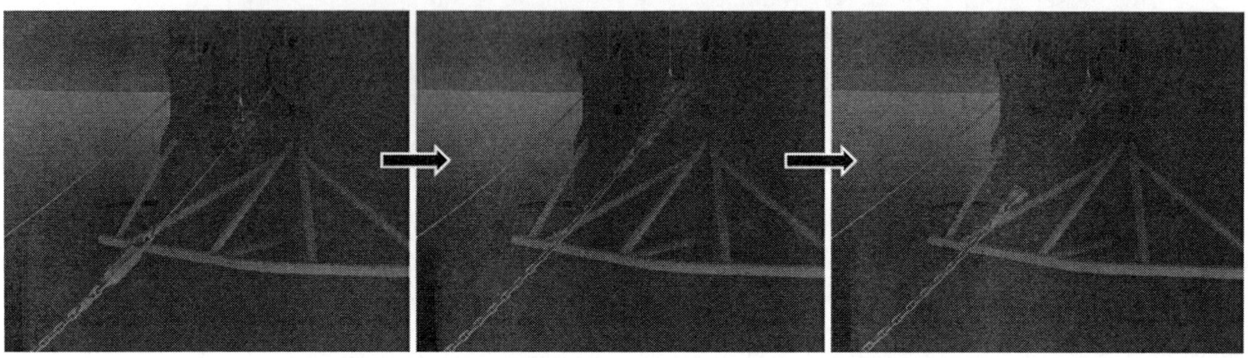

Figure 8: RAR Plus Mechanical Backup Release Procedure

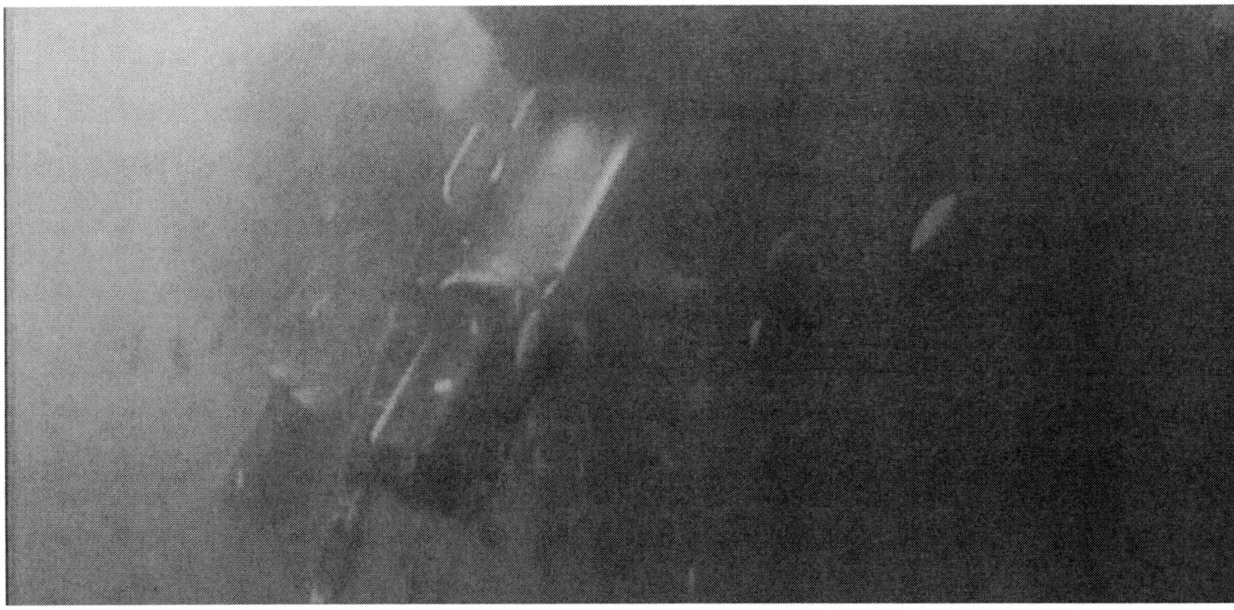

Figure 9: RAR Plus Units Nested with Trigger Sleeve Against Rig Fairlead and Ready for Rig Transit

The RAR Plus™ can also be equipped with an acoustic telemetry system with integrated sensors providing real-time continuous monitoring of local line inclination, depth, and strain. The sensors inside the RAR Plus and topside processing unit will communicate through acoustic modems and algorithms to display both direct-measured and indirect-calculated line tensions onboard the rig in real-time. This data is independent of the rig's existing load cell tension readings. A user-friendly graphical user interface (GUI) will display valuable data for onboard personnel to use during mooring activities and for the duration of the drilling campaign. The data stream can also be integrated with 3D catenary monitoring systems, providing extra risk mitigation for mooring in congested fields. For redundancy, all data will be saved in both the topside processing unit and locally within the RAR Plus™ memory for any desired post-analytical purposes. The telemetry system is optional and can be easily removed for applications that do not require enhanced risk mitigation measures.

The track record of deployments and releases conducted is an important factor to consider for any critical technology. The design and operation principles of the internal release mechanisms between the RAR Plus (Table 2) and RAR 6500 (Table 3) are similar, hence the track records can be considered complementary.

Table 2: RAR Plus Track Record

RAR Plus Track Record					
Year	***Vessel Type***	***Location***	***System***	***Objective***	**Release Triggered Successfully?**
2018	Semi-sub	Gulf of Mexico	Moored 9-line	Rig move efficiency	Yes, 9 lines individually
2018	Semi-sub	Gulf of Mexico	Moored 9-line	Rig move efficiency	Yes, 9 lines individually
2018	AHTS	Norway	1-line with AHTS	Release tests	Yes, 2 times
2018	Semi-sub	South Africa	Rig tow line	Static tow emergency release	Job in Progress

Table 3: RAR 6500 Track Record Since 2012

RAR 6500 Track Record					
Year	*Vessel Type*	*Location*	*System*	*Objective*	**Release Triggered Successfully?**
2012	Semi-sub	Alaska	Moored 12-line	Ice avoidance	Not required
2012	Drillship	Alaska	Moored 8-line	Ice avoidance	Yes, 8 lines individually
2013	Semi-sub	Alaska	Moored 8-line	Ice avoidance	Not required
2014	Semi-sub	Russian Arctic (Kara Sea)	Moored 8-line	Ice avoidance	Not required
2015	Semi-sub	Alaska	DP/Moored 8-line	Ice avoidance	Yes, 8 lines individually
2015	Drillship	Alaska	Moored 8-line	Ice avoidance	Yes, 8 lines individually
2015	Semi-sub	Gulf of Mexico	DP/Moored 8-line	Cyclonic storm avoidance	Yes, 8 lines individually
2016	Semi-sub	Gulf of Mexico	DP/Moored 8-line	Rig move efficiency	Yes, 8 lines individually
2016	Semi-sub	Gulf of Mexico	DP/Moored 8-line	Cyclonic storm avoidance	Yes, 8 lines individually
2017	Semi-sub	East Canada	DP/Moored 8-line	Ice avoidance	Yes, ALL lines simultaneously
2017	Semi-sub	Gulf of Mexico	DP/Moored 8-line	Cyclonic storm avoidance	Not required
2017	Semi-sub	Gulf of Mexico	Moored 9-line	Rig move efficiency	Yes, 9 lines individually

For the rig move efficiency case, the well duration, rig rates, AHTS rates, and weather patterns for the region and time of year should be considered. The objective of a remote mooring release is to disconnect the rig and move away from location faster than a traditional disconnect with one or more AHTS vessels. The time savings can be significant when you consider the spread rate of a typical rig move operation (including rig, survey equipment, multiple AHTS vessels for line handing and tow operations). It should also be noted that reduced time in AHTS vessel operations yields reduced exposure for personnel working, thereby lowering Health, Safety, and Environmental (HSE) related risk. Another important advantage is that a faster disconnect reduces the weather window needed for a rig disconnect operation. This can save significant downtime when vessels would otherwise need to shut down and wait for the weather to pass before resuming operations. The number of AHTS vessels required for a rig disconnect can be reduced significantly while still achieving faster disconnect times. With the right release tool that is capable of nesting for immediate transit, AHTS operations for a rig disconnect can be eliminated altogether.

Table 4 presents the results of a case study for a 9-line moored rig disconnection that occurred in the Gulf of Mexico using the RAR Plus system. This rig did not have DP capability, so AHTS vessels were still required to be on location for heading control and tow operations. Still, significant savings of ~$1million were realized by reducing the number of AHTS vessels by one and reducing overall disconnection time, which saved rig time and AHTS time. It should be noted that only one line was released at a time; further time savings could be realized by conducting multiple line group releases or by releasing all lines simultaneously. Historically, the idea of releasing all lines at once has been reserved for emergency cases only, as most vessels cannot operate more than one or two winch lines at a time, but some drilling contractors have recently expressed interest in using the "release all" command of the RAR Plus on their DP-capable MODUs. Furthermore, if this were a rig with DP capability, then AHTS vessels would not be required for disconnection operations, which yields further savings.

Table 4: Rig Disconnection Case Study for 9-line Moored Rig Without DP Capability in the Gulf of Mexico

9-line Moored Rig Disconnection (no DP Capability) – Gulf of Mexico					
Without Line Release			*With RAR Plus Release*		
	w/o delays	w/delays	w/o delays	w/delays	Time Savings
Avg. Disconnect Time Per Leg	8.2 hrs	15.0 hrs	0.9 hrs	1.7 hrs	89% time reduction
Avg. Total Rig Time for Disconnect	2.7 days		0.7 days		Saved 2.1 rig days or 76%
Avg. Total AHTS Time for Disconnect	5.4 days		0.7 days		Saved 4.8 AHTS days or 88%
	Rate ($/day)		# Days Saved		Est. $ Saved
Rig Spread	$200,000		2.1		$412,000
AHTS Vessel	$77,500		4.8		$370,000
Tow Vessel	$32,700		4.1		$135,000
			Est. Total $ Saved		**$917,000**

RISK CONSIDERATIONS

More than 23 Moored Mobile Offshore Drilling Unit (MODU) failures occurred in the 2004 & 2005 tropical cyclone seasons. Efforts were made to develop a generally accepted rapid MODU mooring risk assessment methodology to prevent these failure events from occurring again. There has now been a rapid site-specific methodology in use for the last decade which looks at risk events up to 120-miles from location and individually analyzes their risk.

Risk is a function of probability and consequence. The major component of probability for a MODU risk event is mooring failure which is estimated by a limit state analysis. The limit state analysis takes into consideration metocean for the full tropical cyclone season. Once mooring failure occurs, the probability is determined by the distance and size of the risk event. Based on the type of risk event, several other components are taken into consideration. Each risk event also has a corresponding consequence value, the goal of this value is to capture the repair cost to the damaged asset as well as the cost of deferred or interrupted production during repair time (Alonso & Zimmerman, 2010). Risk results are displayed on a typical Whitman style plot, shown in Figure 10; the consequence of the event is represented by the X-axis while the probability of the risk event occurring is represented by the Y-axis. A medium risk region is shown on the plot that represents an equivalent societal risk to those accepted in multiple other industries. Any risk event above the yellow region is considered "High Risk" while any event below is considered "Low Risk".

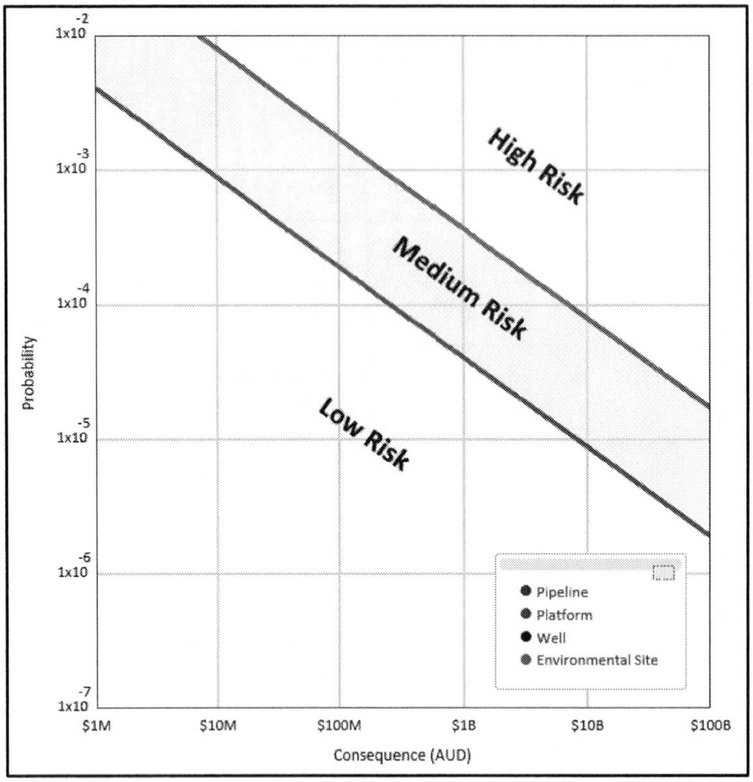

Figure 10: Whitman Style Plot

Risk events considered include both surface and subsea events. Among the surface risk events are interaction and collision events. An interaction risk event is the risk of a drifting MODU causing damage to surface facility assets such as mooring systems or minor structure damage. A collision risk event is the risk of a drifting MODU experiencing hull-to-hull collision and causing complete facility loss. As a result, the probability of an interaction event is greater due to the larger footprint of the facility's assets versus the facility's hull. The consequence value for interaction would include the repair cost and time which is considerably lower than the cost and time to replace the facility. Among subsea risk events are well/manifold and pipeline events. A well/manifold risk event is the risk of a drifting MODU dragging mooring components or anchors over a well or manifold. Due to their relatively small size, well/manifold events have a low probability, and their consequence is based on the cost to re-drill and replace the well. Much like the well/manifold risk event, a pipeline risk event is the risk of a drifting MODU dragging mooring components or anchors over a pipeline. The probability of a pipeline risk event is based on the length of the pipeline within the examined radius of the location; a pipeline that spans a larger area becomes a greater target for a drifting MODU. The consequence value of a pipeline event includes the cost of repair and the deferred or interrupted production during repair time. Since both well/manifold and pipeline risk events involve dragging equipment, one must also consider the equipment type and anchor selection when calculating the probability of either event (Alonso, Sharples, & Zimmerman, 2007).

Though this methodology has worked well for moored systems, it does not take into consideration releasable mooring systems. The major component of probability for a MODU risk event is mooring failure; however, when we consider releasable mooring systems, there is an additional probability of disconnecting from the mooring system. The probability of mooring failure now only comes into play if unable to disconnect from the mooring system. This additional probability creates a shift in the probability of risk events. The ability to disconnect from the mooring system is dependent on the weather conditions at the location. During the rig's operation, a transition time (T-Time) is calculated based on how fast the rig can secure the well and disconnect from the mooring system. This T-Time will vary greatly depending on the type of operation in progress. As the probability of severe weather conditions increase, the longer it takes to transition after a tropical cyclone threat is detected, so T-Time plays a significant role in the ability to disconnect. Based on historical tropical cyclone data, we can estimate the site-specific annual probability of occurrence of a tropical cyclone producing severe site-specific winds at estimated T-Times. Though disconnecting from the mooring system eliminates the probability of mooring failure from the methodology; we still need to consider that the rig could find itself in severe weather if it is not able to outrun the tropical cyclone.

The probability of severe weather conditions during transit can be estimated using historical tropical cyclone data. As an example we can look at a DP MODU that has a transit speed of 5-knots, with a T-Time of 50 hours when a storm is first observed. Local wind speeds are calculated for the first 50 hours of the storm at that location. Once the 50 hours is reached, the rig will start to transit at 5 knots into the direction of lower wind speeds. This assessment will allow us to find the best case local wind speed during transit for that DP system. Figure 11 displays the local wind speeds if the DP MODU stayed on location, while Figure 12 shows the DP MODU with a 5-knot transit speed after the 50-hour T-Time. As seen is the figures, this example is of a tropical cyclone that comes close to the location and produces 90+ knot site-specific winds. When a transit speed is considered, the local winds are reduced to ~44 knots. This same exercise can be done for each storm on record to estimate the probability of severe weather during transit. The reason for the lowered site-specific winds is the DP's ability to move away from the path of the storm and avoid the most severe consequences. The distance from the storm is shown by a black line in the following plots.

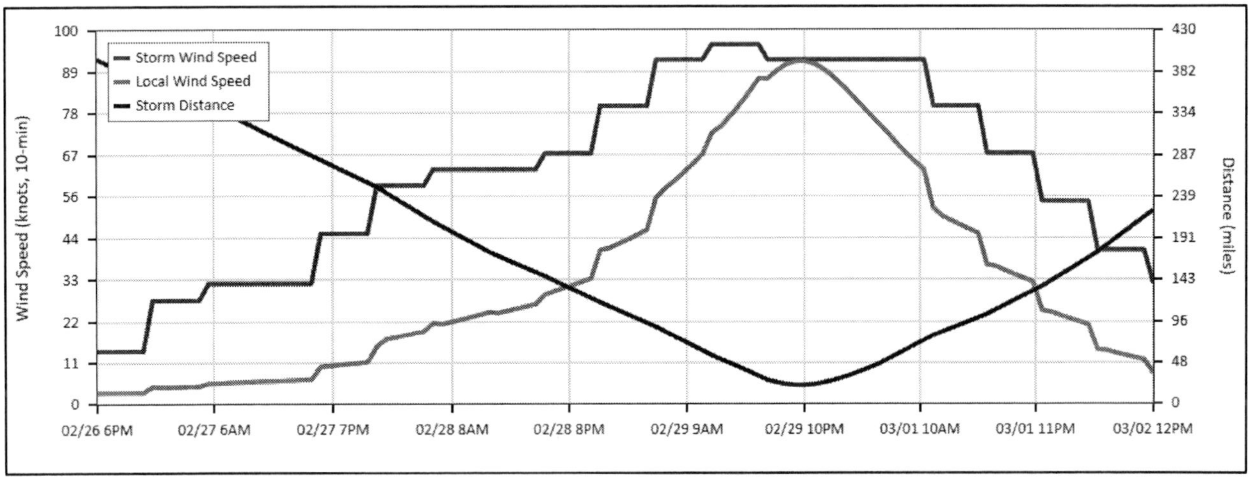

Figure 11: Wind Speed Estimate - 0-knot Transit Speed

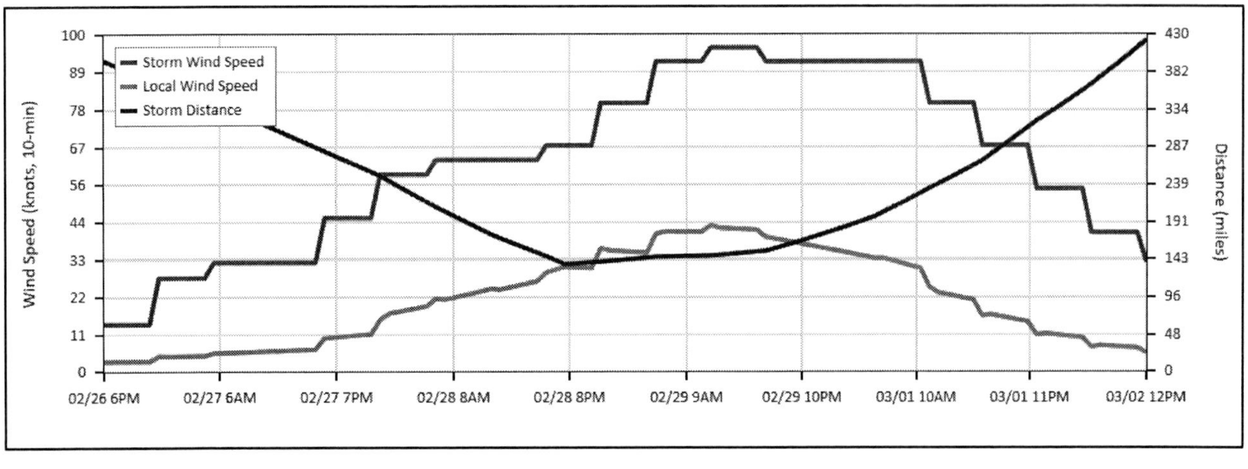

Figure 12: Wind Speed Estimate - 5-knot Transit Speed

A typical moored MODU risk result summary plot can be seen on Figure 13. This example is for an 8-line rig that meets a 10-year return period at an intact FOS of 1.2. That return period correlates with a 60.5-knot, 1-min wind speed at this location. Pipeline risk events are shown by blue circles, well risk events are shown by black circles, and platform risk events are shown by red circles. You will notice that some red circles are connected by a line; these two circles represent the interaction and collision risk events for one platform. Multiple pipeline and platform events are considered high risk in this example since they are above the acceptable risk area shown in yellow. If we calculate the probability of severe weather conditions during transit for this rig, we find that there is a 6.25E-2 annual probability of occurrence. That assumes a 50-hour T-Time and a 5-knot transit speed. Incorporating this probability, we can see the resulting risk result summary in Figure 14. Overall probabilities reduce, and as a result the risk profile improves. All risk results will vary based on location and rig parameters; however, this example highlights the changes one might see when looking at moored rigs with releasable mooring systems. Though a releasable mooring system presents new scenarios when it comes to risk, tools to estimate risk in these scenarios are available.

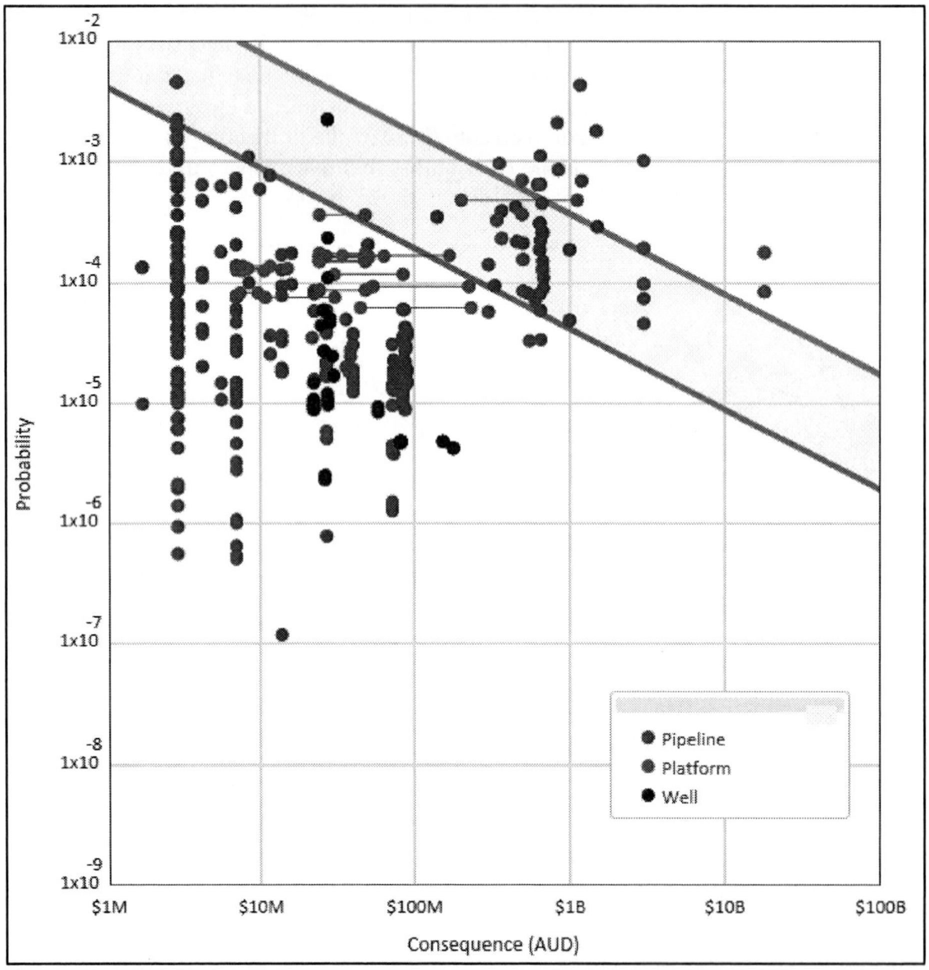

Figure 13: Risk Result Summary: 10-year RP @ Intact FOS of 1.2

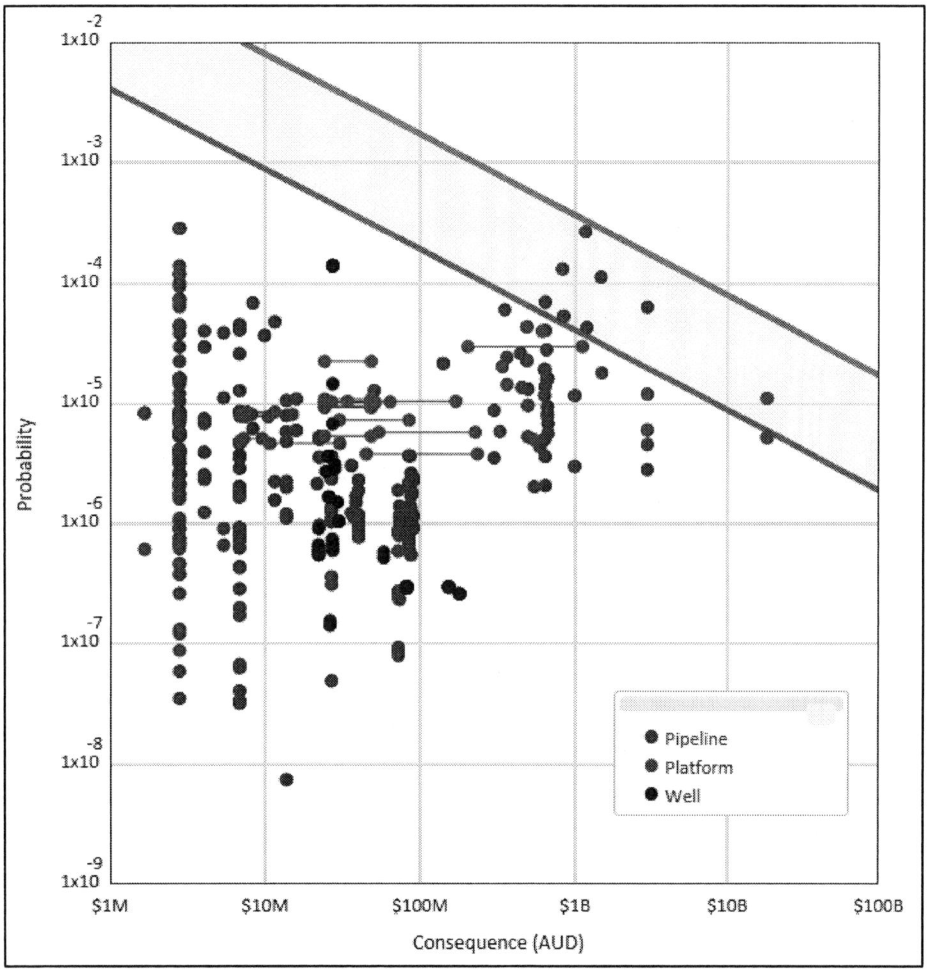

Figure 14: Risk Result Summary: Released from Mooring System

CONCLUSIONS

Industry consensus was that after Hurricanes Katrina and Rita in 2005, any future GOM (or other regions subjected to TRSs) FPSOs should be disconnectable, if only to negate the possibility of being an in-place target for other adrift facilities or MODUs. Of the two FPSOs operating in the USGOM (the *BW Pioneer* at Cascade/Chinook and Shell's *Turritella* at Stones), as of 2018, both are disconnectable (Parshall, J. 2016). With the lift on oil export bans in the US in 2017, the disconnectable FPSO may become an even more attractive future prospect in the USGOM.

During 2017 and 2018, roughly 75% of all moored drilling operations in the USGOM utilized the RAR 6500 or RAR Plus acoustic quick releases. In 2018, RAR Plus units were deployed in the USGOM, the North Sea, and off the coast of South Africa. Synonymous with the ideologies being applied to LTMs in regions subject to sudden extreme conditions, such as ice floes and TRSs, the desire for speed off location during an emergency is ever-increasing. Risk aversion aside, as the trend for operators to look for new means to cut exploration and production costs continues, reducing critical path down time of moored MODUs by utilizing acoustic quick releases during rig moves, especially in regions such as the North Sea, where much of the MODU work is in brown fields, is an increasingly popular concept.

As with previous decades of mooring evolution, the future trends towards efficiency, and smarter and faster systems, with increased capabilities, that are installable by a wider array of vessels.

ACKNOWLEDGMENTS

The authors would like to thank Peter Dorey and InterOcean Systems, Inc. for their support in writing this paper.

REFERENCES

Aker Marine Contractors. (2000, 3 1). Trial Results from Deepwater Anchors, Polyester Moorings. *Offshore, 60*(3).

Alonso, D., Dr. Sharples, & Zimmerman, E. (2007). MODU Mooring Risk Assessment JIP. Houston, TX.

Alonso, D. & Zimmerman, E. (2010). Risk Management of Moored Mobile Offshore Drilling Units in the Western Pacific Ocean. OCEANS'10 IEEE. Sydney, NSW, Australia

API RP 2SK: Design and Analysis of Stationkeeping Systems for Floating Structures. (2005, Oct). *3rd.* American Petroleum Institute.

Aubeny, C. (2018). *Geomechanics of Marine Anchors.* Boca Raton, FL: CRC Press: Taylor & Francis Group.

Flory, J. F., Banfield, S. J., & Berryman, C. (2007). Polyester Mooring Lines on Platforms and MODUs in Deep Water. *Offshore Technology Conference.* Houston, TX: Offshore Technology Conference.

Lasch, J. E., Pearlman, M. D., & Riotte, E. C. (1980). Special Mooring Systems Using Remote Controlled Quick Disconnects for Beaufort Sea Drilling Operations. *Offshore Technology Conference.*

McCarron, W. O. (2011). *Deepwater Foundations and Pipeline Geomechanics.* Fort Lauderdale, FL: J. Ross Publishing.

National Commission on the BP Deepwater Horizon Oil Spill. (2010). The History of Offshore Oil and Gas in the United States. *Staff Working Paper No. 22.*

Noble Denton Europe Limited. (2006). Floating Production System, JIP FPS Mooring Integrity. *Research Report 444.* Aberdeen: Health and Safety Executive.

Parshall, J. (2016, Aug 1). Twenty-Year Effort Brings Only Two FPSO Vessels to US GOM. *Journal of Petroleum Technology, 68*(8).

Priest, T. (2007). *The Offshore Imperative: Shell Oil's Search for Petroleum in Postwar America.* Library of Congress Cataloging-in-Publication Data.

Priest, T. (2014, 10 17). Offshore at 60: The Blue Water Breakthrough. *Offshore, 74*(10).

Robb, S. (1975). Drilling Rig Anchor Release System - A New Concept for Hostile Weather Mooring Operations. *IEEE OCEAN.*

SBM Offshore. (2018, Sept). Turret & Mooring System Supply Record.

Schempf, F. J. (2007). *Pioneering Offshore: The Early Years.* Tulsa, OK: PennWell Custom Pub.

Sincock, P., & Sobdhi, N. (1993). Drag Anchors for Floating Systems. *Health and Safety Executive - Offshore Technology Report.*

Whisenhunt, G., Barrow, J., Pasternak, J. (2017). Saving Rig Time and Mitigating Risk Through "Quick-Release" Mooring Systems. SNAME Offshore Symposium. Houston, TX

Wilde, B. (2009, Oct). Torpedo pile anchors enter the GOM. *E&P.*

SNAME
THE INTERNATIONAL COMMUNITY
FOR MARITIME AND OCEAN PROFESSIONALS

ROBUSTNESS PHILOSOPHY FOR FLOATING PRODUCTION SYSTEMS

BRIAN JAMES CHEATER
SENIOR CONSULTANT, GUSTOMSC

Proceedings of the 24th Offshore Symposium, February 20th 2019, Houston, Texas
Texas Section of the Society of Naval Architects and Marine Engineers

ABSTRACT

Robustness is a system property that reflects its resistance to the initiation of hazardous events and their progression beyond defined limit state thresholds. Under scoped perturbations a robust system should not display disproportionate response or cascading failures. Three levels of robustness are identified: Operational (R1), Survival (R2) and Reserve (R3) that span responses in state space ranging from the safe operating limit (SOL) up to system failure at the ALS limit. Systems engineering methods can be used to define system boundaries then map sub-systems and their interactions. HAZIDS can be used to identify critical failure modes for engineering analysis. Risk methods can be used to assess the acceptability of the robustness measure.

Keywords: Robustness, SNAME, Offshore, Symposium

INTRODUCTION

Robustness is a system property that reflects its resistance to the initiation of hazardous events and their progression beyond defined limit state thresholds. A robust system maintains essential minimum features under scoped perturbations. Furthermore, when subjected to a perturbation (even one just outside the intended design envelope), a robust system should not cascade out of control or experience disproportionate responses; robust systems should not sit on the edge of a design cliff.

Robustness is not a new concept and many measures exist in codes and good practice but these cannot be considered complete:

1. They often arise in response to specific accidents. For example, brace redundancy requirements arose after the Alexander Kielland. Consequently, they may reflect the sub-set of experienced events but may not capture the full population of relevant events.
2. New designs may not fall within the range of applicability of existing codes.
3. Operational and design conditions are often treated separately but they may be tightly coupled. For example the severe effect that a ballasting error can have on longitudinal strength. Consequently, the potential for design and operational measures to influence and thereby help or hinder each other is missed.
4. They may be considered in isolation by individual disciplines consequently significant interactions can be missed or completely inconsistent robustness goals and requirements may arise.

5. They may arise in response to perceived uncertainty in the expected environmental conditions, the design requirements or the operating conditions. However, these may be only loosely defined or goal oriented.

This document aims to address many of these existing issues by:

1. Providing examples and ways of thinking of what robustness means
2. Providing a structured definition of robustness.
3. Introduce a systems engineering method to identify requirements and a risk-based method to assess adequacy.

UNCERTAINTY

Robustness is a required hedge against uncertainties (or unknowns) such as:

1. System parameter
2. Stochastic Environment
3. Operational Actions and Errors
4. Component Failure
5. Construction Errors
6. Degradation and Damage
7. System Interactions

Unfortunately, there may or may not be enough data to characterize these unknowns. More troubling still, there may or may not be enough experience to have even identified that an unknown even exists consequently, there may always be the risk of an unprecedented event. This paper will not open an epistemological discussion but Figure 1 below does attempt to identify how engineering practice maps out to these four basic states of knowledge and Figure 2 shows how they may be dealt with in practice.

		++ Group Awareness	- -
	Certainty / Identification	Certain (Known)	Uncertain (Unknown)
Self Awareness ++	Identified (Known)	Known-Known (identified knowledge) Things we are aware of and understand. Mitigate with design reviews and code certification (Class/CVA) that implementation is correct	Known-unknown (identified risk) Mitigate with expert consultation and R&D as needed. Find upper/lower bounds or regions of stability in state-space
		DETERMINISTIC MODELS; RANDOM PROCESS MODELS	CHAOTIC PROCESS MODELS
Self Awareness - -	Unidentified (Unknown)	Unknown-known (untapped knowledge) Things we understand but do not use or are not consciously aware of. Intuition and Experience. Engineering judgement. Mitigate with peer review and compare/contrast to past projects.	Unknown-Unknown (unidentified risk). Things we are neither aware of nor understand. Mitigate with "catch-all" margins based on hard experience (actuarial statistics or tests), risk ranking and benchmarking against similar attempts in the past. Potential "big-data" applications to discover new trends and relations.
		No Basis for a Direct Engineering Model, May use Indirect "black box" model with calibration coefficents and control parameters when actuarial statistics or model test data exist	

Figure 1: Categories of Uncertainty (Modified from Kim, 2012)

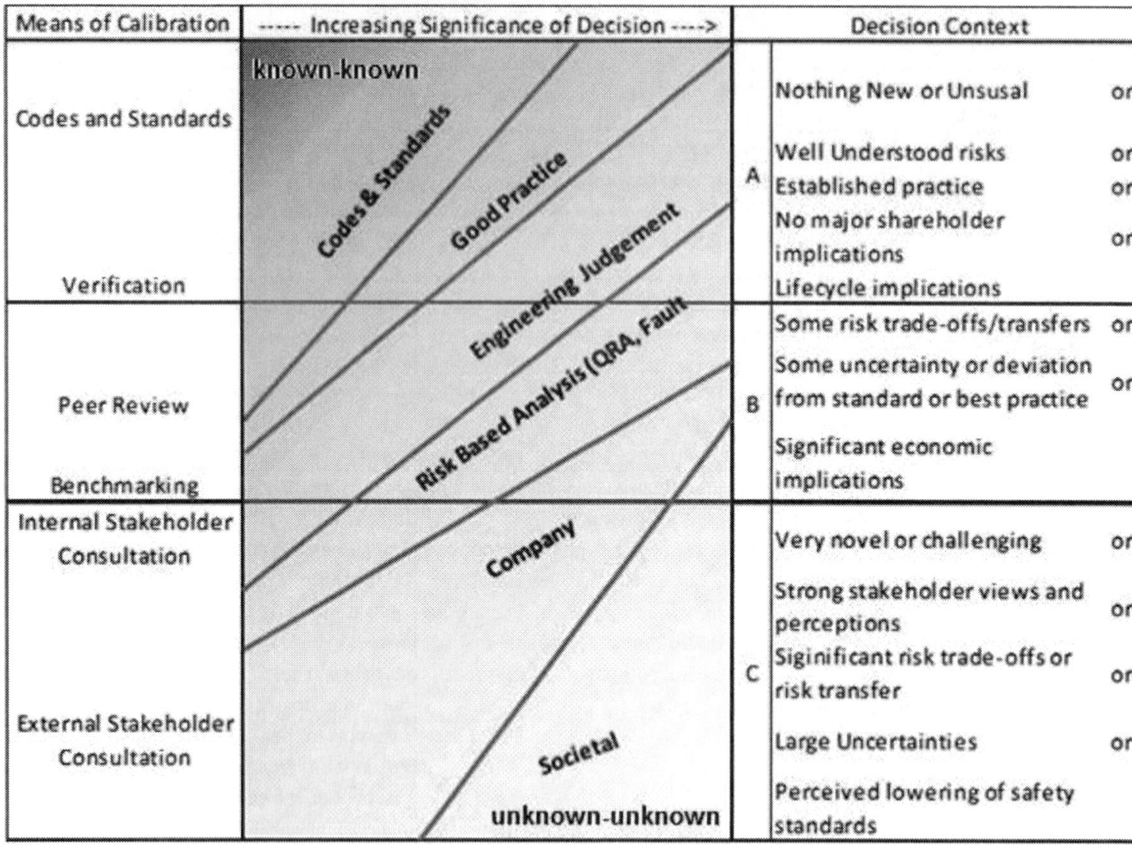

Figure 2: Risk Context Model

How we manage the risk of uncertainty depends on the significance of the decision and the context within which we are making that decision.

For example, a TRL-7 (field proven) off-the-shelf solution being used as before would tend toward a "known-known". A known-known is trivial from a design perspective and verification can default to established rules and norms. Using standardized off-the-shelf designs with a set of specifications and functional requirements that conform to the original design intent promote this.

An unknown-known may result in an intuition-based decision that may result from a mistaken assumption. In good practice, one should always challenge what one assumes is known.

A useful example of an unknown-known is something that is well understood in a different engineering discipline but not part of regular practice for the disciplines tasked with a design. Elastomers and O-ring design to prevent the Challenger disaster were within the state of the art for arctic drilling and production oilfield operations but relatively unknown to aerospace (Morandi, 2018).

For offshore systems, many parameters are treated in a stochastic manner but can be bound with varying degrees of certainty—these are "known-unknowns" which may have lower bounds or upper bounds such as a characteristic extreme value defined for a given return period.

For offshore systems, "unknown-unknowns" are those with no precedent or expectation that they could "reasonably" occur. These are not reasonably foreseeable since they are unknown, but it is important to be imaginative and consider physically feasible possibilities to reduce the pool of unknown-unknowns; this pool increases in size the more complex or more tightly coupled a system becomes.

There is usually no direct management mechanism, particularly prevention, in place for unknown-unknowns except for inherent conservatism that may reside in certain margins or practices that are already in place. For example, Class Rules for hull girder bending moment were developed long before the tools existed to adequately model potential strength limitations in severe weather conditions. However, empirically derived rules were developed over time based on the real performance of successful and unsuccessful designs. Using these rules within the limits of the context in which they were developed resulted in adequate structures with enough margin to serendipitously cover other local strength modes and interactions that may not have been explicitly clear. As first principles have been increasingly adopted over empirical rules, an increased number of previously shielded "unknown unknowns" have become exposed and problematic.

However, a new concept would be expected to require an extensive qualification process complete with risk assessment, model testing and possibly even development of a field prototype.

For one-off designs with high stress factors or novel features, the risk management process is challenging. But, formally applying risk management and systems analysis tools would map out and reduce the extent of the unknowns. Adding robustness measures such as targeted margin and failure mode decoupling further promotes survivability. All of which helps to build a case that reasonable risks have been accounted for.

EVENTS AND RISK

An event is the manifestation (physical occurrence) of an uncertainty within the boundaries of a system.

An event may be the product of a single uncertainty or a group of uncertainties that combine into a state-space vector which can be applied to an engineering model of a system.

The risk associated with the manifestation of an event is the probability of the event occurring multiplied by the consequences of the occurrence.

The probability of an event occurring must be greater than the project defined risk tolerance to be considered significantly foreseeable for project design purposes.

HAZARD

A hazard is defined here as a risk that:

1. Is significant (Exceeds the project defined risk tolerance).
2. Has harmful consequences.

In other words, a hazard must have a significant probability of occurring and a significant consequence should it occur and those consequences must be harmful.

The first goal of risk management in the design process should be to minimize the available pool of potential hazards (significant risks with the potential to cause harm) from which a hazardous event can initiate. Those risks remaining, must be further addressed in design by introducing the appropriate number and type of safety and containment barriers to manage the risk as simply and cost-effectively as possible.

The goal of hazard management in operation should be to fully maintain detection, safety, and containment systems, as designed, and to run the plant within Safe Operating Limits. Should detection or barriers become unavailable for a time (repair or maintenance required), operations and engineering should develop and implement alternative approaches to maintain sufficient control of risk for the interim. If operating requirements change, again, operations and engineering will work together to assess new or changed risks and modify the physical plant or operating procedures as necessary to manage risks appropriately while delivering the new requirements.

Engineering and operations must work together in a complementary fashion to manage risks appropriately in design, delivery, and ongoing operations of assets. The relationship between uncertainty, risk, and hazardous events is illustrated in Figure 3 below.

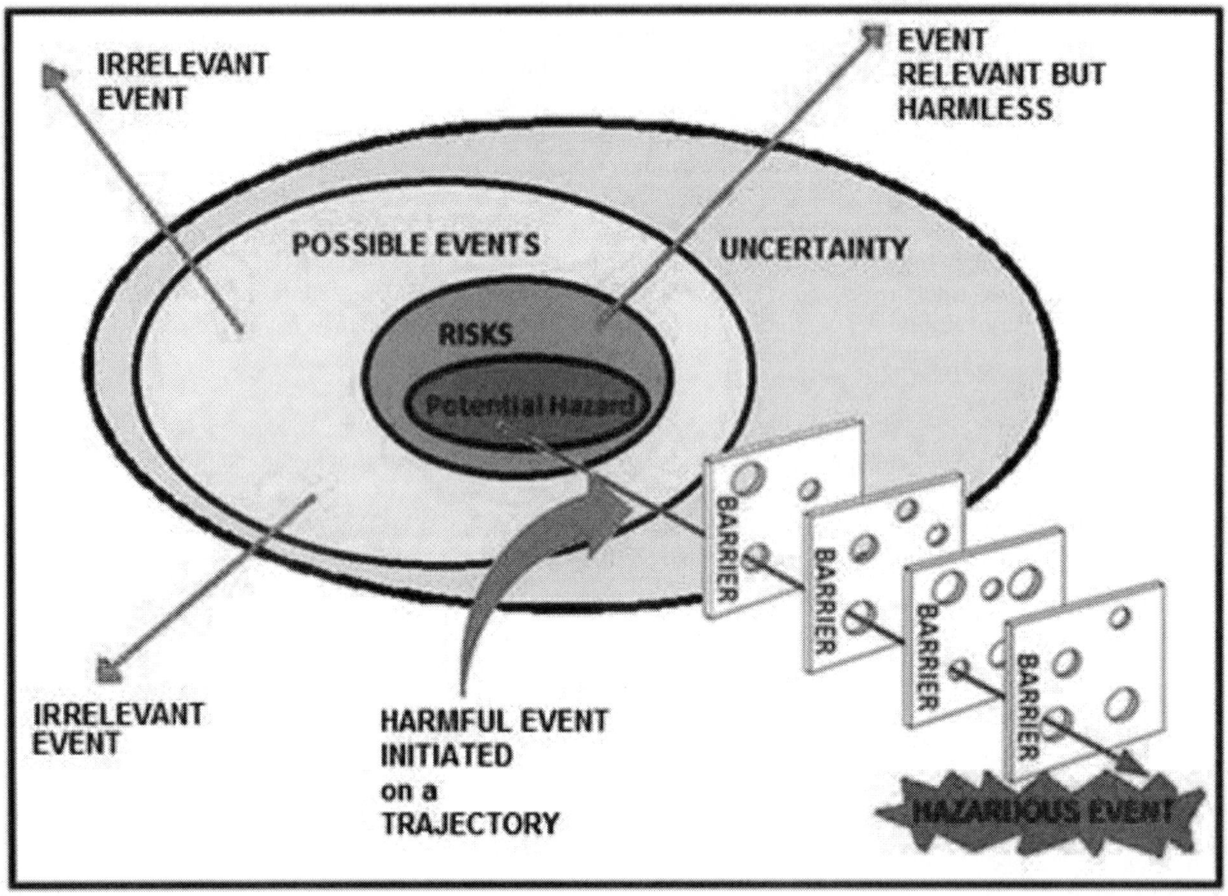

Figure 3 - Schematic Relationship between Uncertainty, Risk and Hazardous Event

ROBUSTNESS

Offshore systems generally involve non-linear stochastic processes similar to those described by Dvijotham in his analysis of power flow. For a given set of state variables (**X**) these processes may possess multiple local solutions and positions of equilibrium which in the most general sense may or may not be stable. Most processes encountered may be stochastic or chaotic but local linearity (strong or weak) may often exist and indeed is often designed for. For example, at low angles of heel with wall-sided structures, the hydrostatic restoring moment monotonically (and linearly) increases with angle of heel but at higher angles this becomes a highly non-linear problem. By setting minimum GM requirements, the system can be kept linear with a stable response (heel) curve. This process may now be said to possess a known local solution and position of equilibrium for given sets of state variables (**X**) which may exist inside a region of feasible solutions (**F**).

The feasibility region is the set of state variables (**X**) that are physically possible and which have solutions that satisfy performance requirements. These requirements may be expressed as a factor of safety, a limit state (SLS, ULS, ALS), a probability of failure, and an ALARP (as low as reasonably practical) threshold or similar. Following on from the example above, the feasibility limit on GM may be determined by the maximum permissible angle for crane operations, or it may be deck immersion or the onset of second-order roll effects. Other examples exist with structural reliability (See Figure 4 below).

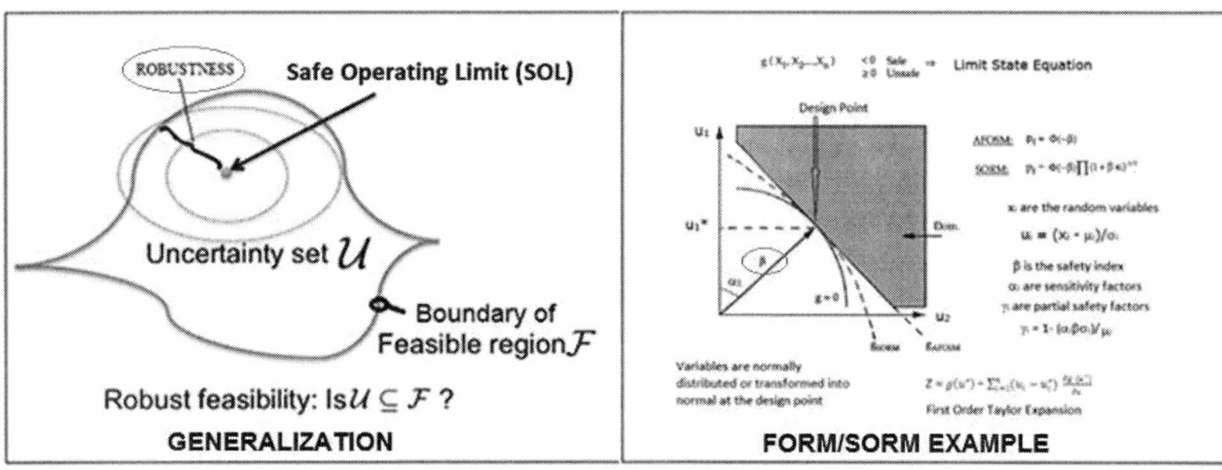

Figure 4-Interpretation of Robustness in State Space

Robustness is formally defined here as the spread (in state space) between the safe operating limit (SOL) and the edge of the feasibility space of a system. For offshore systems this spread may in some cases be expressed as the spread between the different ISO Limit states**Error! Reference source not found.Error! Reference source not found.Error! Reference source not found.Error! Reference source not found.** which serve as threshold markers between different states. In structural reliability calculations this spread may be expressed as the safety index 'β'.

For offshore systems, the safe operating limit (SOL) may be a defined alarm point, such as a maximum operating condition, or it may be a defined range of parameter values within which the system is expected to operate normally and without alarm. This should be defined by the project based on system specifics and operations philosophy.

Three levels of robustness are illustrated in Figure 5 and Figure below and defined as follows:

Operational robustness **(R1)** is for events up to the safe operating limit. It is operational flexibility against shutdown that allow the operational team sufficient time and scope to respond to changes in the process, system behavior or the external environment. This provides for response to alarms and correction of events which are amenable to control efforts such as throttling an engine. For events not amenable to control or exceeding the safe operating limits, the system is put into a stand-by or restricted mode such as a storm-safe condition. Time, space and range of choice typically serve as the first set of robustness barriers.

Survival robustness (R2) is system reserve against damage that allows the facility (put in restricted mode) to survive (without damage) a design level event that cannot be avoided. Such an event may, for example, be a controller failure which allows pressure to build, or it may be a 100-year storm. Provided the failure mode has been identified and mitigated, the design allows for these events to occur without permanent damage through the use of automatic safety systems or design margin in the form of safety factors (i.e. ULS). Code factors of safety typically serve as the second set of robustness barriers.

Reserve robustness (R3) maintains essential performance features (such as life safety) during abnormal or accidental events that are beyond or outside the established design condition. Controlled absorption and dissipation (or diversion) of damage typically serves as the third set of robustness barriers. It is system reserve against failure that allows the facility to endure accidental or abnormal events. This may be a fire and blast or a wave impact. These events may result in permanent debilitating damage to the facility while maintaining defined "essential" performance requirements such as preserve life, prevent pollution, and hold station. Some companies may extend this to include mitigation of economic consequences (i.e. repairable damage) based on their own cost/benefit assessment that satisfies their business needs. Unless there is a clear company specification of "repairable damage" for a particular component or system, these are not strength conditions where the system is designed to remain elastic, so the system is not expected to return to its original state. Debilitating damage that renders the unit permanently inoperable may occur, but the overall system must protect safety of life until evacuation is possible and minimize any discharge of pollutants into the environment. Finally, the unit should not go adrift which would cause it to be a hazard to itself and others. R3 applies regardless of whether control measures or automatic safety systems were applied.

Robustness measures R1 and R2 are often implicitly covered in normal design for failure modes which have been identified and mitigated by the code. This means that even for designs based on minimal functional requirements and which are code based, there is significant robustness inherited from the codes used: this is especially true of experience-based codes such as classification society Rules. But, regardless of any existing measures, care and effort should be expended to identify risks and failures modes outside the scope of any codes and additional robustness measures introduced if necessary.

Critical reserve robustness R3 may be specially identified. On mooring systems this may include design requirements that maintain integrity of the system even in one-line damage conditions. On floating systems, this may include design requirements that maintain afloat stability even in the event of one (or more) compartment damage. Dynamic positioning systems may require the consideration of damage to any component or the loss of any single compartment to flooding or damage depending on the class of the DP system. These examples represent cases where R3 reserve robustness requirements have been identified as critical in general and have been explicitly added to the Codes or Rules by the relevant Authorities. R3 may be specially added to the Basis of Design by owners with specific concerns or they may be discovered during the design process on a project as discussed in this paper.

		Design / Performance Checks	
Performance Level	Designation	API World	ISO World
Limit deflections and vibrations	A	Operational	SLS
Intact without damage	B	Extreme Storm	ULS
Repairable damage	C	As Specified	As Specified
Impending collapse	D	Collapse	ALS

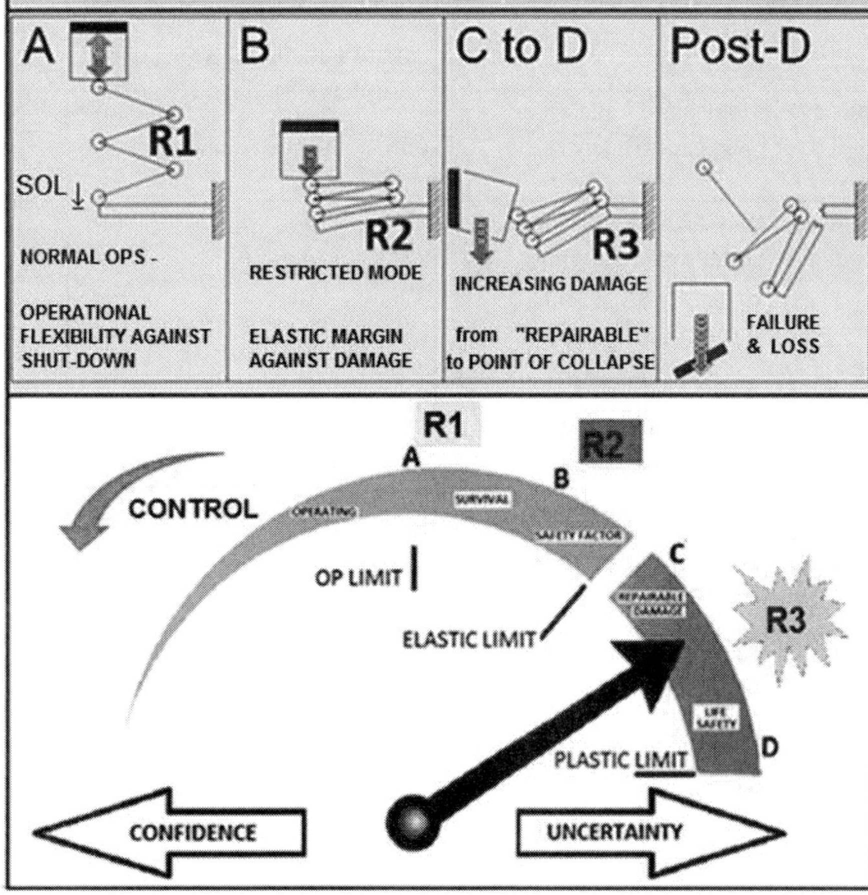

Figure 5-Robustness and Performance Levels

The level of robustness required by the designer against an event depends on the class of the event as summarized in Fig 6 and Table 1 below, which shows the relationship between the event class, the robustness spread, and the threshold being protected against expressed here as one of the ISO limit states: serviceability (SLS), ultimate (ULS), accidental (ALS).

Examples of robustness are shown below in Figure 7 – 10.

Table 1 Event Classes and Mitigations

Event Class	Mitigate Against.....	Robustness Spread	ISO Threshold
Un-eventful normal operation in anticipated operating context	Loss of Functionality or Control	R1	SLS
Expected hazardous events which are anticipated, prepared and planned for (e.g. expected extreme design storm)	Damage to the Facility	R2	ULS
Unexpected hazardous event which can be foreseen or has relevant precedent but little to no warning is possible: (a) natural (tsunami, earthquake, squalls, rogue waves) (b) accidental (fire and blast, component failure, error) (c) abnormal storms exceeding expected characteristic extreme values	Escalation to Catastrophic Loss	R3	ALS
Unforeseen or Unprecedented hazardous event	Mitigation impossible or unreasonable	N/A	N/A

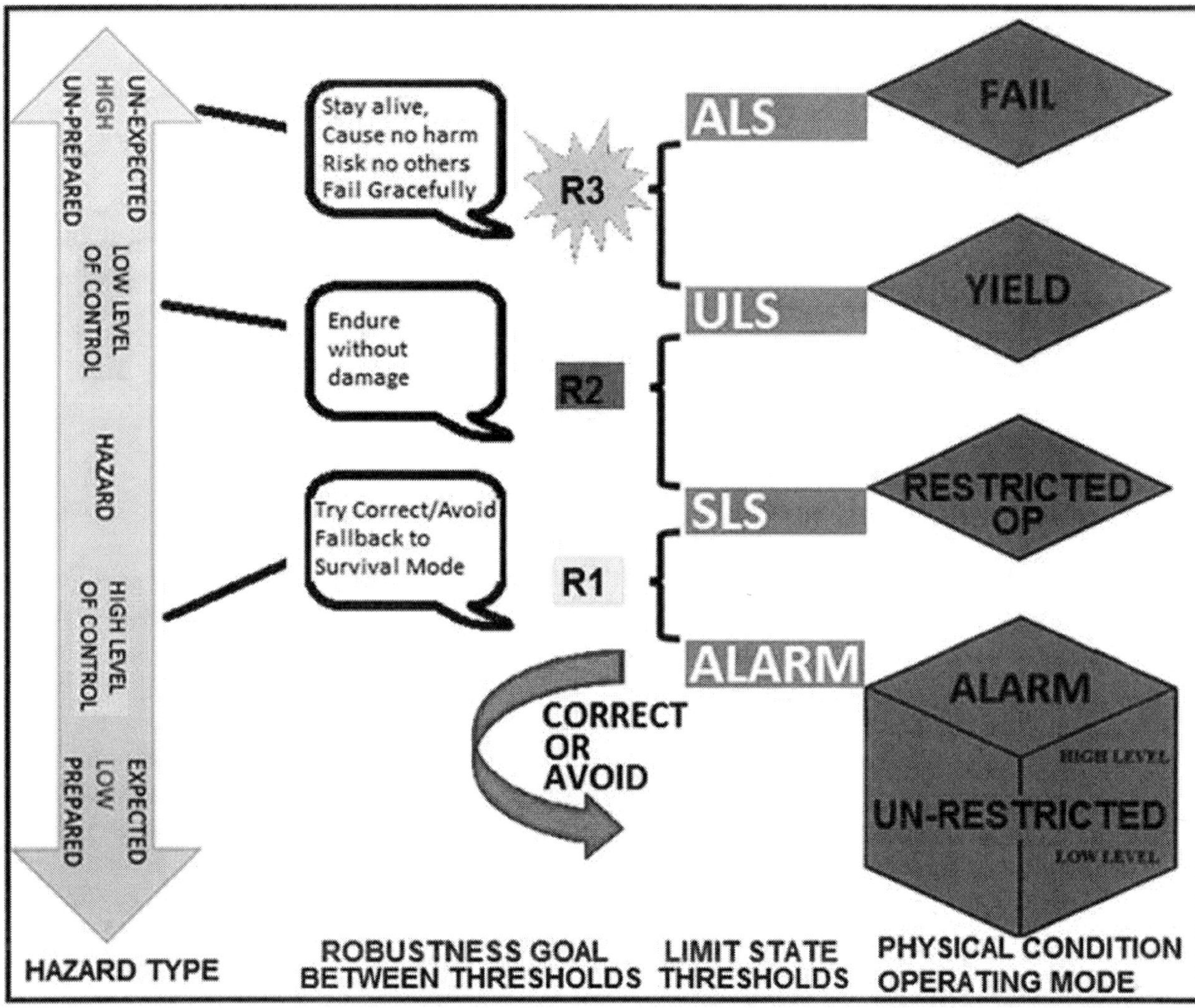

Figure 6 - Robustness Context and Levels

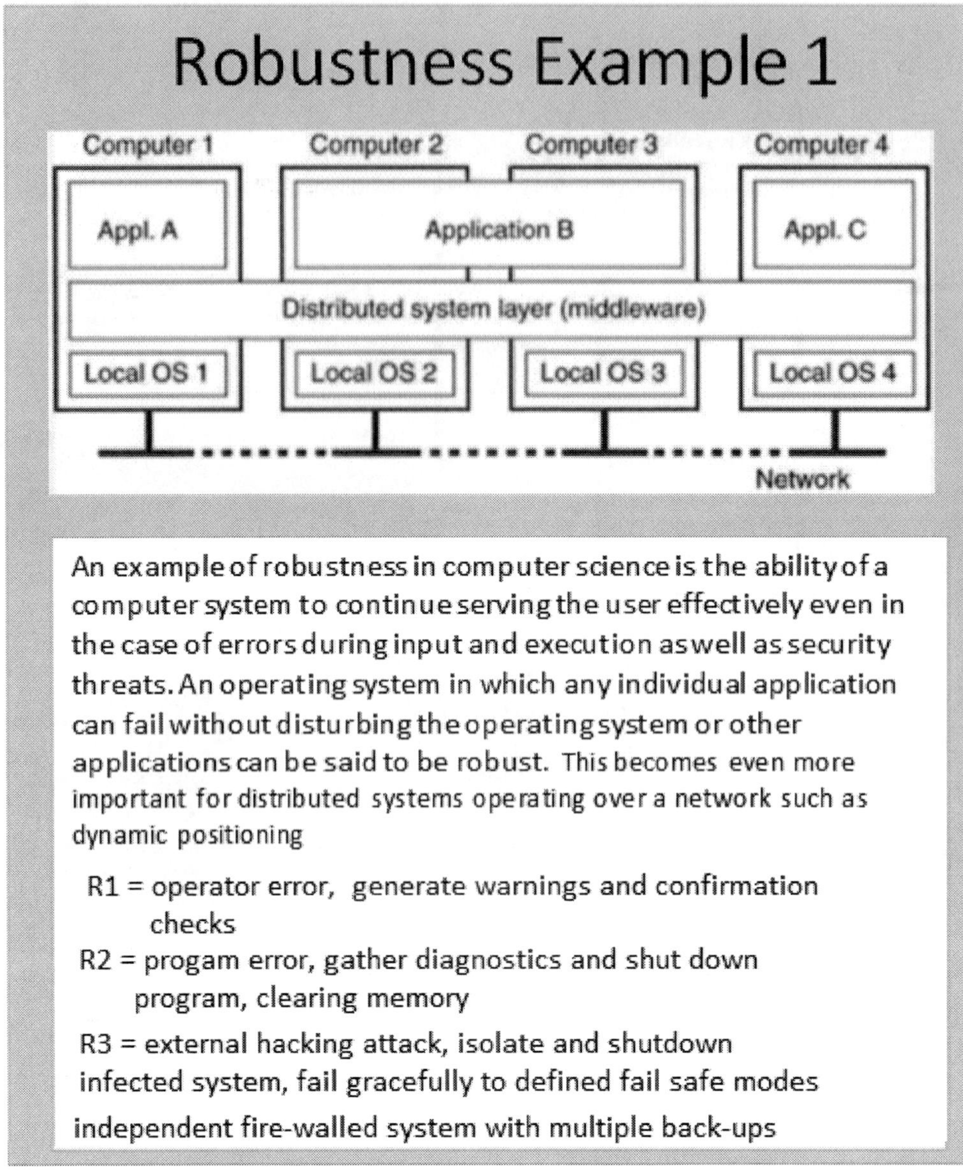

Figure 7 – Robustness Example 1

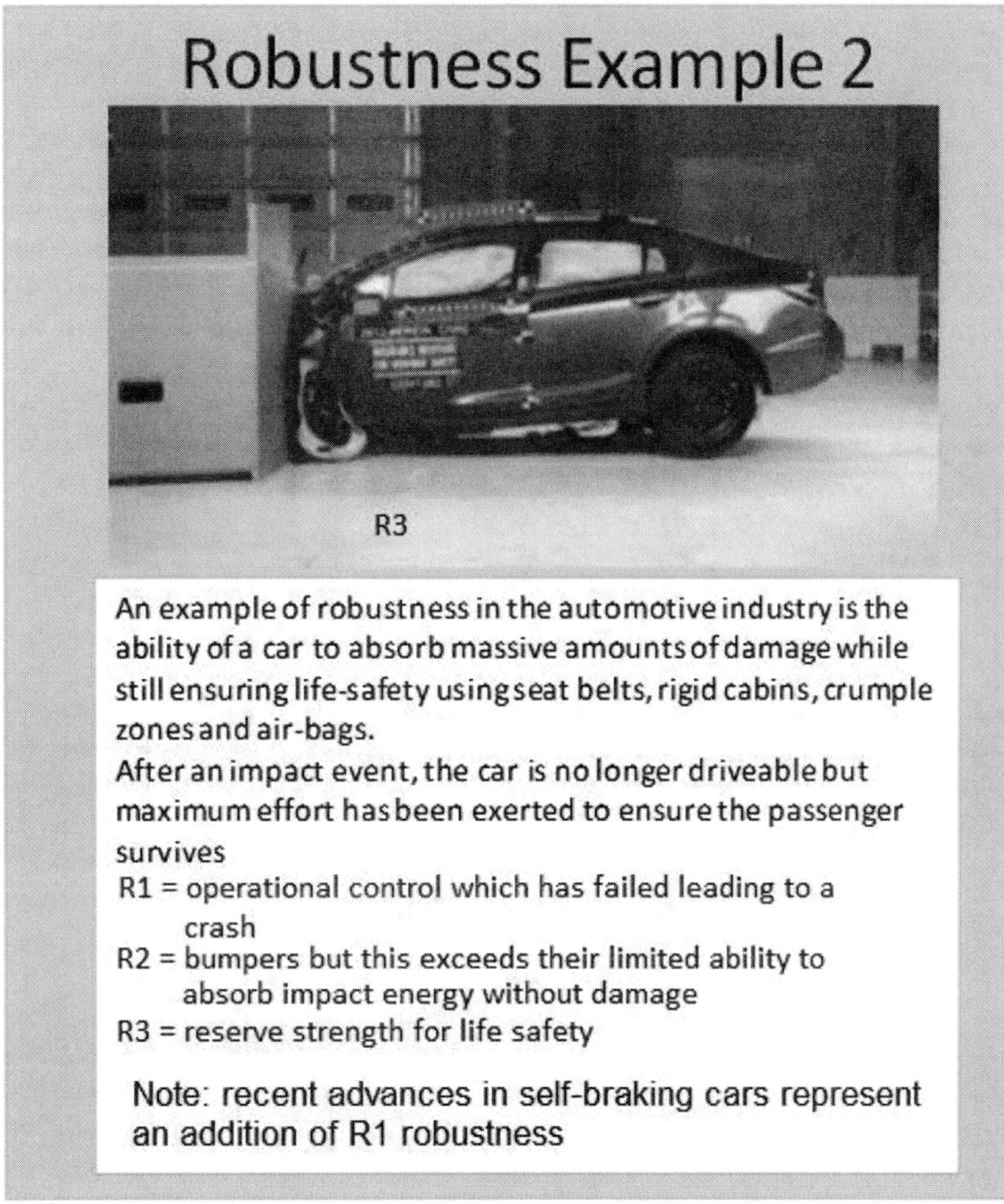

Figure 8 – Robustness Example 2

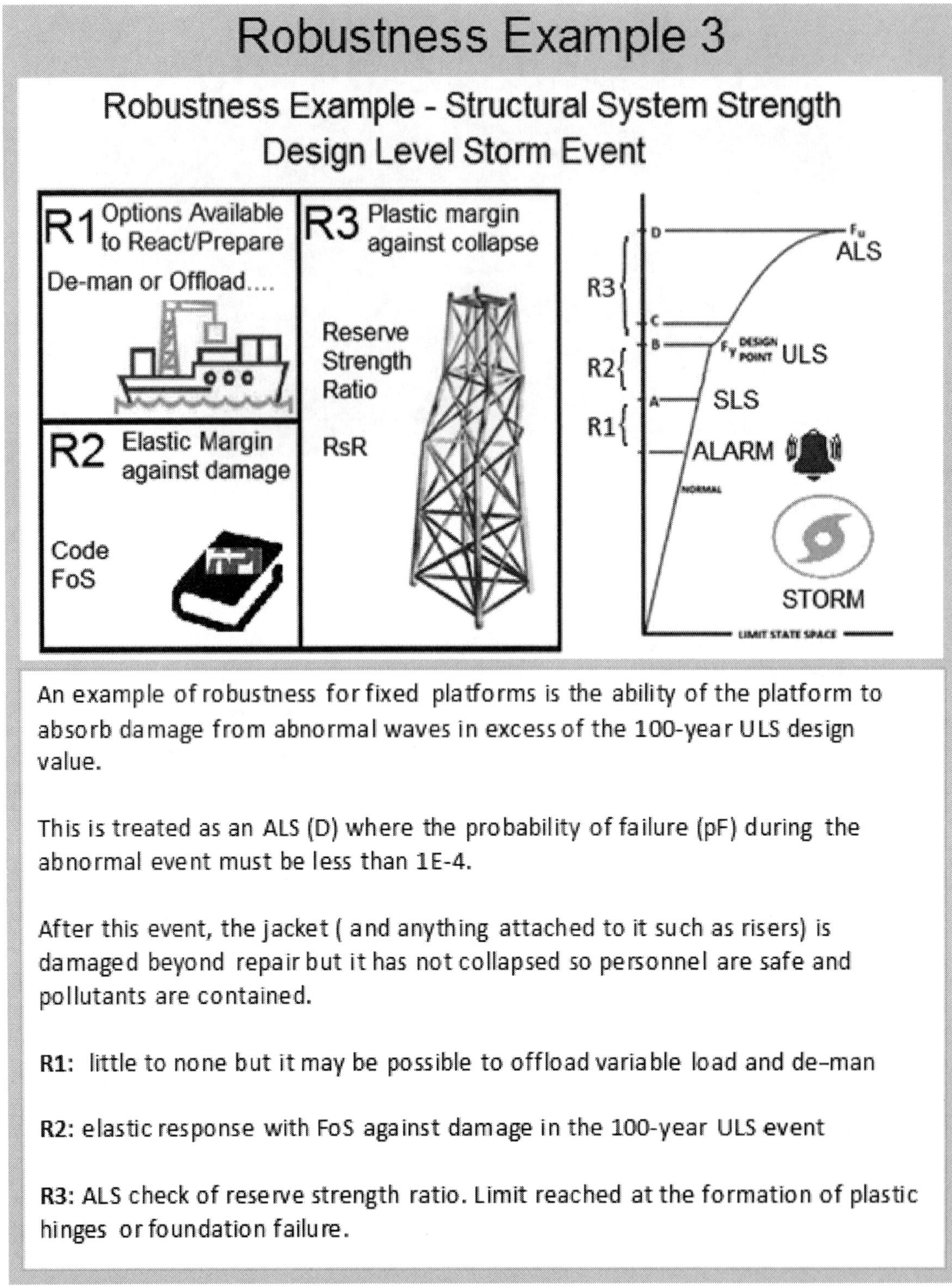

Figure 9 – Robustness Example 3

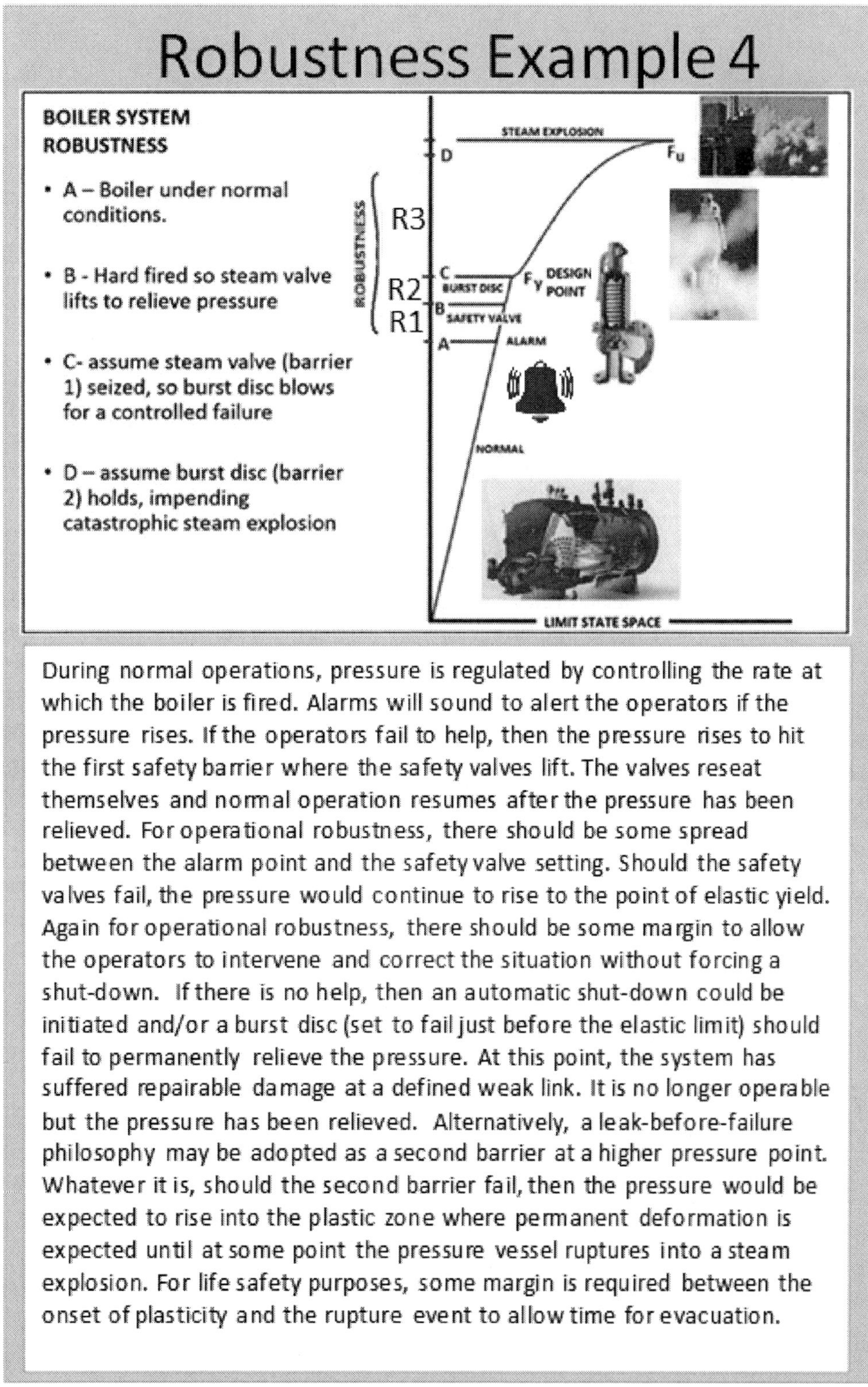

Figure 10 – Robustness Example 4

ROBUSTNESS PROBES

Systems engineering principles can be used to define system boundaries, sub-system interfaces and interactions.

In responding to reasonably foreseeable events, a robust system should not cascade out of control or produce a response or a consequence that is disproportionate to the cause. A robust system maintains essential minimum features under scoped perturbations.

Robustness probes are intended to "stress-test" the design to ensure there are no sudden cliffs the design will fall off with marginal increases above the safe operating limit (SOL) or the design conditions.

Robustness probes may be required if (among other things):

1. Regulations explicitly dictate it.
2. The design is novel or scaled up beyond available in-service experience, hence extra assurance is required by project stakeholders or a compatibility check with existing regulations (which do not quite fit) is required.
3. There is significant uncertainty in the loading or operations warranting probing the limits and the sensitivities of the design.
4. A failure mode lies outside the design envelope but is identified in a project HAZID and it is deemed important enough to require an understanding of the inherent design margins for internal risk assessment purposes.
5. The designer wishes to identify design break-points and hard limits that must not be exceeded in order to assess margin sizes.

A probe is non-binding "for information only" unless explicitly required by Code or noted otherwise by the client in the basis of design. Assessment using a robustness probe may be against:

1. Straight-forward requirements given upfront by the Company or direct regulatory requirement such as the Norwegian requirement for positive airgap at the lifeboat stations in a 10,000-year storm. In this case, the design probe becomes a design requirement equivalent to a simple code check (i.e. the unit has the required airgap; if not, it doesn't, then it must be corrected).
2. An internal risk assessment per Company policy. In this case, the design probe is treated as "for information only" unless the Company risk assessment dictates that it is required. In this case, they implement the findings using a variation order to amend the design.
3. An internal risk assessment is required per Designer practice. In this case, the designer will make their own assessment based on individual company practice and history.

A successful assessment would ensure:

1. Minimum essential features and performance are maintained.
2. The system response is not disproportionate to the perturbation.
3. The system response is not accelerating away from a stable solution.
4. The system response continues to remain meta-stable at each perturbation; i.e., the system may jump but not enough to represent a state change.
5. Failure modes are decoupled (where possible) and do not cascade.

In practice satisfying these may require design changes such as adding margin or redundancy, or decoupling failure modes, or changes in operating procedures such as going to a storm-safe survival mode, evacuation or reduced manning, shutting down production then flushing and isolating risers.

A robust system is locally stable in state-space and does not fail suddenly or catastrophically when subjected to uncertainties. To test for this, the parameter vector (X) defining the safe operating limit (SOL) in state-space is subjected to perturbations in its applicable set of uncertainties ($\mathcal{U}$) and the system response evaluated.

A robustness probe is a virtual tool that provides the means for testing the response of a system to a set of perturbations (ΔX) around a safe operating limit (SOL) as illustrated in Figure 11 below.

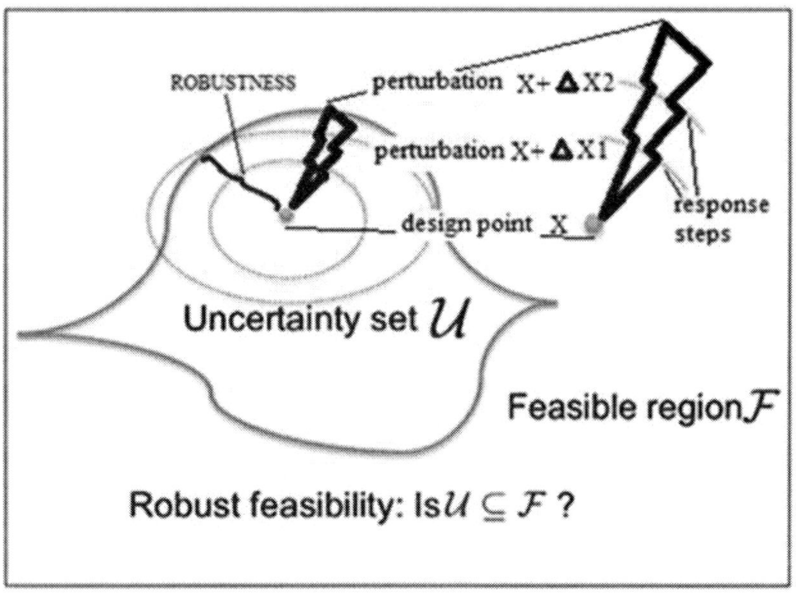

Figure 11 - Robustness Probe Schematic

Fortunately, as noted by Dvijotham, for real physical systems, there is structure to both **F** and **X** that can be exploited to solve these problems efficiently.

For floating offshore systems this structure manifests itself through geometric constraints, engineering models, existing failure modes and physical limitations on the permissible range of values that key parameters may assume.

This structure provides a mechanism to scope the perturbations. The perturbations should be:

1. Limited to parameters which are subject to uncertainty. Where the values of the parameters are certain or where upper/lower bounds exist and would produce economical solutions, then no perturbation or probing is necessary.
2. Limited to parameters which are relevant to the failure mode in question.
3. Limited only to independent parameters. Dependent parameters should be identified and expressed in terms of independent parameters and eliminated to maintain a sharp focus on critical items.
4. Mutually exclusive, jointly exhaustive, and span the relevant parts of state space.
5. Sized based on regulatory requirements.
6. Sized based on physical limitations.
7. Sized based on risk tolerance to the specific failure mode or hazardous event in question.

The range of parameter variation should be based on a combination of physical limitations, engineering models and principles, engineering judgement, operational experience, risk tolerance, and regulatory requirements. These considerations provide insight into how the parameters can vary and what combinations are most significant. This keeps the focus only on what is most important and prevents losing sight of parameters that could become important. The probes should not default to simple systematic variation of parameters such as using a simple matrix of parameter variations. Enormous volumes of data can easily be generated with existing software, however, sheer volume can be overwhelming and can mask the few critical cases. Future application of machine learning may help but, for the moment, effort should be made to focus and restrict the number of load cases to manageable quantities that encompass the likely critical limits or combinations.

Probes should only follow the failure modes with significant consequences. A significant failure mode may for example include a wave impact event on a deck box. An insignificant failure mode (depending on company practice) may be grating washed off by localized wave run-up. Insignificant failure modes should not be pursued.

The failure modes identified should:

1. Be based on the physical and logical layout of the system in question.
2. Arise through a systematic process. This may be an internal company method or a formal method such as Failure Modes Effects Cause Analysis (FMECA) or similar held by qualified practitioners. See ISO 31010 for a comprehensive list of formal methods.
3. Be non-trivial and have significant consequences in the event of a failure.
4. Be independent of each other.
5. Be simple and non-divisible.
6. Be traced back as close to root causes as possible.
7. Identify dependent systems for consequence analysis (trace back up other branches as required).

Formal methods which trace causal relationships, such as the bow-tie method, can be used to identify barriers which are missing or redundant.

Applying the above methodology, it should be possible to map-out the region of stability in state space relevant to a particular failure mode.

A feasible solution (**F**) cannot violate any physical principles and it satisfies the appropriate performance requirement. This requirement may be expressed as a limit state, target reliability, or individual risk per annum (IRPA) depending on the context. Both conditions are required to create an acceptable solution. For example, a capsized unit may be a physical possibility, but is not an acceptable outcome.

A perturbation that results in a state change is a design "break point". For example, the hull girder bending moments at the elastic and plastic limits are both design "break-points" as they represent changes in state.

At a systems level it may be possible to correct or improve break-points. However, some such as hull girder bending moment are intrinsically present; i.e., it will always be possible to find some load that will cause plastic collapse. These can be improved at economic cost but never avoided completely.

A "hard limit" is a "break-point" that cannot be corrected and which must be avoided by adding margin, decoupling or for those which can be subject to some measure of control, imposing operational limits with appropriate operation team warnings.

An example of a robustness probe is presented and discussed below (see Figure 12) for an evaluation of CoG location.

State space parameters are longitudinal center of gravity-LCG and transverse center of gravity -TCG (i.e. state space = physical space for this example).

An elastic system is one which can be trimmed (with increasing difficulty) for an envelope of CoG locations using water ballast only. A plastic system can be trimmed only with a combination of water ballast and redistribution of variable weights such as deck or operating loads. At some perturbation in LCG/TCG, it is no longer possible to trim the unit at all – this is the plastic limit for this system.

During construction, excursions into the plastic region may possibly be absorbed by redistributing lightship items or adding solid ballast at the expense of payload. Whether or not such measures are acceptable is a project specific decision that will influence how tightly the margins are set and obligations are contractually passed in weight budgets from owner to yard before the start of construction.

Designer and owner weight budgets would define the limits and the margins applied based on the sensitivity of the design to shifts in CoG and the potential for scope changes.

This envelope is the plastic limit for this failure mode and represents a hard limit the designer and project team should protect against in construction.

The alarms in this case would be yard produced weight reports tracking the location of the CoG at regular intervals. These would have to be monitored and acted upon as necessary by the project team.

ROBUSTNESS PROBE EXAMPLE 1

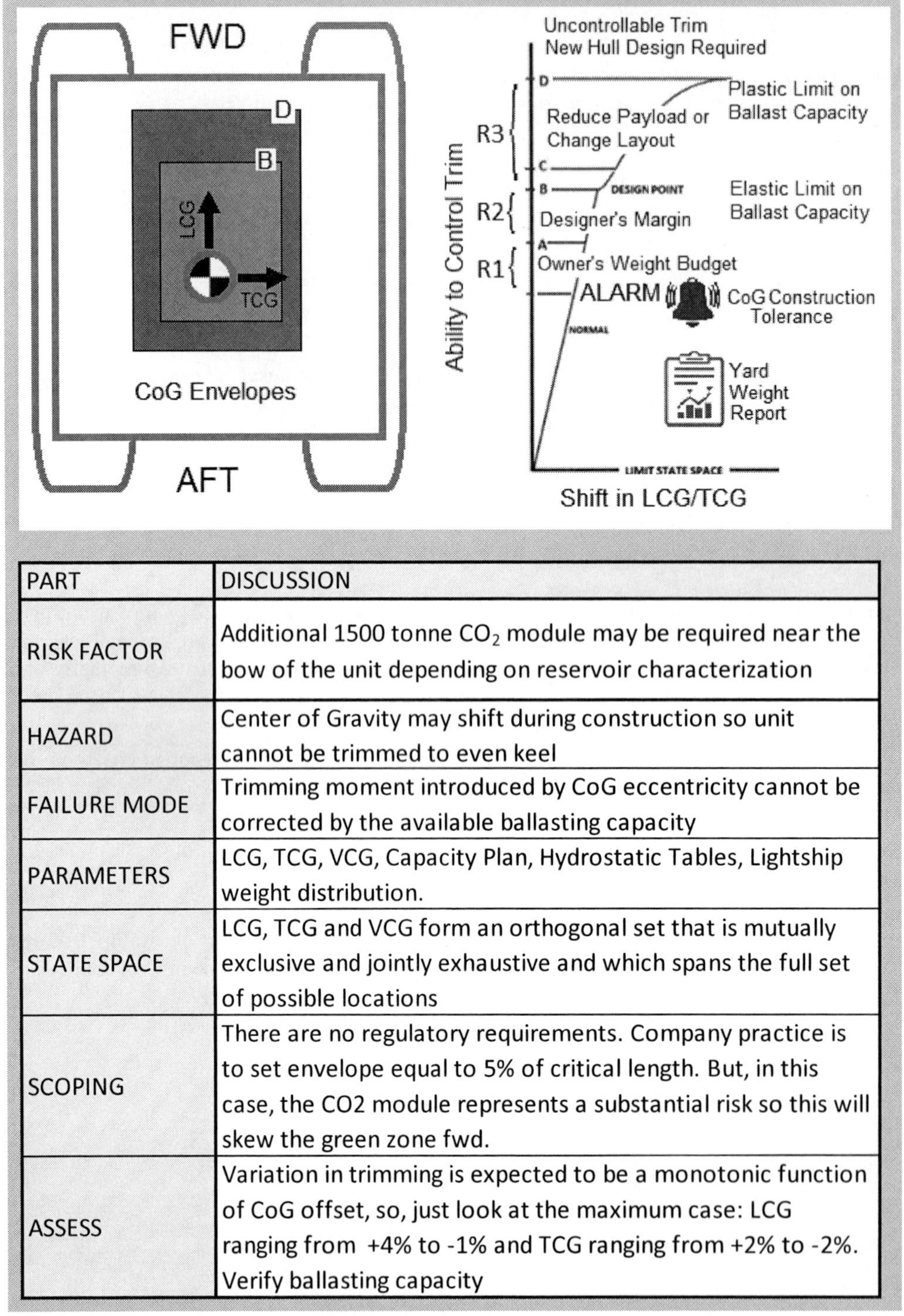

PART	DISCUSSION
RISK FACTOR	Additional 1500 tonne CO_2 module may be required near the bow of the unit depending on reservoir characterization
HAZARD	Center of Gravity may shift during construction so unit cannot be trimmed to even keel
FAILURE MODE	Trimming moment introduced by CoG eccentricity cannot be corrected by the available ballasting capacity
PARAMETERS	LCG, TCG, VCG, Capacity Plan, Hydrostatic Tables, Lightship weight distribution.
STATE SPACE	LCG, TCG and VCG form an orthogonal set that is mutually exclusive and jointly exhaustive and which spans the full set of possible locations
SCOPING	There are no regulatory requirements. Company practice is to set envelope equal to 5% of critical length. But, in this case, the CO2 module represents a substantial risk so this will skew the green zone fwd.
ASSESS	Variation in trimming is expected to be a monotonic function of CoG offset, so, just look at the maximum case: LCG ranging from +4% to -1% and TCG ranging from +2% to -2%. Verify ballasting capacity

Figure 12 - Robustness Example - "Center of Gravity"

Conclusions

Robustness is an emergent property of a system that cannot be achieved by simply designing to higher strength levels. It requires a considered look at the applicable failure modes and the parameters that feed into them. These failure modes should be decoupled from each other as far as possible and barriers should be erected in both design and operations against the manifestation of significant failure modes. The design should also be stress tested for reasonable perturbations of the applicable parameters to ensure local stability.

A systems engineering approach that provides a formal method of identifying and clarifying requirements and the boundaries between them is essential. This approach should cut across discipline lines to treat the entire unit holistically.

Acknowledgements

I gratefully acknowledge the contributions of Jim Brekke, Peter Wallace, Alberto Morandi, Arun Duggal, Flavia Rezende, Rex Song, Piotr Szalewski, Hongbo Shu, David Petruska, Amal Phadke, Steve Leverette, Andrew Kilner, Wei Ma, David Wisch, David Knoll, and Paul Erb.

References

AISC. (1998). Load & Resistance Factor Design: Manual of Steel Construction. Chicago: American Institute of Steel Construction.

Charles, P. (1984). Normal Accidents: Living with High Risk Technologies.

Dvijotham, K. (n.d.). Robustness of Infrastructure Networks. IEE.

HSE. (2001). Reducing Risks, Protecting People. UK Health and Safety Executive.

HSE. (2006). Offshore Installations (Safety Case) Regulations 2005 - Regulation 12 Demonstrating compliance with the relevant statutory provisions. UK Health and Safety Executive.

IACS. (2015). IACS UR S11 "Longitudinal Strength Standard" Rev 8. International Association of Classification Societies.

IMO MSC Circ 64. (n.d.). Guidelines for vessels with Dynamic Positioning Systems. International Maritime Organization.

INCOSE-TP-2003-002-04. (2015). SYSTEMS ENGINEERING HANDBOOK - A GUIDELINE FOR SYSTEM LIFE CYCLE PROCESSES AND ACTIVITIES (4th edition). San Diego: Wiley.

INCOSE-TP-2010-006-02.1. (2017). Guide for Writing Requirements. San Diego, CA: International Council on Systems Engineering.

ISO . (2007). ISO 19902:2007 - Petroleum and natural gas industries - Fixed steel offshore structures - 1st Ed. International Standards Organization.

ISO 19901. (2005). "Petroleum and natural gas industries - Specific requirements for offshore structures". Brussels: International Standards Organization (ISO).

ISO 19905-1. (2012). Site Specific Assessment of mobile offshore units Part 1: Jack-ups. International Standards Organization.

ISO. (2009). ISO 3100 Risk management - Principles and Guidelines. Geneva, Switzerland: International Standards Organization.

ISO. (2013). ISO 19900:2013 - Petroleum and Natural Gas Industries -- General Requirements for Offshore Structures. International Standards Organization.

ISO. (2015). ISO 2394 - General Principles on Reliability for Structures - 4th Edition. International Standards Organization.

ISO. (2016). ISO 19905-1:2016 - Petroleum and natural gas industries -- Site-specific assessment of mobile offshore units -- Part 1: Jack-ups. International Standards Organization.

ISO-17776. (2000). ISO 17776 Guidelines on tools for hazard identification and risk assessment. Geneva Switzerland: International Standards Organization.

ISO-31010. (2009). ISO 31010 Risk Management - Risk Assessment techniques. Geneva, Switzerland: International Standards Organization.

Kim, S. (2012). Characterizing unknown unknowns. Project Management Institute (PMI) Global Congress 2012 - North America.

Moan T, J. J. (1999). Probabilistic Inspection Planning of Jacket Structures. Offshore Technology Conference. Houston: OTC.

Moan, T. (2014). Safety Assessment of Offshore Structures. OSRC - Offshore Structural Reliability Conference. Houston.

Morandi,A. (2018) A Brief Review of Robustness Efforts in the Offshore Industry. 23rd SNAME Offshore Symposium. Houston.

Hybrid Induction – Oxygen Cutting Process for Faster, More Efficient, and Less Costly Scrapping of Off-Shore Petroleum Platforms

Jerald E. Jones
President, EnergynTech, Inc.

Valerie L. Rhoades
Vice President, EnergynTech, Inc.

James R. Dydo
President, Gatekey Engineering

Proceedings of the 24th Offshore Symposium, February 20th 2019, Houston, Texas
Texas Section of the Society of Naval Architects and Marine Engineers

ABSTRACT

A new steel cutting process has been invented and developed, by EnergynTech, Inc., which significantly increases the cutting speed of steel, and reduces the labor cost through automation of the operation. The specialized automated equipment can be used in a scrapping shipyard for dismantling ships or off-shore structures which are moved to a shipyard. In a mobile or portable implementation can be used on-board for scrapping operations. This new process was developed in a U.S. Navy funded Small Business Innovation Research project, the goal of which was to increase the productivity of the scrapping of decommissioned nuclear submarines and aircraft carriers, and reduce the environmental impact of that activity.

Keywords: SNAME, Offshore, Symposium, cutting, steel, automation, hybrid, induction, oxygen, opacity, PSCAA

Introduction

The Ship/Submarine Recycling Program (SRP) is the process that the United States Navy uses to dispose of decommissioned nuclear vessels. SRP currently takes place only at the Puget Sound Naval Shipyard (PSNS) in Bremerton, Washington. The salvage workers cut the submarine into pieces and remove the piece with reactor core. Scrap metal from the remaining hull pieces is cut into crane liftable pieces, then moved to an indoor cutting area where the large pieces are cut into smaller pieces. Scrap metals are sold to private companies or reused.

The primary metal cutting technology used today for shipbreaking is Oxy-fuel torches. This is also the main generator of visible particulate matter (PM), which is measured by "opacity". Opacity is the smoke generated by burning of the torch gas, the iron and carbon in the steel, and the steel surface coatings at/near the kerf (including paint and a rubber-like special hull treatment and its adhesive). The PM emissions linger in the air long enough to be detected and reported by observers using the Ringelmann Method. The fuel that causes 90% of the dark smoke is the surface coatings. A minimum of 9" of each side of the torch path must be cleared to bare steel—inside and out—to avoid these substrates from burning and contributing to opacity. Reducing the opacity emissions will have added safety and health benefits. On November 4, 2010, Occupational Health and Safety Association Instruction CPL 02-00-136 was replaced by CPL 03-00-012 with significant changes to increase the overall worker safety during shipbreaking operations.

The objective of this project was to develop an innovative new metal cutting system for submarine and aircraft carrier hull disposal that achieves competing requirements for environmental compliance; safety and health requirements; scheduling, manpower and time constraints; while achieving increased cost efficiencies to translate into life cycle cost reductions.

Steel, on a microscopic scale, is held together by the metallurgical bond between iron atoms. This is a very strong force – which provides steel with excellent mechanical properties for use in many applications – including shipbuilding. That strong metallurgical bond is also the reason that steel is difficult to cut. A cutting process for heavy section steel used for nuclear submarine and aircraft carrier hulls must break that bond. Three approaches are available: 1) mechanically break the bond by forcing the atoms apart; or, 2) lower the bond strength by a moderate increase in temperature so that the material separates more easily by mechanical cutting; or, 3) increase the temperature of the steel to the melting point and remove the material in a narrow strip along the cut line. The physical cutting method for approaches 1 and 2 is to scrape pieces of metal using a saw blade or grinding medium and remove a small strip of metal forming a cut line. Both of these processes, for thick steel plate can be very slow, but produce the best quality cuts. For thicker metal parts where the cut quality is of lesser, or no, importance approach 3 is most frequently accomplished using an oxy-fuel torch or a plasma torch. It is possible to use a laser cutting system, but for scraping large metal structures such as submarines, aircraft carriers, or offshore structures laser cutting systems would not be suitable because of the rugged, outdoor environment.

The approach of the project was to develop an Hybrid Induction-Oxygen (HI-O) cutting process. Steel, if heated to 1600° F (autoignition temperature), and subjected to a directed flow of pure oxygen, will spontaneously combust (i.e. oxidize), which produces significant heat. Oxy-fuel cutting, used at PSNS today, has small flames which maintain the surface of the steel at autoignition temperature and a small oxygen nozzle makes the cut. The heat released by the oxidation, continues to melt the metal providing a hot liquid surface for the continuing oxidation reaction, and the oxygen stream strips the oxide particles off of the surface thereby maintaining a clean liquid metal surface.

The approach of the project is to substitute induction heating instead of the oxy-fuel flame which the authors have done with other processes[2,4,6]. The flame is a very inefficient heat source for cutting. First, the flame is continually being cooled by the flow of room temperature fuel gas and oxygen into the flame, so those gasses must be heated to the ignition temperature in order for the flame to continue to burn. Where the flame impinges onto the surface of the metal, the flow of the heated combustion gasses after impinging onto the surface of the metal, is that it is re-directed and flows back into the surrounding atmosphere. So, a significant portion of the heat is used up heating the atmosphere in the near vicinity of the steel, and not heating the steel. Finally, the science of heat flow shows that the surface of the metal is a significant barrier to the flow of heat. Near the surface of the steel plate, there is a relaxation of the metallic bonds, which reduces the efficiency of heat transfer – that is, the metal surface, metallurgically acts as a thermal insulator to prevent the heat from flowing into the interior of the metal.

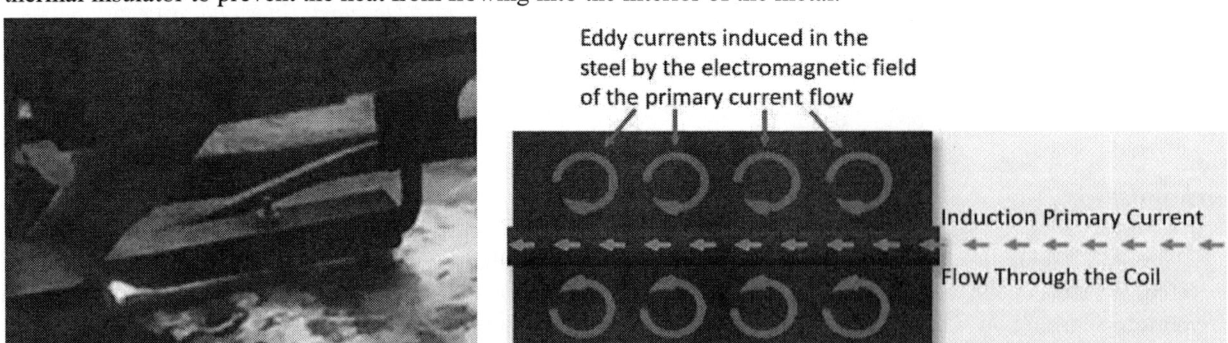

Exhibit 1. (Left) photograph of the induction coil heating the metal surface under the coil. (Right) Schematic representation of the Induction Heating of the metal part.

The induction coil is powered by the Miller ProHeat which is power rated for 35 kW. So, the induction coil provides heating of 1,400 W/mm², while the oxy-fuel flame is approximately 49 W/mm² – 28 times as much heating.

In addition, the induction heating process is 90% - 95% efficient, that is 90+% of the induction power is converted to heat in the steel, while the oxyfuel flame efficiency is much lower, perhaps 30% depending on conditions and the operator. So, the Hybrid Induction-Oxygen process can potentially cut at 50+ times the speed of the oxyfuel process.

This translates to a much faster cutting speed. The limit on the speed of oxy-fuel cutting is the flame energy. on the speed of HI-O cutting it is the rate of the oxidation reaction. For example, the oxy-fuel cutting charts show 10 in./min. for this thickness of steel. To date, in the laboratory, HI-O cutting has demonstrated cutting at 100 in/min. 10X the speed of Oxyfuel cutting. The Phase I project proved that the theory could be made into a practical cutting apparatus. The Phase II project has shown that a portable robot can be used to cut submarine hull steel at very high speed, the authors have developed other mobile/portable robots for manufacturing processes[1,5].

The result of the project is a new cutting process which successfully met or exceeded all of the project goals set by the Navy. These same project goals (with the possible exception of the environmental constraints) would be the same for scrapping of off-shore structures. The project goals were:

1. Semi-robotic control;
2. Able to cut high-tensile steel;
3. Portability: preferably in "man-handle-able" modules;
4. Power: hydraulic/electric/pneumatic;
5. Control: hydraulic/electronic at cutter;
6. Person drivable, preferably not requiring programming for each cut;
7. Cut: Quality of cut and accuracy of cut are not an issue because the cut pieces are scrap;
8. Speed -- minimal accepted lineal cutting speed of 10 inches per minute on 2" thick steel sheet, with the same speed or better on thinner work pieces;
9. Submarine and ship hulls are not flat plate and have curvature;
10. Flexibility in the cutting path so that it could be steerable around obstructions;
11. Cut kerfs (width) is required to be ½" minimum ¾-1" preferred;
12. Thickness of metal from ½" to 4";
13. Should not require access to both sides of panel;
14. If a cutter is used, cutter should be designed for dry or very minimal coolant/lubrication;
15. Cutter system must be suitable for use in year-round, open-weather, salt waterfront environment;
16. Resultant opacity below the limit set by Puget Sound Clean Air Act (PSCAA) and also lower than the norm of oxy-propane torch cutting;
17. Safety: cut large sheets of plate at high speed within OSH noise limits, prevent repetitive motion, and prevent fatigue injuries.

METHODOLOGY

The first part of the project was to design the induction heating system. EnergynTech has invented and developed other Hybrid Induction processes, including Hybrid Induction Arc Welding. It was determined that a similar coil could be used for cutting. However, there were two additional considerations. First, although the welding environment is harsh and the light gauge copper tubing used for the coil must be protected from the welding arc radiated heat, the potential of hot liquid metal spray from the cutting process posed a much larger damage threat. So, coil protection became an important consideration in the cutting head design. Second, was the coil length. Although the project goal set by the navy was a cutting speed of 10 inches per minute, the internal goal set by EnergynTech was 100 inches per minute. So, the coil design was determined, in part, by the rate of heating of the coil. The easy to use, and semi-portable Miller Proheat 35 induction power supply was selected for the cutting system. Therefore the induction power was limited to 35 kW. The heating time is proportional to the coil length divided by the travel speed, limited by the maximum inductive power available.

A Finite Element Analysis (FEA) heat flow model was developed for the project by Gatekey Engineering, Inc. The model was in two parts:

1. Develop a model of the induction electromagnetic field to determine how the induction heating system affects the metal which is to be cut and then to allow the heating model to provide a precise result for the thermal effects.

2. Develop a heat flow model, where the power inputs are derived from the induction electromagnetic field model. Identify the conditions where the metal surface reaches the autoignition temperature for cutting to occur. Use the model to determine the preliminary operating conditions for the cutting process

The mathematical modeling[7] of the time for the metal to reach the autoignition temperature of 1600° F (1144° K) which is needed for the oxygen lance to cut the steel is shown in Exhibit 2. It is important to note that the time (t_{ig}) is inversely proportional to the Heat Flux (q'') raised to the second power. Consequently, the heat flux has a significant influence on the speed of the cutting process. The Heat Flux of the induction system and the oxy-fuel flame are:

Oxy-fuel: $q'' = 48.88$ MW/m^2
Induction: $q'' = 1400.88$ MW/m^2

Induction Heating

$$t_{ig} = \left(\frac{\pi}{4}\right)(k\rho c)\left[\frac{T_{ig} - T_o}{q''}\right]^2$$

t_{ig} = Time to reach autoignition temperature, Speed,
κ = Thermal conductivity (W/(m-°K)),
ρ = Density (kg/m³),
c = Specific heat capacity ((J/(kg-°K)),
T_{ig} = Autoignition Temperature (°K),
T_0 = Ambient Temperature (°K),
q'' = Heat flux (W/m²)

Exhibit 2. Induction heating analysis – to calculate the time needed to reach the autoignition temperature of the steel, used to calculate both the induction electromagnetic field effect and the heat flow analysis.

Even if the two heating processes had the same efficiency, the value of q'' for induction is 28 times that for oxy-fuel. So, the oxy-fuel heating system takes, theoretically, 784X as much time as induction to heat the metal surface. This, of course ignores the fact that the electromagnetic field from the coil is radiating in all directions around the coil. However, an electromagnetic flux concentrator is used to re-direct the field, which substantially reduces the losses in the electromagnetic field. Laboratory tests conducted at EnergynTech of actual production coils shows that the heating of the steel surface by the induction system is at least 20 times the speed of an oxy-fuel torch.

The cutting speed is very important in scrapping of ships and off-shore structures because of the shear volume of metal that must be cut. For a submarine, for example, there is over one-half million square feet of hull which must be cut into small pieces.

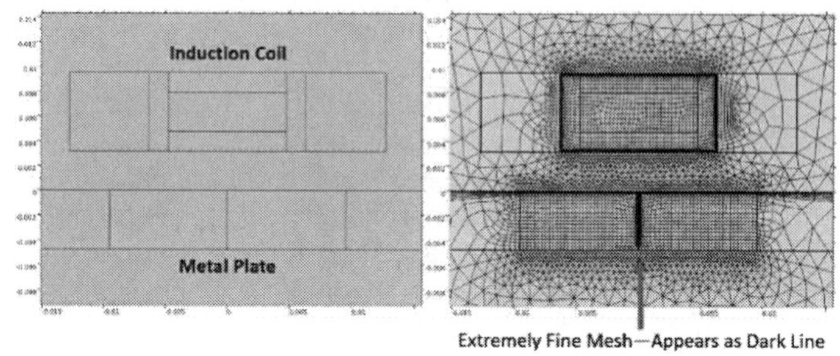

Exhibit 3. Finite Element Analysis schematics showing: (Left) the rectangular copper tube which was the induction coil at the top and the metal plate being heated at the bottom; and (right) the Finite Element mesh structure which was used to calculate both the induction electromagnetic field effect and the heat flow analysis.

Exhibit 4. For cutting a submarine hull, over one-half million square feet of hull steel must be cut into small enough pieces to be transported to a foundry and melted.

RESULTS

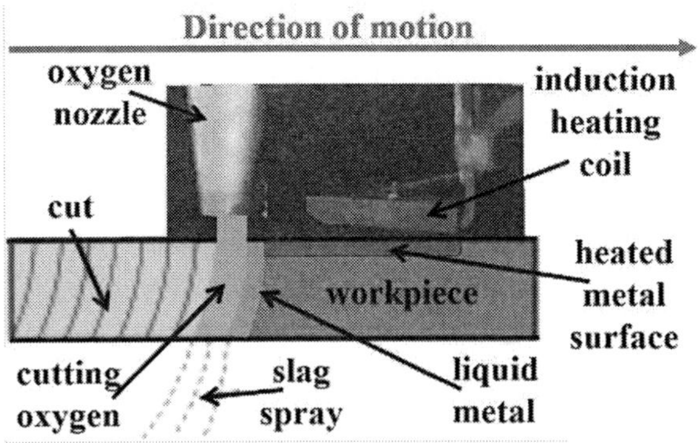

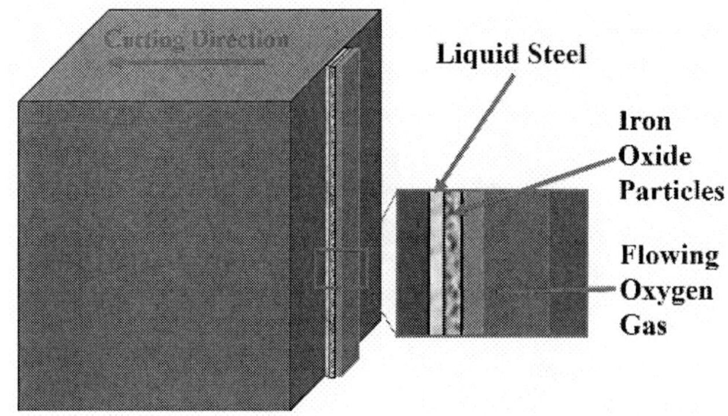

Exhibit 5. The Hybrid Induction Oxygen cutting process: (left) schematic cross section diagram of the cutting operation; and, (right) a diagram showing the liquid steel, oxide particle layer, and the flowing oxygen stream inside the cut.

The two primary criteria for the project were: 1) maximize the speed of cutting; and, 2) minimize the burning of the coating on the steel surface which causes fume which violates the opacity requirement of the Puget Sound Clean Air Act (PSCAA). For the Hybrid Induction – Oxygen cutting process the heat from the oxidation of the iron, melts the surface of the metal just ahead of the cutting process (exhibit 5). This has two results: 1) the liquid melted metal provides the "fuel" for the generation of heat because the liquid is much more easily oxidized; and, 2) the formation of a layer of iron oxide particles on the surface of the liquid metal. Thus, there are two primary objectives for the oxygen stream. The first is to remove the iron oxide particle layer from the liquid surface, so that oxygen can reach the liquid metal. So, the oxygen stream must have sufficient velocity in order facilitate sufficient transfer of momentum to the oxide particles to remove them from the liquid metal surface. The second objective for the oxygen stream is to supply a sufficient quantity of oxygen to oxidize the all of the liquid metal on the cutting surface. For this project, a standard convergent/divergent oxygen nozzle, made for large machine oxy-fuel cutting systems, was used.

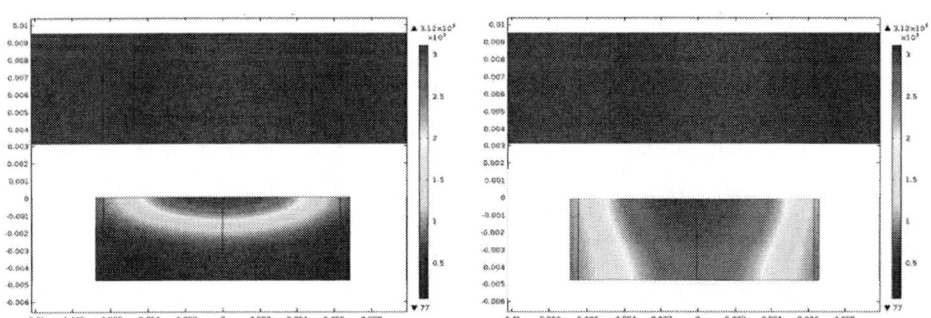

Exhibit 6. Heat flow analysis shows the heat profile on the metal top surface as the coil passes over and heats the metal surface.

The speed of cutting is limited by the heating rate, as shown in exhibit 2. The heat flow analysis provided information for the parameters (e.g., coil length, induction power, etc.) for the induction heating. The analysis showed that the metal temperature was well past the required 1600° (light green in exhibit 6) in one second of heating. The coil length is 4 inches, so the cutting speed can be at least 240 inches per minute. This is 24 times the 10 inches per minute criteria which the Navy had requested as the requirement for a successful project.

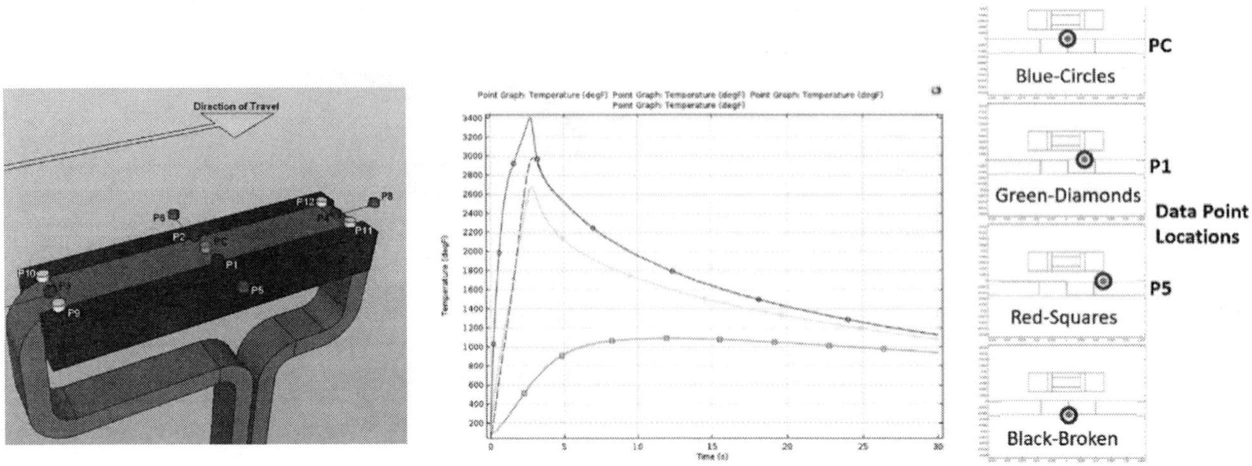

Exhibit 7. (Left) Schematic diagram of the induction coil upside down, with points (top of the cylinders) on the metal surface; (Right) temperature as a function of time at those locations on the steel plate.

Because the heat is primarily generated by the oxidation of the iron, the only heating criteria is to have the steel surface at the autoignition temperature when the oxygen lance arrives. The heat flow analysis shows that the steel directly under the center of the coil reaches that temperature in less than one second, confirming that the cutting speed can, theoretically, be maintained at 240 inches per minute.

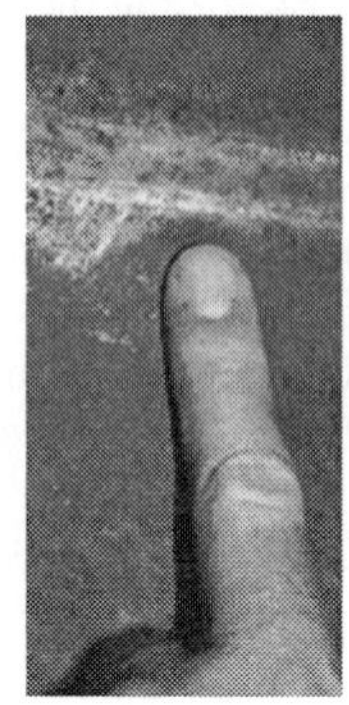

Exhibit 8. A human sensor at position P5 on the plate, just a few seconds after the induction coil passed, during a heating test. This is confirmation that the coating will not burn at that location.

The other criteria of importance is meeting the Puget Sound Clean Air Act (PSCAA) requirement of less than 20% opacity caused by fumes – which are primarily from the burning of the hull plate coating. Using Oxy-fuel cutting, the requirement is that the coating must be removed from the cut line 9 inches on either side of the center. This is a tremendously labor intensive process and adds significant cost to the scrapping of the ship and delays the schedule of the recycling of the submarine hull metal. The other approach is to dismantle submarines inside tensile fabric tents equipped with fans to capture fugitive PM emissions – these tents are nearly 5 stories tall and are very cumbersome to lift and position as well as they interfere with crane operations to remove cut metal pieces.

The Temperature-Time diagram shows that on material 1/8″ from the edge of the cut (Exhibit 7, red line), the metal surface barely reaches 1000° F so that very little, if any, of the coating reaches the temperature which generates the black smoke, and does not stay at the temperature long enough to generate enough fumes to violate PSCAA. While cutting at higher speed is important, the ability to maintain the fume level below the allowable maximum, of 20% opacity, is critical to the operation of the shipyard and the ship recycling program.

The cutting speed is very important in scrapping of ships and off-shore structures because of the sheer volume of metal that must be cut. For a submarine, for example, there is over one-half million square feet of hull which must be cut into small pieces.

Exhibit 9. The Hybrid Induction Oxygen cutting process cutting submarine hull at 70 inches per minute.

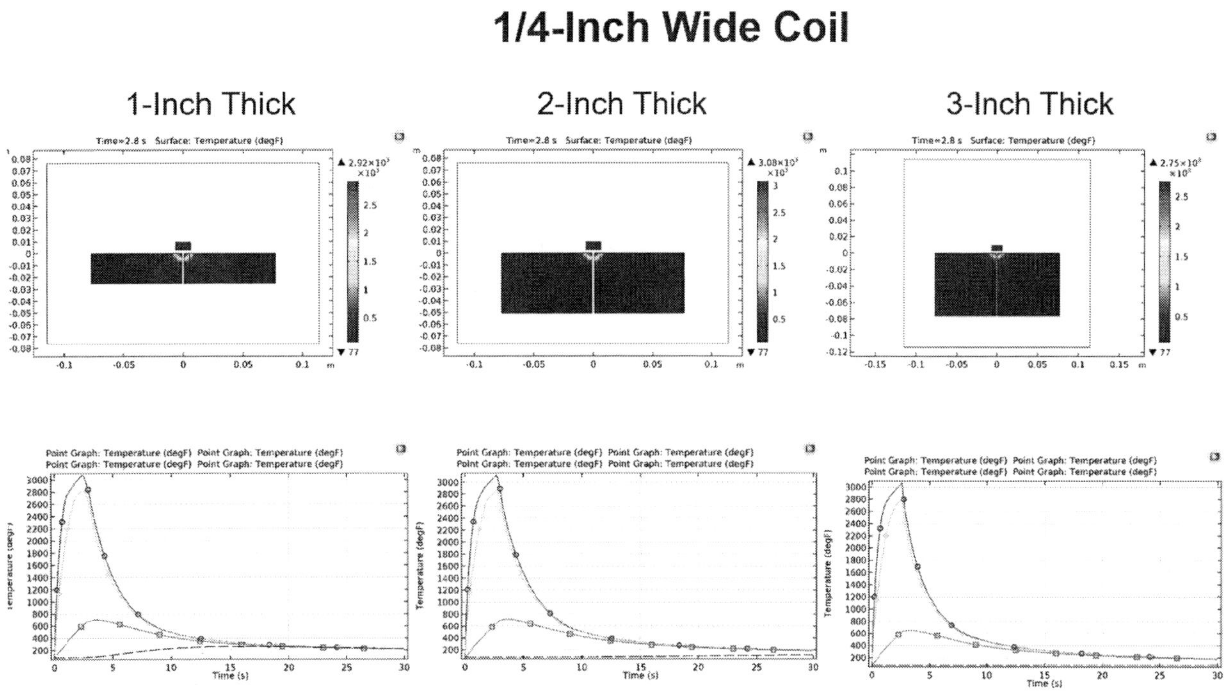

Exhibit 10. The Hybrid Induction Oxygen cutting process heat flow model analysis (same color coding as Exhibit 7): This analysis demonstrates the ability of the HI-O cutting process to cut very thick material at high speed. While the heated area gets somewhat smaller with increasing plate thickness, (left – to – right: 1″, 2 ″, and 3″ plate thickness) the material directly under the coil still reaches the autoignition temperature (1600ºF) in less than one second (blue line in the plots).

Testing of the HI-O cutting system showed that at cutting speeds up to 20 inches per minute (double the speed requested in the Navy Solicitation), the system produced fume which was observed to have much lower opacity than the allowed amount of 20%. However, at speeds of 40 up to 70 inches per minute, while the system produced good cuts, the fume opacity exceeded the allowable level. Consequently, a study was undertaken to understand the cause of the fume. The fume which had been cited in the Solicitation, which was caused by the burning of the coating on the steel surface was not the origin of the observed fumes for the HI-O process, since there was no evidence of the coating being overheated.

An experimental cutting system was set up with a welding fume collection system. Then, cuts were made on submarine hull plate. The fume collection system captured the fume and it was passed through a filtration system. The fume material was captured on the filter (Exhibit 11). Two observations were important to note. First, the material was a dry powder which if lightly tapped would fall off of the filter; and, Second, he powder was a reddish brown much like the color of rust.

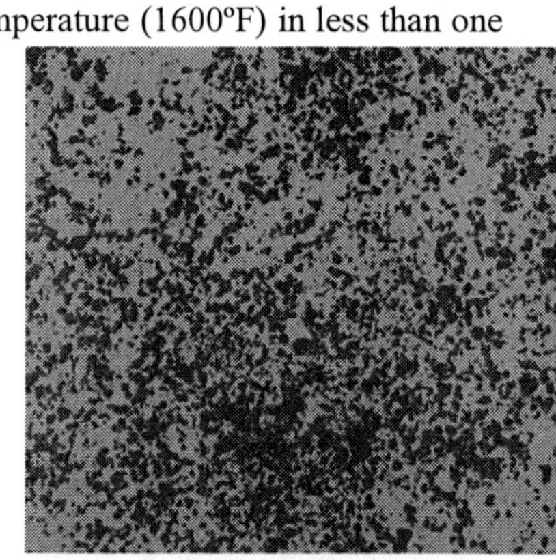

Exhibit 11. Fume collected on a filter from the Hybrid Induction-Oxygen cutting process. The fume is made up entirely of fine, dry, iron-oxide powder

After some analysis, it was determined that this fume material was iron oxide which had been formed by the oxygen stream as it impinged on the liquid metal surface. This was not the fume which had been mentioned in the SBIR Solicitation. The fume generated by the burning of the coating was black, dense, and sticky – that is, it could not be easily removed from the filtration material. This powder is formed due to turbulence in the oxygen stream, which breaks up the large particles (dross). This fine dry powder fume can be collected in simple type vibration self-cleaning filters. Thus, this new process has shown the ability of cutting below the allowable maximum fume opacity at double the speed requested by the Navy. In addition, the new process demonstrated the ability to cut at a speed up to 7 times that required by the solicitation. So, this new process can cut hull steel in the dry dock without violating the Puget Sound Clean Air Act (PSCAA), and can also substantially increase the cutting productivity during the cutting of the crane liftable pieces into pieces which are then shipped to the foundry – that is accomplished inside buildings with ventilation and filtration systems. The big advantage is that this fume will not clog and destroy filters.

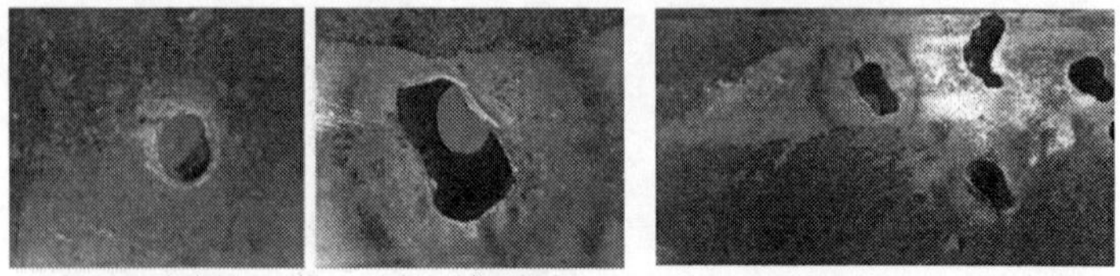

Exhibit 12. The demonstration of the multi-dimensional control robotics, capable of rotating the torch through a 90 degree angle to pierce the steel, then finishing the Pick Hole for the Crane..

The cutting system, for making long (i.e., 16 foot) cuts could be a robotic gantry, or a Mobile Cutting Robot. The Phase II project designed and assembled an 8 foot test gantry system which was specifically designed for use outdoors in a sea coast environment[8].

The robotic system is produced from anodized aluminum and stainless steel to be corrosion free. The controller was demonstrated to be capable of producing cuts and following the curved surface of the submarine hull[3]. Testing was done to demonstrate the automating surface following system, it was able to sense the surface topography along the cut line and follow the surface to maintain the tilt of the cutting head the coil stand-off to cut a non-uniform curved surface. Additionally, the robotic system is versatile enough to pierce the steel, and to cut the "pick holes" which are used to attach the large cut pieces to the cables of the cranes to hold them during cutting and remove the cut pieces once they are completely cut.

EnergynTech has produced mobile/portable robots as well as fixed automation systems[1,4,5] for specialized applications. At present, any system produced would not be a standard commercial robotic product. Since the chemistry of the cutting process is identical to the Oxy-fuel cutting process, any iron based alloy, which can be cut with that process, should also be able to be cut with the HI-O process.

CONCLUSIONS

1. The Hybrid Induction – Oxygen (HI-O) Cutting process was developed and then demonstrated capable of cutting submarine hull plate at speeds up to 7X the current cutting rate using the Oxy-fuel process. The Ozy-fuel charts show 10 in./min., the project goal was to cut at that speed, but a test of the HI-O process at the shipyard demonstrated cutting at 70 in./min. This would still require the fume control, but would substantially increase cutting productivity.
2. HI-O cutting can reduce the fumes produced by the cutting process to a level below the fume opacity limit of the Puget Sound Clean Air Act. At 20 in./min. the average opacity was 6.1%, well below the 20% maximum allowable.
3. The portable robotics developed for the HI-O process provides Semi-Robotic Control which can follow a complex curved shape surface and so can be used for automated cutting of submarine hull plate without any programming by the cutting personnel.
4. The HI-O process can not only cut all types of low and medium strength steel, it can also cut high-tensile strength steel.
5. The HI-O process and the robotic system is suitable for use in year-round, open-weather, salt waterfront environment; and was tested in that environment.
6. Software and hardware was designed for the HI-O Cutting System which was demonstrated to be capable of piercing the steel (not needing to start from an already cut edge), cutting Pick-Holes for the crane, and the ability to cut around and avoid obstacles in the submarine hull.
7. The HI-O process can produce cuts at very high speed (i.e. up to 70 inches per minute) which have a sufficient wide kerf that hull pieces can be removed by crane and not "jam" in the hole during removal.
8. The HI-O process can produce cuts at very high speed (e.g., it was demonstrated in the shipyard at 70 inches per minute) in thick steel up to 4+ inches thickness. Thicknesses of 4" were cut easily – the maximum thickness has not been tested.

9. The HI-O process can produce cuts at very high speed within OSH noise limits, prevent repetitive motion, and prevent fatigue injuries, which can occur using the Oxy-Fuel manual cutting process. Also, the HI-O process reduces the potential hazard of fire, because there is no fuel gas used – only oxygen.

ACKNOWLEDGEMENTS

The authors gratefully acknowledge the support of the Navy Small Business Innovation Research program.

The authors gratefully acknowledge the support of the National Shipbuilding Research Program for their funding of previous Hybrid Induction Arc Welding process development projects – which was instrumental in developing much of the science and technology base used for this project.

REFERENCES

1. Jerald E. Jones, Valerie L. Rhoades, James R. Dydo, Kenneth Fast, Jamie Beard, Andrew Bryant, Radford M. Arner, John H. Gaffney, "Development of a Collaborative Robot (COBOT) for Increased Welding Productivity and Quality in the Shipyard", World Maritime Technology Conference, November 7, 2015.
2. Jerald E. Jones. Valerie L. Rhoades. James R. Dydo. Dennis Fanguy, Craig Long, Radford M. Arner, Mark D. Mann, "Hybrid Induction Arc Welding: A New Process for Reducing Distortion in Stiffener-to-Panel T-Fillet Welds", World Maritime Technology Conference, November 7, 2015.
3. Jerald E. Jones, Valerie L. Rhoades, Mark D. Mann, Todd Holverson, James R. Dydo, "Hybrid Induction Plasma Cutting: An Innovative New Process for High Speed Cutting and Low Emission Cutting, World Maritime Technology Conference, 2014.
4. Jerald E. Jones, Valerie L. Rhoades, Adam N. Cuneo, Mark D. Mann, Radford M. Arner, John H. Gaffney, Todd Holverson, James R. Dydo, Leonard Blocher, Christopher knoble, Jamie Beard, "Hybrid Induction Plasma-Oxygen Cutting, Using A Mobile Robot System for Shipyard Application", Society for Naval Architects and Marine Engineers Maritime Convention, November 2016.
5. Garth E. Turner, Jerald E. Jones, "Portable Automated Plate Straightener", Society for Naval Architects and Marine Engineers Maritime Convention, November 2007 – Ship Production Symposium.
6. Jerry E. Jones, Valerie Rhoades, James R. Dydo, Todd Holverson, Lisa Elles, "Development Ultra-Low Distortion Stiffened Panels for ships using the 'HiDep' Two Sided T-Fillet Welding Process", Society for Naval Architects and Marine Engineers Maritime Convention, November 2018 – Ship Production Symposium.
7. Jerry E. Jones, Valerie Rhoades, James R. Dydo, Todd Holverson, Lisa Elles, "Development of Modeling Techniques for the 'HiDep' Welding Process", Society for Naval Architects and Marine Engineers Maritime Convention, November 2018 – Ship Production Symposium.
8. Jerry E. Jones, Valerie Rhoades, James R. Dydo, Todd Holverson, Lisa Elles, "Hybrid Induction Arc: A New Welding Process for Reducing Distortion and Increasing Productivity" – Shipyard Implementation", Society for Naval Architects and Marine Engineers Maritime Convention, November 2017 – Ship Production Symposium.

THE INTERNATIONAL COMMUNITY
FOR MARITIME AND OCEAN PROFESSIONALS

VIRTUAL MULTIMODAL AUTOMATED OBJECT DETECTION WITH DEEP NEURAL NETWORKS

NIKOLAOS MITSAKOS
DEPARTMENT OF MATHEMATICS, UNIVERSITY OF HOUSTON

SANAT UPADHYAY
INNOVATION CENTER, UNIVERSITY OF HOUSTON AND LOLAARK LLC

MANOS PAPADAKIS
DEPARTMENT OF MATHEMATICS, UNIVERSITY OF HOUSTON

Proceedings of the 24th Offshore Symposium, February 20th 2019, Houston, Texas
Texas Section of the Society of Naval Architects and Marine Engineers

ABSTRACT

In this work, we attempt to illustrate possible applications of automated object detection in video sequences and still images using state-of-the-art artificial intelligence methods, such as convolutional neural networks. These novel tools can be used in various application domains in offshore operations. Neural networks can be trained to detect humans, specific objects and even learn how to conduct pipeline, anchor, tank and other types of inspections where they can significantly reduce the possibility of human error, not by replacing the human operator, but by helping him or her to be more effective and efficient. One of the problems we will discuss is how to harness the power of image pre-processing methods to increase the informative content of inputs, especially in environments where video or still-image visibility may be poor due to weather conditions, underwater turbidity or smoke.

Keywords: Water turbidity, fog, uneven illumination, visibility improvement, object detection, multimodal deep learning, convolutional neural networks, virtual lookout,

1 INTRODUCTION

Artificial intelligence for big data management is gradually becoming a reality of life, although there is a significant need for theoretical work in the understanding of how neural networks with a significant number of hidden layers work. They can be used for decision-making and harness the innate ability of neural networks to handle multiple inputs to generate one output, namely a decision or suggestion. Prior to the AI revolution, the trend to load airplane cockpits with numerous sensors resulted in a number of multiple inputs often in the form of audible alarms or visual displays simultaneously activated. This is one instance, where numerous information input channels have to be simultaneously interpreted by a single operator, namely the pilot, who has to make a decision explaining the causes of these alarms. Aviation psychologists even in the 80s have worked on this problem of the increased monitoring workload of humans and concluded, that big amount of information provided to pilots by monitoring systems and warnings poses as equal problem as limited amount of information would do Endsley (1999). Needless to mention that in aviation multiple alarms are often related to serious flight incidents.

Although not smarter than humans, AI can handle more effectively information from multiple sensors, but to do so, AI systems often require excessive training. We can think of an artificial intelligence system, as an integrated piece of software and hardware which will interpret a condition and assign to it a label, or an interpretation of a situation. This integrated system may function properly only if any situation it encounters fits well within the probabilistic parameters of its training. Anything outside its experience, will only trigger a random response. One way to increase its experience, especially in cases, which are not going to be fit properly within its knowledge space, is to use multiple inputs. One example is ABB's experimental Virtual Lookout, which combines visual camera inputs with LIDAR sensors too measure distance from various visible targets. Hyperspectral imaging is another example of the use of multiple sensors or modalities to help interpret information. This modality has been used for the detection of underwater pipeline corrosion, or from airborne sensors to detect types of marine life or water pollution, e.g. Roper & Dutta (2006), Johnsen et al. (2016). In the jungle of Deep Learning, *multimodal learning* has been used with audio and video with of lips movement for speech recognition Ngiam et al. (2011), Srivastava & Salakhutdinov (2012), and images combined with distance data obtained by RGB-D sensors for classification Eitel et al. (2015).

In this work we use a single input to generate artificially additional inputs, which we fuse, as we would do, had we have more than one physical modality. In no case, do we advocate that additional physical modalities are not useful. Definitely they are, but our purpose is to hint that virtual modalities can potentially help the automated decision process, when the input sensor operates in conditions which may potentially deliver less reliable inputs. To investigate the role of virtual modalities in improving the overall system accuracy, we use a top performer Neural Network, the Fast Recursive NN. In fusing the virtual with the physical inputs, we chose not to change the network's basic elements of architecture.

AI acolytes may argue that when we train systems, we must include examples even from marginal conditions. However, there is always a risk of using an excessive number of training examples to account for a multitude of acquisition conditions which render the system unable to generalize, that is to be unlikely to derive a correct decision, when an input presented to it is not similar to the set of its experiences. The size of the training set depends not only on the number of labels the system must learn to output, but also on the inherent variability within the class of same label inputs and the variability of input acquisition conditions. Taking all three into account it is not difficult to realize that the training set quickly grows immensely.

The remedy we propose for this situation is to help the system improve its generalizability by reducing the set of conditions influencing the needed volume of training examples, and compensate for this reduction by adding the inputs from the virtual modalities. Moreover, the need to have a huge training set requires an equally big data collection enterprise, which may be both costly and time consuming. It is not hard to be trapped into this kind of predicament. For instance, if we want to develop a system to detect metal surface corrosion, we need to present to the system a balanced number of training examples from healthy and corroded surfaces. Here, we need to introduce the term *ground truth* which is widely used when referring to training examples. The key work here is the word "balanced", which requires a possibly time-consuming data collection from corroded surfaces with medium of low levels of deterioration. In such cases, the best approach is to collect data samples of what is designated as normal, and use a NN which scores samples presented to it. Score outputs express the network's belief that an examined surface patch is healthy. In this case, with a properly trained system we can reasonably expect that patches with signs of deterioration will be outliers and therefore will receive low scores, which are often called by the NN community as *confidence scores*.

We begin our presentation with a brief overview of the two virtual modalities we use, followed by a sketchy description of the neural network architecture we use to fuse the physical and virtual modalities inputs. Finally we close with our experimental results.

2 METHODOLOGY

2.1 Digital Image Enhancement and Illumination Neutralization

Digital image enhancement is the process of adjusting the pixel values of a digital image, with the purpose of deriving an image which is visually more appealing and/or more informative. To this end, a broad range of image enhancement techniques have been proposed, starting from the simplest brightness, contrast or sharpness adjustment, to the removal of various types of distortion, such as noise created by the digital camera sensor. In addition, to this distortion type, image enhancement methods, often, focus on specifically enhancing specific image informative content, such as edges (e.g. Canny Edge Detector) or texture. However, there are image acquisition conditions which inhibit the capture of a still image or video footage with enough information to allow these image enhancement methods to fulfill their role. Examples of such conditions are low-light, especially at night with insufficient illumination, shadows cast by objects, and the presence of light scatterers, such as haze, fog, water vapors in general, and underwater suspended particles. Object detection by artificially intelligent systems depends on the extraction of features, or more simply, the generation of identifying signatures which indicate the presence of a certain local shape in an image indicative of the presence of a certain object, e.g. a cylindrical surface. As more objects may share the same shape, e.g. in our example both cans and pipes have this shape, the AI system will learn how to piece together these signatures and conclude that *a certain type of an object is present at a certain location of an image* and perhaps quantify its belief on how reliable the detection is. This is a relatively complex method. A system under supervised learning achieves this goal to a certain extend. In the next section we discuss this issue in greater detail.

State of the art object detectors are exposed through supervised training to a variety of imaging conditions. so low illumination, or shadows do not often pose significant problems to them. But when imaging conditions, such as water vapors, haze, smoke or water turbidity, in general conditions conducive to the higher than normal scattering of light rays, detections become acutely uncertain. So, what if there are methods which can essentially "lift the veil of invisibility" and make the object detector to work properly? To make this point more clear, we stress that the extra variability these exterior conditions impose create perturbations to a classifier which do not necessarily fit well within its learned experiences. Consequently, ameliorating their effects on an input image makes the objects in the input more similar to the classifier's experience. In this work we used two such methods, originally designed to neutralize the effects of uneven or low illumination. We use these methods to create supplementary features, as if we incorporate them as separate inputs from additional modalities.

Illumination neutralization algorithms into two main categories Ochoa-Villegas et al. (2015): Those that treat the illumination problem as a *relighting* problem and those that treat it as an *unlighting* one. Relighting methods improve similarity of illumination conditions between gallery images and a probe image Unlighting methods utilize the Lambertian reflectivity principle:

$$Image(I) = Illuminance\ (B) \times Reflectance\ (R). \tag{1}$$

Reflectance of objects in a scene is a quantification of what luminous energy the surface of an object can reflect back regardless of the sources of this energy. So reflectance depends on both geometric and textural properties of surfaces of various objects in a scene. Image edges, in particular, are associated with areas of high curvature of the surfaces of objects in an image as seen from a camera. Illuminance, represents the amount of light that reaches all objects in an image reduced by scattering on the surfaces in the image scene. Hence, illuminance, or simply luminance, does not necessarily capture correctly the effects of light scattering when the space between the scene and the camera contains anything other than clean air or, ideally, vacuum. Consequently, methods aiming to solve (1) are ideally suited for visibility improvement in turbid, or foggy environments.

Unlighting methods aim at solving this under-determined problem by approximating B and R from I. A good classical and with a lengthy history example is *Retinex* Land (1964), Land et al. (1977) modeled after the color perception system human vision. We use a modern and relatively fast variation of it, the **Variational Retinex (VR)** (Fu et al. (2016)). The other method we use is Illumination Normalization (Upadhyay et al. (2018)).

After taking logarithms of both sides of (1), the right-hand side has two terms which are approximated by means of a constraint optimization algorithm. This is the key idea of VR. The assumption for the solution or the constraints more precisely, are that solution should be such that regions with small value changes in R occupy most of the image, and what remains is edges and that the overall difference $I - R$ must be small. The optimization uses two weights or free parameters which control how much each of these constraints weigh in the final outcome. Clearly, the accuracy of the solution is heavily influenced by the choice of these two parameters. Prior knowledge of imaging conditions can possibly help to better choose the values of these two parameters. In our experiments, we used the MATLAB implementation provided by the authors of Fu et al. (2016) to acquire the reflectance estimation of the scenes (Figure 1 (b) and (d)). A major challenge in this implementations is finding optimal values for the two free parameters which control the minimization process.

On the other hand, the **Illumination Normalization (IN)** operator is a deterministic method, so it is faster than VR, thus, suitable for real-time applications of visibility improvement (Upadhyay et al. (2018)). The IN operator uses a multiscale transform to brake down an image into a individual components which, if put together, reconstruct the original. The IN operator takes each individual component and normalizes it in a way that the effect of illuminance at each scale is canceled out (Figure 1 (c) and (f)). The main effect of the transform is that the lower-scale components of this transform become more flat in terms of brightness, whereas the higher scale components which primarily carry the details of texture and edge information become more prominent. As a result, details necessary to object detection become more visible as if the image was acquired with adequate and uniform illumination. In Upadhyay & Papadakis (2019) the last two authors give a set of mathematical arguments proving that the outputs of IN operator contain information on shapes which is essentially the same with the information carried by a well-illuminated image of the same scene.

(a) Raw (original) (b) VR reflectance (c) IN enhanced

(d) Raw (original) (e) VR reflectance (f) IN enhanced

Figure 1: VR and IN image enhancement methods generate images where details, necessary to object detection, can become more visible. Given any raw input (**1st column**), we use the VR method to acquire an estimate of the scene's reflectance (**2nd column**) and the IN method to obtain a derivative image of the scene where the effect of illuminance has been reduced significantly (**3rd column**). In section 2.5 we state our premise that such image derivatives can be perceived as *virtual imaging modalities*, as the informative content they create may contain sufficient complementary information, with respect to the original image. In section 3 we demonstrate how state-of-the-art machine learning algorithms can be trained in tandem on the raw and artificial imaging modalities to learn representations that effectively harness the complementarity of information of virtual modalities (Original images are from the VOC2007 dataset).

2.2 Object detection

In an automatic **object detection task**, provided an input image and a pre-determined set of **target object classes** (e.g., person, car, dog etc), a system is expected to designate the smallest possible rectangular image regions containing an object from the target classes (**detection regions**) in a tight manner and also to predict the correct class label for each detection region. The fact that the number of possible detections per image, as well as the range of classes where those detections fall into, is not fixed, adds an extra degree of difficulty to this, supervised learning[1], task. Indeed, it is common to observe that a detection region contains objects from more than one of the target classes. In such a case, an efficient system should be able to assign all the appropriate labels to this same detection region. In this and the following sections we discuss the use of artificial modalities as inputs in DNN-based object detectors.

With a variety of applications, ranging from security (e.g., face detection and action detection) to industrial quality control (e.g., machine inspection, aerial surveillance and image analysis) and even commercial (e.g., visual search engines, autonomous navigation systems, point and shoot cameras), object detection holds an exceptionally dominant place in the efforts of computer vision researchers' community.

2.3 Modern object detection and classification

Our discussion, from now on, presents design ideas for DNNs which are fairly generic, despite the use of object detection as a case study. The ideas we present can easily be adopted for any multiclass, or one-class classification NNs with images as input data.

Notable early approaches to solve the object detection include the fast Viola-Jones framework Paul Viola (2001) and the combination of Histogram of Oriented Gradients (HOG) with Support Vector Machines (SVM) Dalal & Triggs (2005). During the recent years, artificial Neural Networks as **Deep Neural networks (DNNs)**, and especially their sub-class of **Convolutional Neural Networks (CNNs)**, have proven to be very successful on supervised learning image analysis tasks, including object detection. One of the first DNN approaches proposed the use of CNNs with multi-scale sliding windows scanning the input image Sermanet et al. (2014) which evolved into the various versions of YOLO (Redmon et al. (2015, 2016), Redmon & Farhadi (2018)) and RetinaNet(Lin et al. (2017)). These approaches are classified as **one-stage** methods for object detection. An alternative approach proposed the use of two stages, where regions with high probability to contain objects are singled out first, in what is known as a **region-proposal stage**. Then, a classifier examines each of these regions, to come up with a final decision on whether an object appears in the region and from which target class Girshick et al. (2014), Girshick (2015), Ren et al. (2015). There is a trade-off between processing time versus detection performance in both these approaches and utilizing the whole spectrum of their potential remains a very active domain of research.

2.4 Faster R-CNN

In the original settings of Region-proposal object detection, given an image input and a set of target object classes which the system is trained to recognize, a region-proposal detection system initially generates a feature representation of the input. Then, a binary "objectness" classifier examines this representation, to select **Regions of Interest (RoIs)** on the image, i.e. rectangular image regions where high confidence arises that one object from, at least, one of the target classes, is present. Subsequently, a learned representation for each RoI propagates, independently, into a second classifier that yields two final decisions. Primarily, for each ROI, this classifier gives 'confidence score' per target class, corresponding to the system's belief that an object from the specific target class appears in this ROI. Usually, one more confidence score is estimated, corresponding to class background, i.e. confidence that the ROI does not contain any object. Secondarily, for those target classes that the ROI receives relatively high confidence score to contain a corresponding object, the classifier attempts to adjust the dimensions and location of the rectangular ROI, aiming at a final **detection region** that captures the object in a tight-fit manner.

This, region-based, approach was introduced in Girshick et al. (2014) as **Region proposal CNN (R-CNN)** and quickly evolved into the Fast Girshick (2015) and finally the **Faster R-CNN** Ren et al. (2015). The main innovation of the latter, as compared to its region-based predecessors, is that the system addresses both region proposal

[1] In machine learning terminology, **supervised learning** refers to the process of training an algorithm by providing a 'labeled' training data-set. This means that, each data-set instance during the training is a pair of an input (typically an array of descriptors) and the 'correct output' that the algorithm is expected to return (**supervisory signal** or **label**).

and classification challenges resorting exclusively to CNNs, assisted by ranking operations and thresholding rules w.r.t. estimated 'confidence' values. In short, Faster R-CNN has two main components (Figure 2). The input image

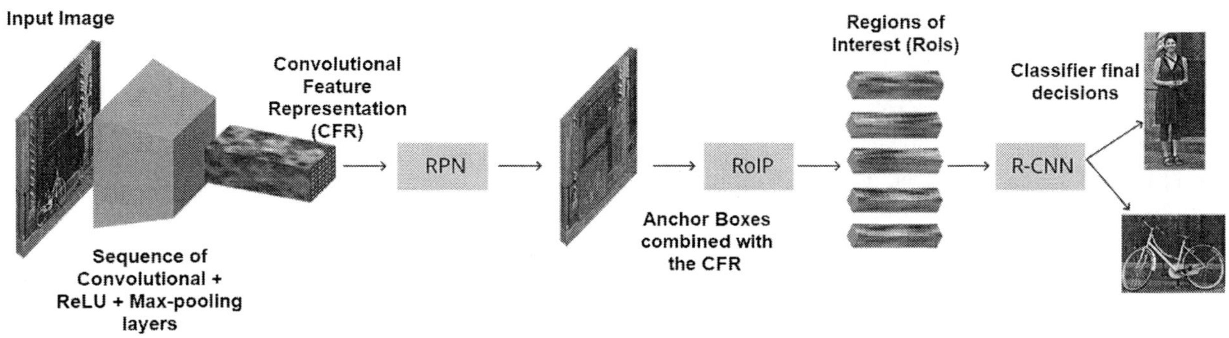

Figure 2: The Faster R-CNN object detection model depends exclusively on convolutional layers to tackle both Regions of Interest (RoIs) detection and Region regression-classification. The input image assumes its Convolutional Feature Representation (CFR), propagating through a sequence of convolutional layers with ReLU activation and periodical max-pooling operations, which reduce the spatial dimensions. The Region Proposal Network (RPN) receives the CFR and again through convolutional layers singles out Regions of Interest (RoIs), i.e. rectangular regions of the image with high 'probability' of an object to exist. Finally, the RoIs get re-sized and vectorized to fit as input to the R-CNN classifier, where through Fully Connected (FC) layers the final classification decision is returned.

propagates through a sequence of (trainable) convolutional layers which transform it into a **(convolutional) feature volume**[2] (3-dimensional array of descriptors), preserving the spatial organization, with periodical sub-sampling operations reducing its overall spatial dimensions. A (convolutional) **Region Proposal Network (RPN)**, utilizes this deep feature extraction to identify rectangular sub-areas of the image that 'probably' contain one or more objects from the target classes (RoIs), completing the first component. For the second component, the feature volume's portion that corresponds to each RoI is extracted (RoI Pooling (RoIP)) and propagates through a regression predictive model, implemented using fully connected layers (R-CNN), that assign the final target class confidence scores (including a 'rejection confidence score') and the final corrections of the RoIs location and dimensions.

One of the key novelties for the generation of RoIs in Faster R-CNN is the introduction of a deterministically constructed set of reference bounding boxes, called **anchor boxes** (Ren et al. (2015)). These boxes have fixed, predetermined, sizes and are (practically) uniformly positioned on the input image. There are many of them and overlap with one another. Any region candidate for becoming a ROI during the RPN stage, is referenced back to a unique anchor box, with a specific center location and dimensions with respect to the original input image I. If the dimensions of a convolutional feature volume, generated from a specific input are $W(width) \times H(height) \times D(depth)$ in pixels, for each anchor location $(i, j) \in [1,2,...H] \times [1,2,...,W]$ the system uses a set of rectangular anchor boxes, $S_{anchors}(i, j)$, with a variety of predetermined dimensions. In our implementation, we use nine anchor boxes per pixel (i, j), one square and two equal rectangles, landscape and portrait, each one of them with three different area sizes.

The Faster R-CNN architecture, and more specifically the VGG-16 based which is one of its most representative implementations, constitutes the basis for our *virtual multimodal* architecture. In the next sections we will review in detail the mechanisms of Faster R-CNN's components and will discuss the changes we propose, aiming in utilizing the artificial multimodal inputs.

2.5 Multimodal Faster R-CNN with artificial modalities

The success of DNNs and CNNs on supervised learning tasks using single modalities, motivated efforts to further improve the performance of such systems by combining multiple modality inputs (**multimodal learning**). (Ngiam et al. (2011), Srivastava & Salakhutdinov (2012), Eitel et al. (2015)). A common element in all these, recent, deep

[2]Often referred as feature representation in the literature.

multimodal approaches is that multiple modalities need to be a-priori available with the data-set. However, there are many situations where only one modality (e.g., monochromatic images) is available. In this section, and our experimental results in section 3, we demonstrate how image enhancement methods, such as VR and IN, derive scene perspectives (see Figure 1) that can complement the, often, limited information in the original (raw) modality. We treat the IN and VR derivatives as virtual imaging modalities, derived by the single available raw image modality.

We introduce a deep feature fusion architecture, the **virtual multimodal Faster R-CNN**, that combines the complementary information found in the original images and the derived modalities, improving detection and classification accuracy. Our proposed architecture is designed and implemented as a region proposal network, based on the popular Faster R-CNN model. However, its main innovation, fusion at the feature representation level using **derived artificial imaging modalities** when only one data modality is available, can be easily extended to multi-scale window networks. This remains part of our ongoing research.

2.6 Generating complementary information from virtual imaging modalities

The algorithmic design we propose in this section, relies on one premise:

> Image derivatives, generated from raw images using powerful image enhancement methods such as VR and IN, can be perceived as **virtual imaging modalities**. The informative content they create may contain sufficient complementary information, with respect to the original image. State-of-the-art machine learning algorithms can be trained in tandem on the raw and artificial imaging modalities to learn representations that effectively harness the complementarity of information of virtual modalities. Moreover, when the conditions during image acquisition had been adverse, e.g., under low/uneven illumination, the artificial modalities can compensate for the loss of information.

Of course, this is also true for the case of non virtual modalities. For example, Lidar and Sonar can reveal the shape of objects even in zero visibility conditions, where image acquisition in the visual spectrum is useless. Successful methods have been developed to combine complementary information from different acquisition sensors, as we discuss in the following sections. In this work we focus on the following problem, which is of pivotal importance in the common case where only one imaging (original) modality is available:

> Can the complimentarity of information from original and virtual modalities be leveraged to improve detection and recognition accuracy of state-of-the-art algorithms?

2.7 Virtual multimodal Faster R-CNN

We adopt the architecture of the very widely used, VGG-16 based, Faster R-CNN object detector. We independently train clones of this model and then transfer trained sections into a fusing architecture, which we design explicitly for accommodating the complementary informative input from IN and VR. We should note, though, that that this strategy can easily be extended to different Faster R-CNN set-ups, and even to other popular, not region-based, object detection architectures (e.g., YOLO (Redmon & Farhadi (2018)) and RetinaNet(Lin et al. (2017))). For simplicity, we describe in detail only the fusing architecture of our **Bi-modal Faster R-CNN** set-up. However, the generalization to accommodating a larger number of, physical or virtual, modalities, such as the **Tri-modal** version that we use in our experiments (section 3), is fairly straight forward.

Starting with the original (**raw**) data-set, which consists of monochromatic (e.g., gray-scale) images (**RD**), we select and apply the image enhancement method of our choice. The derivative data-set (**SD**) will serve as additional, virtual imaging modality. Using the real RD and the derivative SD imaging modalities, we train two independent, but identically structured, **Uni-modal VGG-16 Faster R-CNN** networks, as described in the following paragraph.

Uni-modal VGG16 Faster R-CNN

For each training session (RD and SD), the corresponding image becomes the input of a **Convolutional Feature Representation (CFR)**. This mapping is implemented by a series of 13 convolutional layers, each one followed by Rectified Linear Units (**ReLU**), with the max pooling operation activated after selected layers, as shown in Figure 3. The number of filters increases after every few layers, as is common in CNN architectures. This process transforms

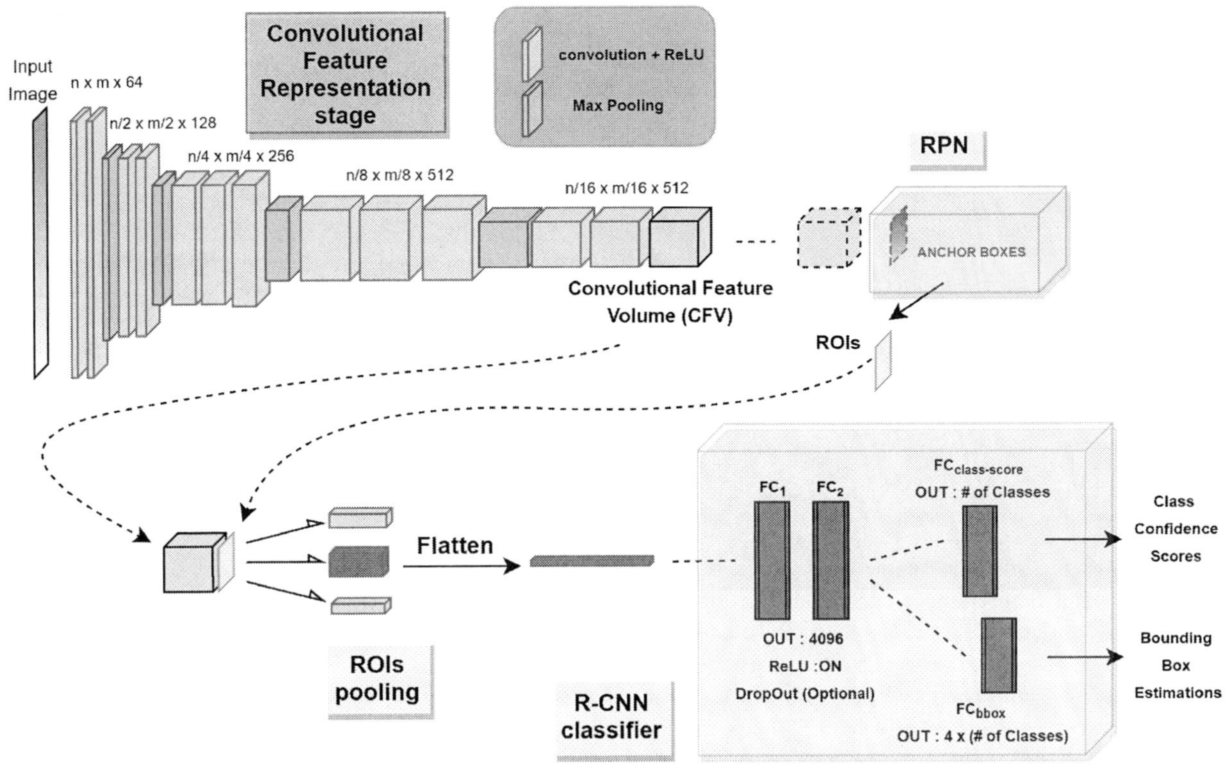

Figure 3: The Unimodal VGG16 Faster R-CNN propagates the image input through a sequence of convolution layers (Convolutional Feature Representation (CFR) stage), periodically reducing the spatial dimensions with Max Pooling, generating the (convolutional) feature volume. This representation enters the RPN which, using the pre-determined anchor boxes as reference regions, selects ROIs, i.e. regions of high 'objectness'. Finally, each ROI is vectorized and pushed into the R-CNN classifier stage where, through a succession of Fully Connected (FC) layers, the system arrives at the final decision: object from which classes appears in the detected region (if any) and final location and dimensions adjustments to fit the object tightly (if any).

the original input into a 3-dimensional array, the (convolutional) feature volume[3]. The max pooling operations (red color layers in Figure 3) divide the CFR component in five sections. The spatial dimensions of the feature map are bisected as the volume of processed data passes forward, from one section to the next. We label each convolutional layer as $convM.P$, where $M = \{1,2,3,4,5\}$ corresponds to the section where the layer belongs and P is its position in the specific section. For example, $conv1.1$ refers to the very first convolutional layer that receives the input image, while $conv5.3$ refers to the last convolutional layer in the CFR component. The coefficients throughout the CFR component, both for the convolutional filters and the biases, are all initialized using the corresponding coefficients of the VGG-16 network pre-trained on the (RGB) Imagenet data-set (Deng et al. (2009)) (see following paragraph for more details on the model parameters' initialization). The spatial dimensions of all convolutional filters (**kernels**) used in the CFR component are the same, 3×3. We allow the training to update all layers' filter coefficients. Furthermore, all convolution operations act on the previous layer's output-array with zero-padding applied as needed, so that the output's spatial dimensions do not drop due to the convolution operation Thus, for an image input of size $n \times m$, the output of the CFR stage, i.e. the output of the $conv5.3$ layer, is a convolutional feature volume with dimensions $(m/2^4) \times (n/2^4) \times 512$.

The convolutional feature volume enters the next stage of the unimodal network, the RPN. In this stage, one more convolutional layer ($RPN_{3\times3conv}$), composed by 512 filters of size 3×3 and followed by ReLU operation, acts on

[3] Often called **feature map** or **feature representation**

the feature representation. Once more, appropriate zero padding is used to retain the map's spatial dimensions. The $RPN_{3\times3conv}$ output enters two independent 1×1 convolutional layers, each with enough channels to output objectness scores (RPN_{class}) and bounding box predictions (RPN_{bbox}) respectively, for all corresponding anchor boxes of the image. More precisely, since we use $|S_{anchors}| = 9$ anchor configurations, the RPN_{class} layer will have $2\cdot9 = 18$ output channels (a pair of confidence scores per anchor representing confidence of being an object or being background) while RPN_{bbox} will have $4\cdot9 = 36$ output channels, corresponding to 4 regression bounding box (**bb**) adjustments for each anchor ($\Delta_{x_{bb-center}}$, $\Delta_{y_{bb-center}}$, $\Delta_{bb-width}$ and $\Delta_{bb-height}$ respectively).

An $(m/2^4)\times(n/2^4)\times512$ feature volume yields a total of $M = (m/2^4)\times(n/2^4)\times9$ candidate ROIs. Each of them has now been assigned objectness scores and bounding box adjustment values. However many of these regions are in practice 'significantly' overlapping due to the anchors' construction. At this point, a process known as **Non-Maximum Suppression (NMS)** takes place. The aim of this process is to eliminate most of the overlapping ROI candidates, allowing those with the highest objectness scores to prevail. The list of candidate regions is re-arranged in decreasing order with respect to objectness score[4] and a predetermined NMS threshold, $NMS_{thres} \in (0,1)$, comes into play. The regions are compared on the basis of their **Intersection over Union (IoU)** ratio; the IoU of two image regions A and B is defined as the ratio of their intersection area over the area of their union $\left(\frac{|A\cap B|}{|A\cup B|}\right)$, measured in pixel units. Each RoI candidate region that demonstrates IoU$> NMS_{thres}$, with a region found higher in the list, gets discarded. Through this process, the RPN yields the final list of ROIs.

ROIs, through their reference anchor and the quadruple of region adjustments, should be regarded as sub-regions of the original, $n\times m$, input image. However, ROIs can also be projected onto the convolutional feature volume, defining feature volume '$3D$-slices'. These $3D$-slices are extracted and interpolation in space is applied, to achieve a uniform size of $7\times7\times512$, completing what is commonly referred to as the **ROI pooling** step. These, new equally sized, **RoI feature volumes** assume now the role of RoI descriptors for the final, R-CNN classification, stage.

Each ROI feature volume that comes out of the ROI pooling process is vectorized and pushed into the Fast-RCNN classifier. Here, two consecutive, **Fully Connected (FC)**+ReLU layers, FC_1 and FC_2 respectively, transform the vector representation, both producing outputs of size 4096. The (random) **drop-out** operation, a very common sparsification (regularization) technique that allows only an arbitrary percentage of the coefficients to be fine-tuned at each training iteration (to reduce over-fitting), may be used here[5]. Note that certain sparcification (a reduction in the number of small in magnitude thought as insignificant values) already has taken place, through the ReLU and the max-pooling operations. This additional sparsification step can compensate for some residual over-redundancy of the feature mapping mechanism, built before the RPN (original sparsification comes from max-pooling operations and the ReLU non-linearity). We note here that, the drop-out operation is commonly regarded as regularization of the back-propagation process. We believe that both viewpoints offer a different, useful perspective.

At last, the feature vector output of FC_2 enters two parallel, independent FC layers, that yield the two final system predictions. The first ($FC_{class-score}$) generates the softmax[6] confidence scores per class, including one background class. The second (FC_{bbox}) yields the final regression bounding box adjustments, intended to make each detection region as tight as possible around the predicted object. Thus, in a detection task where the number of target classes is C, the final output of any original input is a summary 'report' for a number of RoIs, say N, presented in two tables:

$P \in [0,1]^{N\times(C+1)}$: contains the $FC_{\text{class-score}}$ layer output (softmax class scores). Each row corresponds to a RoI. Each column corresponds to a target object class, plus one more column for class 'background'.

$B \in \mathbb{R}^{N\times4\cdot(C+1)}$: contains the FC_{bbox} layer output (box adjustments). Each row corresponds to a RoI. Every next quadruple of entries in a row corresponds to one of the target classes, again with the addition of an extra quadruple for 'background'.

[4]Note that only the confidence score of being an object is used

[5]In our experiments we observed that in this particular situation drop-out does not allow the model to achieve its best performance. Thus, we adopted an early-stopping-like technique, by allowing the training to continue for many epochs after convergence on the training set has been achieved, and then choosing the model that performs best on the validation set as representative of the scheme

[6]Given a vector $u = (u_1, u_2, ...u_K) \in \mathbb{R}^K$ the softmax of u is composed by the elements $s(u)_i = \frac{e^{u_i}}{\sum_{k=1}^{K} e^{u_k}}$.

Initialization

During training all but the predefined, if any, filter entries the system will use will be learned via a process which forces them to change in a way that a certain input from the training set must be matched to a certain output. One classical approach is to give random values to those entries. Assigning values to any filters entries or weights to fully connected layers at step one of the training is called *initialization*. One of the main difficulties when training deep CNNs is that the convolutional layer coefficients, if randomly initialized, tend to over-fit the training data-set. Furthermore, for many computer vision tasks, even data-sets composed by hundreds of thousands of instances can be considered relatively small to represent adequately the input space in its generality. These factors can lead, not only to very long training sessions but, even worse, to models where the loss-function optimization process (see next paragraph) gets trapped inside a 'local minimum', or reaches a 'saddle point', quite far from a model that would generalize efficiently on un-seen data. A widely accepted solution to this problem, akin to the 'human experience', is **transfer-learning**. According to the transfer-learning strategy, new architectures that have to be trained on relatively small datasets, can benefit by 'borrowing' convolutional or FC layers' coefficients, previously trained on much larger data-sets. There is evidence that this transfer of coefficients works even when the tasks are not exactly the same. We use the, openly available, pre-trained VGG 16 coefficients from Simon et al. (2016), to initialize the filter coefficients in all convolutional layers in the CFR stage of our unimodal sessions. Ideally, in our enhanced data SD-unimodal session, we would rather use an initialization using coefficients obtained from pre-training VGG16 with an enhanced version of the ImageNet data. However, due to computational limitations arising from the large size of the ImageNet data-set, this has not been feasible during our experiments.

The coefficients for all the layers following the CFR stage, convolutional and fully connected, are randomly initialized, using normal distributions random inputs, but ignoring the values that are more than two standard deviations (**stddev**) away from the mean. Values outside this interval are discarded and re-drawn. The parameters of the normal distributions we use in our implementation are: *(i)* mean$=0$ and stddev$=0.001$ for the FC_{bbox} layer and *(ii)* mean$=0$ and stddev$=0.01$ for all remaining layers.

Loss function and optimization

Supervised machine learning apparatus is driven by what is known as the **loss function**, that measures the discrepancy between the system's predictions and the correct 'labels' (**ground truth** values). Loss function plays the role of the objective function in an optimization scheme and is evaluated at each training iteration. To control the quality of the approximation of the model's predictive function (here the predictive function is a scoring function), the model aims at optimizing (minimizing) the loss function, by updating the model's trainable variables as needed (**backpropagation**). However, simply minimizing the loss function does not necessarily lead to a good predictive model. During the training phase, therefore also during the loss function minimization process, only instances from the training set are examined. But the evaluation of a model's predictive accuracy is done by measuring its predictive ability on the separate test set, after the model's training has been completed (usually on the basis of loss function convergence). The ability of a model to predict the labels of instances that never appear during its training phase is commonly known as the model's **generalization capability**. Good generalization capability is the key ingredient that needs to complement a successful loss function minimization scheme, in order to obtain an efficient predictive model. These issues need to be very carefully considered when designing the model architecture, the loss function as well as the optimization scheme.

For the Uni-modal Faster R-CNN, we use the loss function defined in Ren et al. (2015). In short, the loss for an image input is defined as the sum of four partial losses, evaluated in it's two prediction stages:

$$Loss_{Total} = Loss_{cls}^{RPN} + Loss_{reg}^{RPN} + Loss_{cls} + Loss_{reg}$$

The first two terms of the sum (**RPN loss**) refer to the classification (log loss) and bounding box regression loss of the RPN respectively. The last two terms (**classifier loss**) steer the final classification and detected region regression loss. The classification loss terms penalize the difference between the systems confidence that a region contains an object and the actual objectness label of the region (0 if the region does not contain object and 1 if it does). Both regression loss terms measure the 'mismatch' of an estimated detection box with the corresponding ground truth box, at some location of the image that contains an object.

A critical factor of the training process is the balanced mix of anchor boxes used to evaluate RPN loss. Anchor boxes can be divided into two categories. Those that do overlap 'significantly' with ground truth box annotation of an object from the target classes, and those which do not. We call the former **foreground anchor boxes** and the latter **background anchor boxes**. Since objects from the target classes in most images occupy a small portion of the total image area, the RoI candidate anchor boxes dealt by the *RPN* that are background anchor boxes highly outnumber the foreground ones. This may bias the model towards the background 'class', if all anchor boxes are allowed to enter the RPN loss. To avoid this, during training, the set of anchor boxes is divided into two groups:

positive: Those that have an IoU greater than 0.5 with a ground-truth region of the image, and

negative anchors: Those that have $0 \leq \text{IoU} < 0.1$ with all ground-truth regions of the image.

Then, a balanced random sample of 256, both positive and negative, anchors is selected. In the evaluation of the RPN loss terms at each training iteration, only the mini-batch anchor boxes are used in the calculation of $Loss_{cls}^{RPN}$ and, from those, only the positive anchor boxes and their 'closest' ground truth regions are used in the calculation of $Loss_{reg}^{RPN}$.

Using the $Loss_{Total}$, Stochastic Gradient Descent (SGD) with momentum (Sutskever et al. (2013)) is used as the **optimization scheme**, with the momentum value set at 0.9. The initial value for the **learning step** is set to 0.001 and decreases in a step-decay manner, multiplied by a factor of 0.1 every 70,000 training iterations.

Bi-modal VGG16 Faster R-CNN

We allow the two independent training sessions, for the RD and VR Unimodal Faster R-CNN respectively, to converge. Then, we transfer the CFR stage coefficients and the $RPN_{3\times3conv}$ coefficients of the best performing agent[7] from each, into the **Bi-modal Faster R-CNN** (Figure 4), which we design and train as follows:

The RD physical modality and the SD virtual modality are used as inputs to two CFR streams, arranged in parallel. Each stream is initialized with the coefficients learned and transferred from the corresponding RD and SD Unimodal sessions, from the previous step. The coefficients of these two, independent, streams will remain un-effected by the (optimization guided) updating process, throughout the whole bi-modal training session. The feature volumes generated by **RD-CFR** and **SD-CFR** enter the RPN stage, where two separate $RPN_{3\times3conv}$ layers receive them, the $RD - RPN_{3\times3conv}$ and $SD - RPN_{3\times3conv}$ respectively, both with the same set-up as in the unimodal case[8], but each initialized using the pre-trained $RPN_{3\times3conv}$ coefficients of its corresponding unimodal peer. The outputs of these two, parallel, convolutional layers, are concatenated (deep features fusion) towards the 'depth' dimension. The rest of the RPN stage maintains the same structure as in the Unimodal case, producing the RoIs of high objectness confidence, and the training remains the same. It is important to note though, that, since the depth dimension of the fused representation has doubled due to the fusion of the two feature volumes, the depth of the 1×1 filters in the two, parallel, convolutional layers that conclude the RPN stage, are also doubled. Now, the Bi-modal model has, hopefully, already used the dual information from the RD and SD fused representation volumes, created in the RPN stage, to better estimate the RoIs. For the R-CNN classifier stage, we perform a second deep feature fusion, by concatenating the, already generated, feature outputs from the RD-CFR and SD-CFR stages, towards the depth dimension. Then we (inversely) project the generated RoIs on this **fused convolutional feature volume (FCFV)**. We note that in this case, the representation 'slices' extracted in the RoIs pooling stage inherit the depth of the FCFR, which is twice the size of that in the uni-modal case. The rest of the Bi-modal architecture proceeds with the same manner as the Uni-modal network. The RoI feature representations are vectorized and enter the R-CNN classifier stage. However, since this time the vectorized RoI array has twice the size of the Uni-modal equivalents, the size of the embedding matrix in the FC_1 layer for the Bi-modal architecture needs to have dimensions $(7 \times 7 \times 2 \times 512) \times 4096$. Once more, for FC_1 and FC_2 layers in the Bi-modal network, one can choose to use Drop-Out, probably to correct the redundancy of information, which may lead to over-fitting. However, in our implementation, in order to avoid the negative implications of fine tuning the model, we do not use drop-out but, once again, follow our Uni-modal 'early-stopping' approach. That is, we train until the model has safely converged and then we run validations using snapshots of the model, recorded at every 5,000 training iterations, choosing the best-performing one as the optimal model. The final predictions of class confidence scores and bounding box estimations enter the same loss functions with those described in the uni-modal case.

[7]We save a 'snapshot' of the models' coefficients every 5,000 training iterations. A validation is performed on the validation dataset using each coefficient's snapshot and the best performing is determined according to our 'over-all classes' evaluation metric, described in section 3

[8]512 filters of size 3×3 followed by the ReLU operation

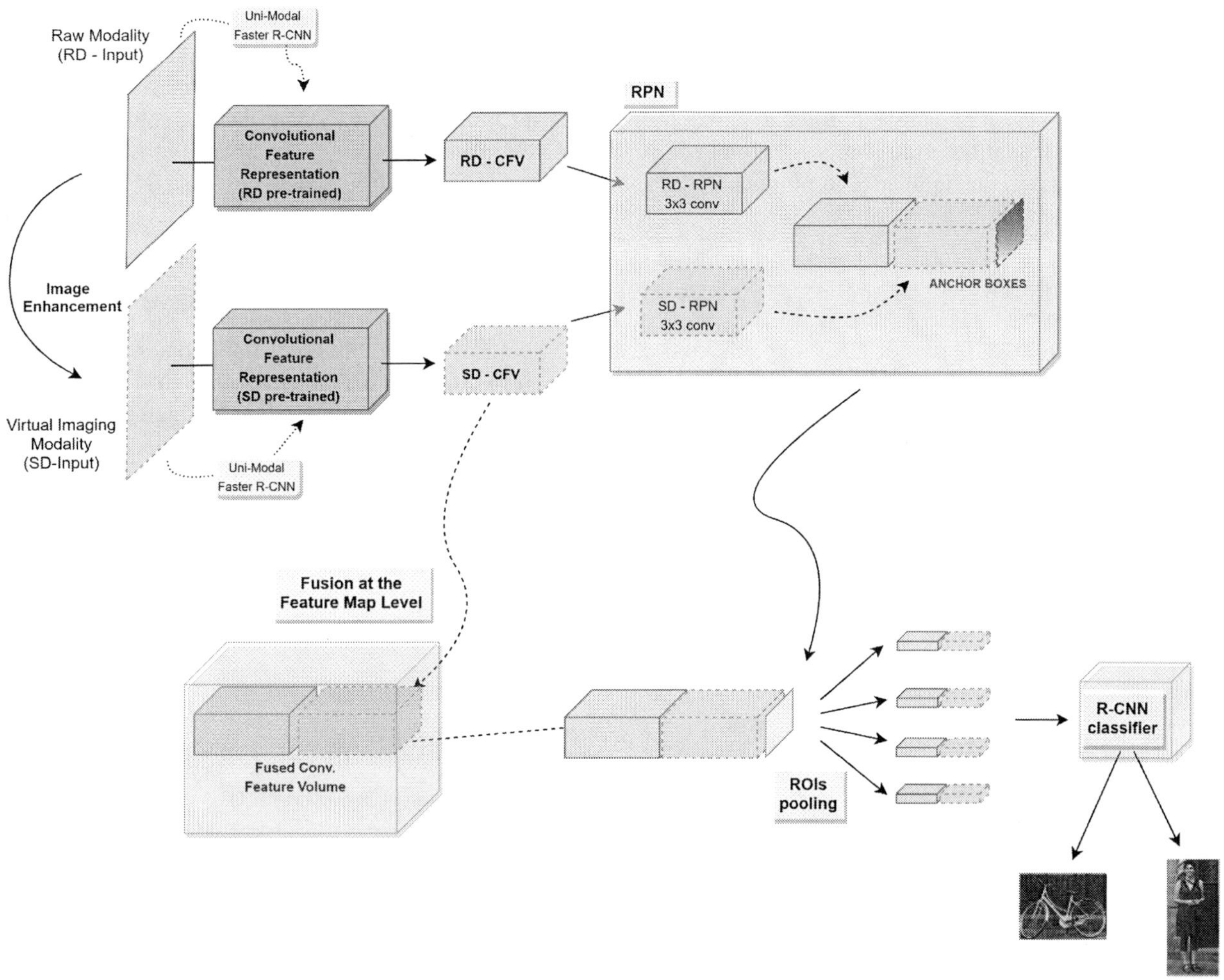

Figure 4: BiModal Faster R-CNN propagates the dual modality input through two parallel streams of identically structured, but independently pre-trained, CNNs. The CNN that receives the Raw (physical) modality (RD) is pre-trained on Raw images while the CNN that receives the virtual modality (SD) is pre-trained trained on processed images. The two streams produce two separate Feature volume representations. The two representations are concatenated (deep features fusion) and the new, deeper, volume enters the Region Proposal Network (RPN), allowing the system to estimate more accurate Region of Interest (RoI) detections. Finally, fused representations of these RoIs enter the classifier leading to more accurate label decisions.

Initialization of the Bi-modal Faster R-CNN

As we already mentioned, all the filter coefficients in the two parallel CFR stages, as well as the filter coefficients in the two parallel $RD-RPN_{3\times3conv}$ and $SD-RPN_{3\times3conv}$ layers, are initialized by transferring the corresponding coefficients of the best performing models from the Unimodal networks. Once more, we emphasize that these coefficients will remain fixed during the Bi-modal network training process, unaffected by the update which takes place after each training 'forward pass', in what is commonly known as the 'back-propagation' step in deep learning terminology. Apart from these, transferred, coefficients, all other coefficients, in all remaining layers of the Bi-modal network, convolutional and FC, are randomly initialized, using the same initialization set-up we used in the Unimodal training sessions. Furthermore, all these coefficients in the remaining layers, are allowed to get updated during each back-propagation step, until the model converges.

Before we close this paragraph, we need to mention one more thing. It would be possible to transfer the independently pre-trained coefficients of the FC_1 layers from the RD and SD Uni-modal networks, and combine them to initialize the FC_1 layer in the Bi-modal network. This could provide further generalizability gains to the Bi-modal network, and in extension to any multimodal network, especially since the size of FC_1 layer increases proportionally to the number of modalities used, but our postulate does not imply increase in the usable complementary information that is proportional to the number of used modalities. This type of modified multimodal Faster R-CNN initialization remains part of our ongoing experimentation.

3 EXPERIMENTS

We test our Multimodal Faster R-CNN implementation on two very popular, publicly available datasets suited for object detection, the **Pascal VOC 2007** (Everingham et al. (n.d.*a*)) and **VOC 2012** (Everingham et al. (n.d.*b*)) challenge databases. More specifically we follow the object **detection task** rules: Predict the bounding box and label of each object from the twenty target classes in an image unknown to the system being tested. As the challenges' creators state: *The main goal in these challenges is to recognize objects from a number of visual object classes in realistic scenes (i.e. not pre-segmented objects).* This can be treated as a supervised learning problem and a training set of labeled images, from twenty different object classes, is provided. The twenty object classes that have been selected for both challenges are:

Person: person

Animal: bird, cat, cow, dog, horse, sheep

Vehicle: aeroplane, bicycle, boat, bus, car, motorbike, train

Indoor: bottle, chair, dining table, potted plant, sofa, tv/monitor

We run two independent experiments, comparing the accuracy of the Unimodal Faster R-CNN with raw image input (benchmark) against our Bimodal and Trimodal Faster R-CNN architectures, that use the IN/VR artificial imaging modalities. We follow the standard experimental set-up for the VOC2007 and VOC2012 detection tasks:

VOC2007 detection: We train on both the *train* and *validation* VOC2007 data-sets, composed by $5,011$ images containing $12,608$ object instances in total. We validate the model on the VOC2007 *test* data-set, composed by $4,952$ images containing $12,032$ object instances in total.

VOC2012 detection: Due to lack of available annotation for the VOC2012 *test* set, here we train only on the VOC2012 *train* data-set, composed by $5,717$ images containing $13,609$ object instances in total. Then, we validate the model on the VOC2012 *validation* data-set, composed by $5,823$ images containing $13,841$ object instances in total.

3.1 Precision over recall evaluation per object class

In all the Faster R-CNN based network's we use in our tests, a usually large number of RoIs are selected from the RPN classification as significant object-containing RoIs. Each one of these RoIs enter the R-CNN classifier, which predicts the final detection region and gives to it a confidence score per class. It is crucial to understand that this NN is a scoring system and the score is its belief that a certain region selected among those that have been qualified by RPN contains an object from a certain class. In Figs. 7,8, and 9 the green detection boxes are such regions but those displayed are a minuscule fraction of all anchor boxes qualified by both RPN and classifier. Recall, that RPN assigns to each anchor box a similar "objectness score"

Given some **confidence-threshold** value ct_A for class A, a detection region, say BB_{detect} (recorded in the array B), is declared as a **positive detection** of object from class A, if the class A confidence estimated by the R-CNN classifier, say $score_A^{BB_{detect}}$ (as recorded in the array P), exceeds the value of ct_A. If $10^{-5} < score_A^{BB_{detect}} < ct_A$ then BB_{detect} is registered[9] as **negative detection**, i.e. the network's classifier decides that no object from class A is found in BB_{detect}. This rule, yields a classifier for each class and confidence-threshold value.

Now, for evaluation purposes, a positive detection region $BB_{pos-detect}$, for some ct_A, is considered a **True Positive (TP) detection** if it satisfies both following rules:

(i) $BB_{pos-detect}$ has an IoU$>$ 0.5 overlap with a class A ground truth box, say GT_A

(ii) GT_A ground truth region is not overlapped (under the same IoU criterion) by another positive detection box of higher class A confidence

If at least one of (i) or (ii) is not satisfied, $BB_{pos-detect}$ is registered as **False Positive (FP) detection**. It is important to observe here that, the system needs to be 'chary' with its positive decisions. Indeed, in case of multiple 'valid' detections of the same object in an image, all detection regions, except from the most confident one, will be declared as false positives for the evaluation purposes[10] .

To obtain a comparable measure for the accuracy per class of all Faster R-CNN based networks we use in our tests, we adopt again the **Precision (Prec)** over **Recall (Rec)** standard metric. In probabilistic terms, for any class c and some confidence-threshold ct_c, we define

$$\begin{aligned} Prec_c =& P(\textit{detected region indeed contains object from class c} \mid \\ & \textit{system declared that the detected region contains object from class c}) \end{aligned}$$

which is also known as *positive predictive power* in biomedical tests. On the other hand, recall is the *sensitivity* of a system, i.e.

$$\begin{aligned} Rec_c =& P(\textit{system declared a detection region containing object from class c} \mid \\ & \textit{an object from class c indeed exists in this region}) \end{aligned}$$

where $P(A|B)$ represents the conditional probability of event A given B. For the purpose of extracting empirical probabilities of performance, precision and recall are given by the following formulas:

$$Prec_c = \frac{\text{Number of class c TP detections}}{\text{Total number of class c positive predictions}}$$

and

$$Rec_c = \frac{\text{Number of class c TP detections}}{\text{Total number of class c ground truth regions in the (test)set}}$$

Typically, lower precision is expected at higher levels of recall and vice versa. However, in many applications, it is more important to know the precision of a system even at low recall levels, or the recall at low precision levels. Furthermore, it can be of great importance to know what is the best precision one can expect when recall below a specific level cannot be tolerated, or the best recall one can expect when precision below some level is unacceptable. To help understand those concepts we stress that recall is sensitivity and precision is the positive predictive value of the system for screening tests in medicine.

We plot corresponding curves for all significant confidence-threshold values. This curve represents the, per class, accuracy of the multi-class object detection model. Depending on the use of each modality, physical or virtual, we obtain the following networks, of which we compare their individual performances:

Raw: The Unimodal Faster R-CNN network, described in section 2.5, using only the raw grayscale version of the data both for training and testing.

Bimodal.Raw.IN: The Bimodal Faster R-CNN architecture, described in section 2.5, fusing the raw grayscale and the virtual modality induced by IN version of the data, both for training and testing.

[9] It is common for this minimum score threshold value to be set higher than our 10^{-5} choice, so that less regions will enter the evaluation stage as negative, leading to faster evaluation calculations. However, we purposefully decide to set this threshold in such a low value that (almost) all RoIs will enter the evaluation stage, providing us with a more complete picture of the over-all precision/recall trade-off of each tested model.

[10] In statistics language this means that all rank 2 and higher detection regions that effectively capture the specific object will enter the evaluation as false positives

Bimodal.Raw.VR: The Bimodal Faster R-CNN architecture, fusing the raw grayscale and the virtual modality induced by VR version of the data, both for training and testing.

Trimodal.Raw.IN.VR: The Trimodal Faster R-CNN architecture, fusing the physical raw grayscale and both the IN and VR virtual modalities.

Results per class

In Figure 5 we present the precision over recall curves (for a few classes) from the Raw Unimodal Faster R-CNN detector against our Trimodal Faster R-CNN architecture. In most cases, class accuracy improves anywhere from somewhat to significantly, when we use the proposed Multimodal networks (Bimodal or Trimodal Faster R-CNN) relative to the raw-Unimodal model (e.g., for classes *potted-plant*, *bird*, *sheep* and *sofa* in Figure 5(a)-(d)). The gain in precision with the Trimodal architecture is anywhere between 0.02% up to 18.16% in most recall levels. On the other hand, there are object classes, e.g., the class *bus* (Figure 5(e)) where the Trimodal network results in a lower precision than its raw-Unimodal peer. We also find object classes, such as *dining-table* (Figure 5(f)), where performance is mixed

We observe a similar behavior with the VOC2012 database. Thus, in order to reach a conclusion on whether an overall accuracy gain arises from our Multimodal architecture in both these tasks, relatively to the Raw-Unimodal model, we propose to use a descriptive augmented metric, to represent performance of each model over all object classes.

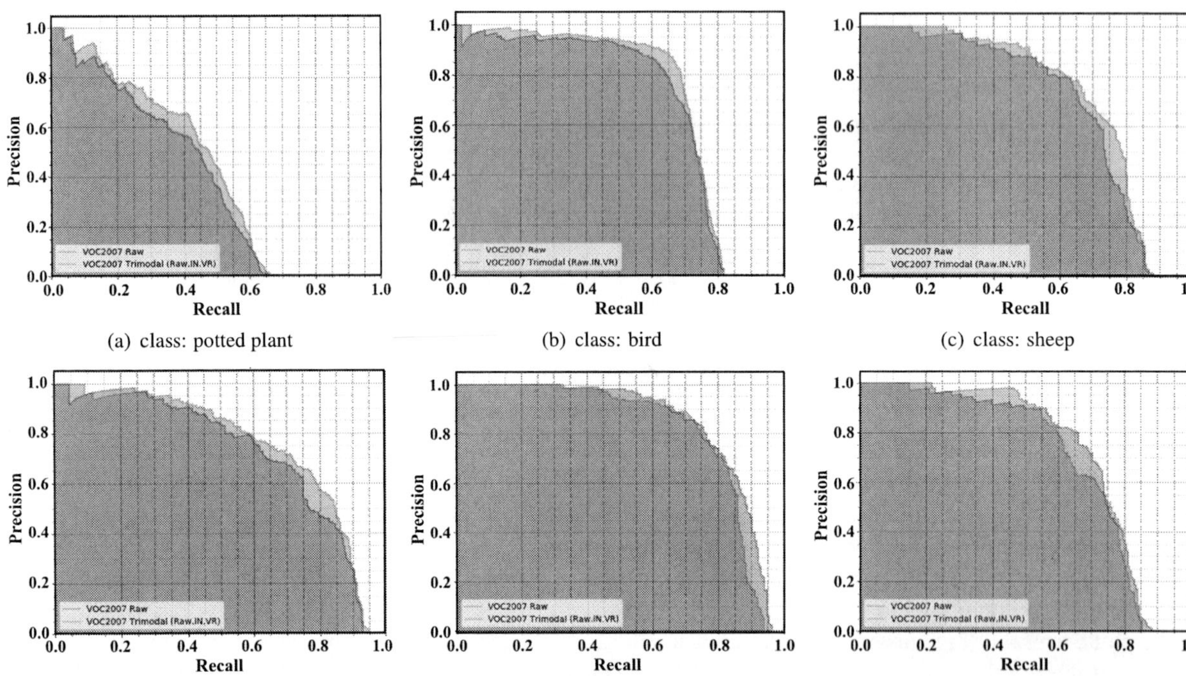

(a) class: potted plant (b) class: bird (c) class: sheep

(d) class: sofa (e) class: bus (f) class: dining table

Figure 5: Precision over recall, from VOC2007 object detection task, for classes *potted plant* (a), *bird* (b), *sheep* (c), *sofa* (d), *bus* (e) and *dining table* (f). Both models, the Raw Unimodal Faster R-CNN (blue) and the Raw.IN.VR Trimodal Faster R-CNN (orange), have been trained on the corresponding version of the VOC2007 *train+validation* dataset. Their performance is evaluated on the corresponding VOC2007 *test* data-set. In most object classes, e.g., (a)-(d) here, we observe a significant increase of precision when we use our Raw.IN.VR-Trimodal model. However, there are cases, e.g. (e) here, where the virtual modalities based Trimodal architecture demonstrates lower precision, and others where performance is mixed (f).

Each model's performance can be studied on a three-dimensional space, with Recall (**r**), Precision (**p**) and Object Class (**c**) defining the three dimensions. The range for the first two dimensions, r and p, is the interval $[0,1] \subset \mathbb{R}$.

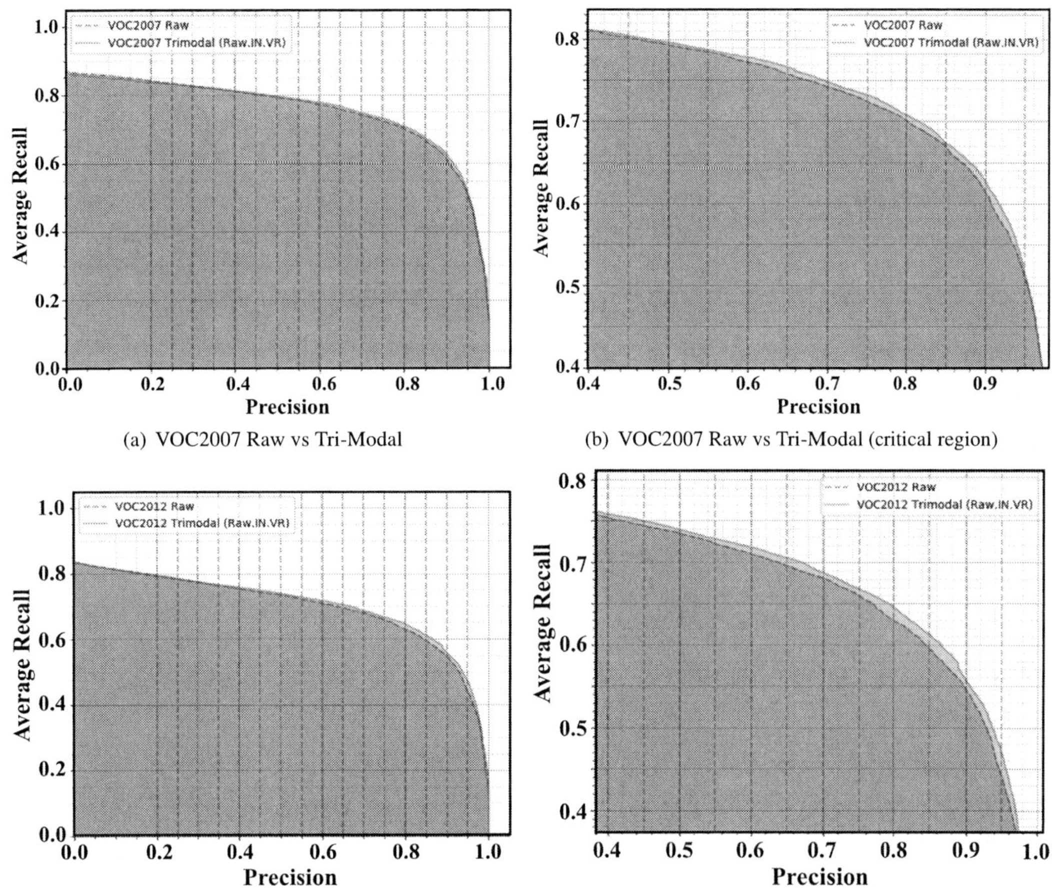

(a) VOC2007 Raw vs Tri-Modal

(b) VOC2007 Raw vs Tri-Modal (critical region)

(c) VOC2012 Raw vs Tri-Modal

(d) VOC2012 Raw vs Tri-Modal (critical region)

Figure 6: All-class average recall *(AR)* plotted as a function of precision *(p)* for the VOC2007 (**top row**) and VOC2012 (**bottom row**) multi-class object detection tasks. The blue curve is drawn for the Raw-Unimodal Faster R-CNN benchmark. The orange curve is drawn for the proposed Trimodal Faster R-CNN architecture, where we combine deep features from the Raw and both virtual IN and VR modalities. Our Raw.IN.VR-Trimodal detector achieves significant increase in the (expected) recall on the largest part of the precision range. In the VOC2007 case, apart from a low region below 37% precision, the Raw.IN.VR-Trimodal detector achieves an increase of the expected (overall object classes) recall from 0.09% to 1.42% (at 90% precision). In the VOC2012 case, the proposed Raw.IN.VR-Trimodal architecture achieves significant increase in the (expected) recall on the whole precision range, from 0.15% up to around 1.76% (at the neighborhood of 85% precision), relatively to the Raw-Unimodal benchmark.

Dimension c is discrete and assumes any value from the set $\{1,2,...,C\}$, where C is the number of target object classes. In our VOC2007 and VOC2012 tasks, $C = 20$. For any model for which the precision recall values (curves) have been evaluated for all object classes, we define the function $r(c,p)$ as

$$r(c,p) = \text{recall value at precision level p for class c}$$

Note here that, precision does not necessarily cover all the available $[0,1]$ range, however there exists a meaningful correspondence between all achieved precision ranges between the different classes. Now, we define the **Average Recall (AR)** as

$$AR(p) = \sum_{c=1}^{20} emp(c)r(c,p) \tag{2}$$

where $emp(c)$ equals the relative frequency of appearance of object class c in each task's data-set, i.e. the ratio of class c ground truth regions in the data over the total number of ground truth annotations in the data. Thus *AR* is the *average empirical expectation of recall r as a function of precision p.*

We use the *AR* function to evaluate the (expected) over-all performance of the Faster R-CNN based models. We plot the *AR* over precision curves from different models on the same graph, to compare their performance in terms of expected over-all recall at any level of precision. In both cases, on the VOC2007 and on the VOC2012 databases, for the task of multi-class object detection, we have observed a significant improvement in the AR, the class-average recall, for the largest part of the precision range, when we engage the virtual modalities with any of our Multimodal architectures, relative to the Raw-Unimodal benchmark model. However, the boost in expected recall is higher when we use our Raw.IN.VR-Trimodal Faster R-CNN architecture (Figure 6). More precisely, in the case of the the VOC2007 database, apart from a low region, below 37% precision, using our Raw.IN.VR-Trimodal detector we achieve an increase of the all class-average recall from 0.09% up to 1.42% (at 90% precision). In the case of the VOC2012 database, the respective increases are observed over the whole range of precision, and range from 0.15% up to 1.76% (at 85% precision).

4 CONCLUSIONS

Multiple physical modalities provide complementary information for artificial intelligence tasks. They work in such a way that when the information flow of one modality does not contribute sufficiently to a decision making process, additional modalities with play a complementary role which, will hopefully decrease disambiguation of this task. If only one physical modality can generate inputs, then there is only one way to proceed; maximize the extraction of information from the single physical modality inputs. In this work we propose that one specific pathway to achieve this goal is by creating virtual modalities, which generate artificial inputs from the single available input. The single physical input and the artificially generated derivative inputs are jointly fed to a Neural Network classifier. We do this in two ways, either by expanding the feature space or by utilizing the structure of the Neural Network classifier, which now receives the single output along with the virtual inputs and generates an expanded feature vector for any non-virtual input.

Within this framework, we have been able to provide preliminary experimental evidence to support the claim that the IN and VR operators increase the ability of the deep object detection NN to assign less ambiguous confidence scores, without imposing significant changes in the network architecture other than those necessary to accommodate the additional artificial modalities. The reader can see the result of this disambiguation in Figures 7,8 and 9.

Our results are preliminary since we have not ascertained the statistical significance of our findings, mainly, due to our intention to adhere as much as possible to the evaluation protocols of the Pascal VOC object detection competitions. The computational overload required for this task was another contributing factor for this decision. Moreover, the precision-recall curves of the R-CNN classifier performance show mild improvement with the fusion of the features of the virtual modalities we use. Of course, one may wonder whether the same or better improvement can be achieved with changes in the networks architecture. Although, this is quite possible, the improvement of discriminability seems to suggest that the artificial modalities inputs may even yield additional improvement. However, the huge variety in network configurations and design can lead to huge volumes of experiments that one cannot possibly perform in order to establish, beyond any doubt, the usefulness of the artificial modalities along with a physical one. Moreover, the identification of the correct parameters for the artificial modalities is another issue we did not touch upon in a systematic way (sensitivity analysis). In fact, in Fig. 9 the reader can see that in adverse atmospheric conditions VR needs other than its default parameters. Despite the number of unanswered questions our study raised we believe that the size of the two databases we used for our test experiments provide sufficient evidence that the virtual modalities can help deep NN classifiers improve performance, not only when inputs are acquired under marginal conditions, but also when all is, or looks, normal. Furthermore, we achieved this goal without changing their prototypical design of the NNs we use.

Overall, our present results suggest that the fusion of physical and derived artificial modalities can potentially improve the accuracy of AI systems.

(a) Raw Only

(b) Trimodal (Raw.IN.VR)

(c) Raw Only

(d) Trimodal (Raw.IN.VR)

(e) Raw Only

(f) Trimodal (Raw.IN.VR)

Figure 7: Each detection system draws a 'labeled' rectangle, enclosing in a tight manner any object (from the learned classes) they detect. The number on the side of the label represents the confidence, of the system, that the rectangle contains an object from the specific class (in the range $[0, 1]$, where 1 corresponds to certainty). In many occasions, the use of complementary information from the artificial IN and VR imaging modalities within our virtual multimodal Faster R-CNN architecture leads to higher detection capability (**1st column**), relatively to the unimodal system (**2nd column**) where only the raw image modality is used, especially when poor visibility obstructs the systems decision. System's detections with confidence below 0.4 are not displayed.

5 ACKNOWLEDGMENTS

This work was partially supported by NSF awards DMS-1720487 and DMS-1320910 and by a University of Houston Seed Funding for Advanced Computing (SeFAC), 2017 "Overcoming Poor Illumination for Real-time, Large Scale

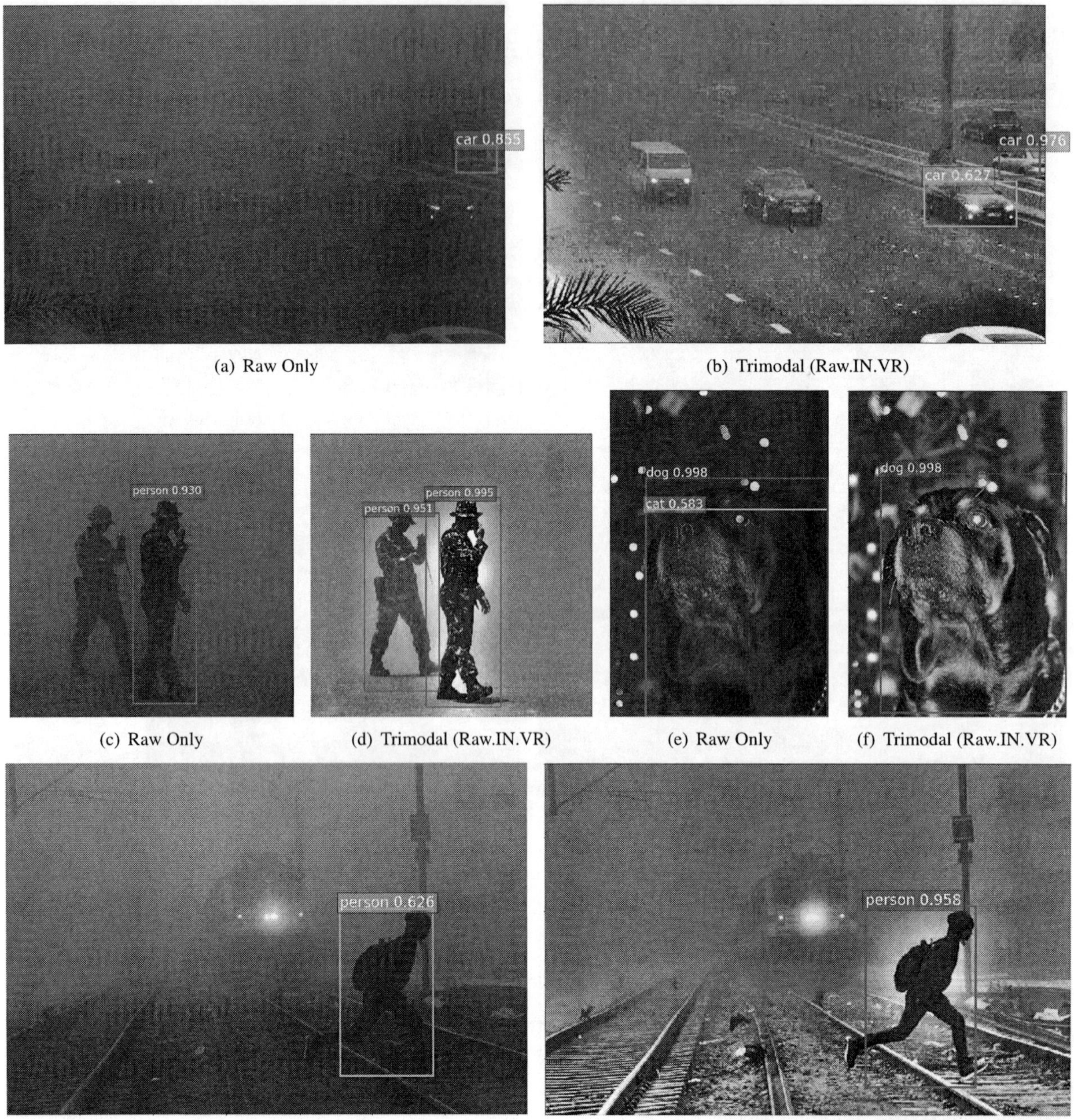

(a) Raw Only (b) Trimodal (Raw.IN.VR)

(c) Raw Only (d) Trimodal (Raw.IN.VR) (e) Raw Only (f) Trimodal (Raw.IN.VR)

(g) Raw Only (h) Trimodal (Raw.IN.VR)

Figure 8: As in Figure 7, system detections appear as 'labeled' rectangles, enclosing any object they detect. Again, the number on the side of the label represents system's confidence that the rectangle contains an object from the specific class (in the range $[0, 1]$, where 1 corresponds to certainty). Again we observe that the use of complementary information from the artificial IN and VR imaging modalities within our virtual multimodal Faster R-CNN architecture can lead to higher detection capability, relatively to the unimodal system where only the raw image modality is used. Higher detection capability of the Trimodal Raw.IN.VR system here translates into either correcting false detections of the unimodal Raw-only system (e.g. in (f)), or new accurate detections (e.g. in (d)), or improving significantly the system's confidence on accurate detections (e.g. in (h)).

Face Recognition Applications".

(a) Raw modality (b) IN modality (c) VR_1 modality (d) VR_2 modality

(e) Unimodal detection (Raw) (f) Trimodal detection ($Raw.IN.VR_1$) (g) Trimodal detection ($Raw.IN.VR_2$)

Figure 9: Optimizing the parameters of the image enhancement methods, used to derive the artificial imaging modalities, can further improve the multimodal system's accuracy. **Top row:** The raw (physical) modality (a) next to the derived (artificial) imaging modalities (b)-(d). Both (c) and (d) are generated by the VR method, but using different parameter settings. **Bottom row:** Improvement in detection performance, compared to the Raw-Unimodal Faster R-CNN detection attempt (e), is achieved only when Var_2 modality replaces Var_1 in the virtual Trimodal model (g).

References

Dalal, N. & Triggs, B. (2005), Histograms of oriented gradients for human detection, *in* 'Proceedings of the 2005 IEEE Computer Society Conference on Computer Vision and Pattern Recognition (CVPR'05) - Volume 1 - Volume 01', CVPR '05, IEEE Computer Society, Washington, DC, USA, pp. 886–893.
URL: *http://dx.doi.org/10.1109/CVPR.2005.177*

Deng, J., Dong, W., Socher, R., Li, L.-J., Li, K. & Fei-Fei, L. (2009), ImageNet: A Large-Scale Hierarchical Image Database, *in* 'CVPR09'.

Eitel, A., Springenberg, J. T., Spinello, L., Riedmiller, M. A. & Burgard, W. (2015), 'Multimodal deep learning for robust RGB-D object recognition', *CoRR* **abs/1507.06821**.
URL: *http://arxiv.org/abs/1507.06821*

Endsley, M. R. (1999), *Handbook of Aviation Human Factors*, Lawrence Erlbaum Asscoiates, chapter 11.

Everingham, M., Van Gool, L., Williams, C. K. I., Winn, J. & Zisserman, A. (n.d.*a*), 'Pascal visual object classes challenge 2007 (voc2007) complete dataset'.
URL: *http://host.robots.ox.ac.uk/pascal/VOC/voc2007/*

Everingham, M., Van Gool, L., Williams, C. K. I., Winn, J. & Zisserman, A. (n.d.*b*), 'The PASCAL Visual Object Classes Challenge 2012 (VOC2012) Results', http://www.pascal-network.org/challenges/VOC/voc2012/workshop/index.html.

Fu, X., Zeng, D., Huang, Y., Zhang, X.-P. & Ding, X. (2016), A weighted variational model for simultaneous reflectance and illumination estimation, *in* 'Proceedings of the IEEE Conference on Computer Vision and Pattern Recognition', pp. 2782–2790.

Girshick, R. B. (2015), 'Fast R-CNN', *CoRR* **abs/1504.08083**.
URL: *http://arxiv.org/abs/1504.08083*

Girshick, R., Donahue, J., Darrell, T. & Malik, J. (2014), Rich feature hierarchies for accurate object detection and semantic segmentation, *in* 'Proceedings of the 2014 IEEE Conference on Computer Vision and Pattern Recognition', CVPR '14, IEEE Computer Society, Washington, DC, USA, pp. 580–587.
URL: *https://doi.org/10.1109/CVPR.2014.81*

Johnsen, G., Ludvigsen, M., Sørensen, A. & Aas, L. M. S. (2016), 'The use of underwater hyperspectral imaging deployed on remotely operated vehicles-methods and applications', *IFAC-PapersOnLine* **49**(23), 476–481.

Land, E. H. (1964), 'The retinex', *American Scientist* **52**(2), 247–264.

Land, E. H. et al. (1977), *The retinex theory of color vision*, Citeseer.

Lin, T., Goyal, P., Girshick, R. B., He, K. & Dollár, P. (2017), 'Focal loss for dense object detection', *CoRR* **abs/1708.02002**.
URL: *http://arxiv.org/abs/1708.02002*

Ngiam, J., Khosla, A., Kim, M., Nam, J., Lee, H. & Ng, A. Y. (2011), Multimodal deep learning., *in* L. Getoor & T. Scheffer, eds, 'ICML', Omnipress, pp. 689–696.

Ochoa-Villegas, M. A., Nolazco-Flores, J. A., Barron-Cano, O. & Kakadiaris, I. A. (2015), 'Addressing the illumination challenge in two-dimensional face recognition: a survey', *IET Computer Vision* **9**(6), 978–992.

Paul Viola, M. J. (2001), 'Rapid object detection using a boosted cascade of simple features', *Proceedings of the 2001 IEEE Computer Society Conference on Computer Vision and Pattern Recognition, 2001.* **1**, I–I.

Redmon, J., Divvala, S., Girshick, R. & Farhadi, A. (2016), You only look once: Unified, real-time object detection, *in* 'The IEEE Conference on Computer Vision and Pattern Recognition (CVPR)'.

Redmon, J., Divvala, S. K., Girshick, R. B. & Farhadi, A. (2015), 'You only look once: Unified, real-time object detection', *CoRR* **abs/1506.02640**.
URL: *http://arxiv.org/abs/1506.02640*

Redmon, J. & Farhadi, A. (2018), 'Yolov3: An incremental improvement', *CoRR* **abs/1804.02767**.
URL: *http://arxiv.org/abs/1804.02767*

Ren, S., He, K., Girshick, R. B. & Sun, J. (2015), 'Faster R-CNN: towards real-time object detection with region proposal networks', *CoRR* **abs/1506.01497**.
URL: *http://arxiv.org/abs/1506.01497*

Roper, W. E. & Dutta, S. (2006), Oil spill and pipeline condition assessment using remote sensing and data visualization management systems, *in* 'Freshwater Spills Symposium, Natural Disasters, Human Error, and Equipment Failure-Causes for Major Inland Oil Spills and the Resulting Multifaceted Response'.

Sermanet, P., Eigen, D., Zhang, X., Mathieu, M., Fergus, R. & Lecun, Y. (2014), Overfeat: Integrated recognition, localization and detection using convolutional networks, *in* 'International Conference on Learning Representations (ICLR2014), CBLS, April 2014'.

Simon, M., Rodner, E. & Denzler, J. (2016), 'Imagenet pre-trained models with batch normalization', *arXiv preprint arXiv:1612.01452* .

Srivastava, N. & Salakhutdinov, R. R. (2012), Multimodal learning with deep boltzmann machines, *in* F. Pereira, C. J. C. Burges, L. Bottou & K. Q. Weinberger, eds, 'Advances in Neural Information Processing Systems 25', Curran Associates, Inc., pp. 2222–2230.
URL: *http://papers.nips.cc/paper/4683-multimodal-learning-with-deep-boltzmann-machines.pdf*

Sutskever, I., Martens, J., Dahl, G. & Hinton, G. (2013), On the importance of initialization and momentum in deep learning, *in* 'Proceedings of the 30th International Conference on International Conference on Machine Learning - Volume 28', ICML'13, JMLR.org, pp. III–1139–III–1147.
URL: *http://dl.acm.org/citation.cfm?id=3042817.3043064*

Upadhyay, S., Mitsakos, N. & Papadakis, M. (2018), 'Real time visibility improvement for underwater video', *Proceedings of the 23rd Offshore Symposium, February 2018, Houston, Texas* .

Upadhyay, S. & Papadakis, M. (2019), A nonlinear multiscale method for illumination normalization. under revision, Applied Comp. Harmonic Analysis.

THE INTERNATIONAL COMMUNITY
FOR MARITIME AND OCEAN PROFESSIONALS

A DISTRIBUTED VORTICITY MODEL FOR FLOW PAST A 3-D HYDROFOIL

CHUNLIN WU
GRADUATE STUDENT, THE UNIVERSITY OF TEXAS AT AUSTIN

SPYROS A. KINNAS
PROFESSOR, THE UNIVERSITY OF TEXAS AT AUSTIN

Proceedings of the 24th Offshore Symposium, February 20th 2019, Houston, Texas
Texas Section of the Society of Naval Architects and Marine Engineers

ABSTRACT

A spatially distributed vorticity model is proposed in this paper to simulate the 3-dimensional boundary layer flow past a hydrofoil. It is a coupled viscous/inviscid method combining the Boundary Element Method (BEM) and the VIScous Vorticity Equation (VISVE) method, in the sense that the viscous results work as corrections on top of the inviscid results. Using the distributed vorticity model, every panel on the wall can release vorticity into the flow field as opposed to the distinct Leading Edge Vortex (LEV) model with a prescribed detachment point. Whether, where and how much vorticity is produced from the wall are determined by the instantaneous flow state. The distributed vorticity model was applied in the case of 3D hydrofoils at an angle. The vorticity and velocity in the flow field is obtained and analyzed. The results from VISVE are compared with those from Navier-Stokes' methods (N-S). Calculated results provided good agreement, which addressed the detachment problem. As a result, VISVE is proved to be quite reliable and efficient.

Keywords: Vorticity Equation, Hydrofoil, 3-D Incompressible, Vorticity Boundary Condition

INTRODUCTION

In this paper, a 3-D vorticity equation method is introduced to solve the incompressible laminar flow past a 3-D hydrofoil. A distributed vorticity model which couples a finite volume solver for the vorticity equation and the panel method is proposed. Using the distributed vorticity model, every panel on the wall can release vorticity into the flow field as opposed to the distinct LEV model with a prescribed detachment point. Whether, where and how much vorticity is produced from the wall are determined by the instantaneous flow state.

The vorticity methods have received myriad attention in academia since the methods have promising advantages. Compared with directly solving for the velocity, solving for the vorticity leads to a velocity field with higher accuracy since the vorticity is one order higher than the velocity. Notably, while modeling flows past a hydrofoil, the vorticity field only concentrates within the boundary layer near the wall and the wake region. The compact distribution of vorticity near the wall leads to significantly smaller computational domains with a significantly smaller number of cells (concentrated close to the foil and its wake) and thus considerably smaller CPU times since the open boundary conditions can be easily calculated using the Green's function (Tian, 2015). Additionally, since only the region near the wall and the wake is necessary for modeling, the grid generation can be rather simplified and can even be automated. Furthermore, because the pressure term, which does not have evolution equations, was eliminated from the equation due to the curl operation, the number of unknowns reduces by 1. Hence, the spatial concentration of vorticity near the wall results in a spatially compact and numerically efficient numerical method.

In the current method, the vorticity equation is discretized using the Stokes' theorem based on the Finite Volume Method (FVM). Instead of solving for the velocity directly by volume integral, the divergence-free velocity field is calculated by solving a Poisson's equation since the former scheme has N^2 complexity while the latter scheme has almost N complexity. Moreover, a vorticity creation scheme is utilized to calculate the amount of vorticity released from the wall by enforcing the no-slip and non-penetrating boundary conditions on the wall. Unlike the distinct leading edge vortex model which requires a prescribed detachment point, by using the presented vorticity creation scheme, every panel on the wall can release vorticity into the flow field.

Tian and Kinnas (2015) originally proposed the VISVE method and conducted comprehensive studies about this method. Then, Wu (2016) improved the solver and gridding algorithm such that it is able to take non-orthogonal grids to model the unidirectional flow around a cylinder in 2-D. Further, Li & Kinnas (2017) successfully applied the method to the unidirectional as well as alternating flow around a 2-D cylinder and presented the results in the SNAME Offshore Symposium in 2017. Xing & Kinnas (2018) incorporated the mixture cavitation model into the VISVE method and it was successfully applied to a 2-D hydrofoil. With the successful validations in 2-D, the extension of VISVE method to 3-D hydrofoils was implemented in advection flow, in which only the advection and stretching of the vorticity was considered, by Wu & Kinnas (2017), where the significant savings in computing time of VISVE versus RANS were clearly shown. Reliable results have been obtained in all cases. In this study, VISVE method is applied to the flow past a 3-D hydrofoil and the diffusion of vorticity is included. The velocity field and vorticity field is compared between results from VISVE and those from RANS to assess the accuracy of the method. This is part of a larger project of which the ultimate goal is to accurately simulate the flow past a propeller and predicting the pressure and forces based on 3-dimensional VISVE method.

METHODOLOGY

1. Velocity Decomposition

The velocity field consists of the in-flow velocity and the vorticity induced velocity

$$\boldsymbol{q} = \boldsymbol{U}_\infty + \boldsymbol{u}_\omega \tag{1}$$

where, $\boldsymbol{q}$ represents the total velocity, $\boldsymbol{U}_\infty$ represents the velocity of inflow, $\boldsymbol{u}_\omega$ represents the vorticity induced velocity. The decomposition is unique according to the Helmholtz velocity decomposition. A given vorticity field $\boldsymbol{\omega}$ would induce one corresponding velocity field $\boldsymbol{u}_\omega$.

2. The Vorticity Equation

In the incompressible flow, the Navier-Stokes (NS) equation is written as

$$\frac{\partial \boldsymbol{q}}{\partial t} + (\boldsymbol{q} \cdot \nabla)\boldsymbol{q} = -\frac{\nabla p}{\rho} + \upsilon \nabla^2 \boldsymbol{q} \tag{2}$$

where, $\boldsymbol{q}$ represents velocity, p is pressure, t is time, ρ and υ denote fluid's density and kinematic viscosity, respectively. The vorticity equation could be derived by taking curl operation on both sides of the N-S equation.

$$\frac{\partial \boldsymbol{\omega}}{\partial t} + \nabla \times (\boldsymbol{\omega} \times \boldsymbol{q}) = -\upsilon \nabla \times (\nabla \times \boldsymbol{\omega}) \tag{3}$$

here, $\boldsymbol{\omega}$ denotes the vorticity. Equation (3) is a less frequently seen curl form of the vorticity. This actually could help enforce the continuity of the vorticity in time. The vorticity equation is discretized using the Finite Volume Method based on the Stokes' theorem. By using the Stokes' theorem to discretize, the skewness of the convection and stretching term would be automatically taken care of. The first order upwind, second order upwind and Quadratic Upstream Interpolation for Convective Kinematics (QUICK) scheme (Woodfield, 2004) were hybridly implemented for the vorticity and velocity interpolation. The Euler backward scheme in time is used to discretize the unsteady term.

3. Vorticity-Velocity Solver

One way of recovering the solenoidal velocity from a given vorticity distribution is to conduct direct volume integral with the help of stream-function.

$$\nabla^2 \boldsymbol{\psi} = -\boldsymbol{\omega} \tag{4}$$

The stream-function can be obtained through a direct volume integral

$$\boldsymbol{\psi}(\boldsymbol{x}_f) = -\int_V \omega(x) G(\boldsymbol{x}, \boldsymbol{x}_f) dV + \boldsymbol{\psi}_b \tag{5}$$

here, $\boldsymbol{x}_f$ is the field point where $\boldsymbol{\psi}$ would be calculated, $\boldsymbol{x}$ is the dummy variable going through the computational domain. $G(\boldsymbol{x}, \boldsymbol{x}_f)$ is the Green's function, which is $-1/(4\pi|\boldsymbol{x} - \boldsymbol{x}_f|)$ in 3-dimensional case. $\boldsymbol{\psi}_b$ corresponds to the stream-function of the background inflow.

The equation (5) was conducted by summing up the inductions from all the discretized cells in the computational domain. However, in 3-D problems it is unrealistic to obtain an analytic form of the volume integral of an arbitrary hexahedron. In fact, the cells near the wall where the vorticity concentrates are almost flat in our applications. Therefore, the inductions are approximately evaluated as the source of planar panels. Obviously, the calculation of equation (5) running through the computational domain is about N^2 complicity, which is extremely expensive in 3-D applications. For the sake of better performance, a Poisson's equation is solved to calculate the velocity, which is about N complexity if the solving scheme is efficient enough. In this case, the equation (5) is only applied at boundaries of the domain.

Now, let us introduce an arbitrary velocity field $\boldsymbol{w}$ that conserves the given vorticity field.

$$\boldsymbol{\omega} = \nabla \times \boldsymbol{w} \tag{6}$$

Notice that the introduced velocity field $\boldsymbol{u}_\omega$ may not satisfy the divergence free condition. So, to reinforce the continuity of the velocity field, a correction term the was added to the velocity field $\boldsymbol{w}$.

$$\boldsymbol{u}_\omega = \boldsymbol{w} + \nabla\varphi \tag{7}$$

Taking the divergence operation on equation (7), we end up with the Poisson' equation

$$\nabla^2 \varphi = -\nabla \cdot \boldsymbol{w} \tag{8}$$

where, φ is the velocity potential unknown to be solved.

4. Vorticity Creation

The non-penetrating boundary condition and the no-slip boundary condition require both 0 tangential velocity and 0 normal velocity on the wall but it is impossible to specify both of the two conditions for the Poisson's equation. Therefore, a vorticity creation scheme is proposed to enforce the non-penetrating and no-slip boundary conditions. The vorticity creation is determined by the instantons flow condition.

At time step n, after solving equation (3), the vorticity $\boldsymbol{\omega}^{n*}$ could be used to calculate the vorticity induced velocity, $\boldsymbol{u}_\omega^{n*}$. Considering the inflow velocity as $\boldsymbol{U}_\infty$, the total velocity at time step n reads:

$$\boldsymbol{q}^{n*} = \boldsymbol{U}_\infty + \boldsymbol{u}_\omega^{n*} \tag{9}$$

Obviously, without imposing the boundary conditions on the wall, the velocity $\boldsymbol{q}^{n*}$ would have normal component $\boldsymbol{q}_n^{n*}$ and tangential component $\boldsymbol{q}_s^{n*}$ on the wall. The normal component $\boldsymbol{q}_n^{n*}$ could be eliminated by the BEM solver and produce potential ϕ on the wall, which is then converted to surface vorticity distribution. To be specific the created vorticity strength reads

$$\boldsymbol{\gamma_b} = -\boldsymbol{q}_s^{n*} - \nabla_s \phi \tag{10}$$

where, $\nabla_s = (\partial/\partial s_1 , \partial/\partial s_2)$ is the gradient operator along the tangential direction on the wall. $\boldsymbol{q}_s^{n*}$ is the tangential velocity on the wall. In this presented model, all panels on the wall could produce vorticity and release vorticity into the flow field, which avoids the pre-determination of the detachment point.

5. Pressure Calculation

Since the pressure term was eliminated from the vorticity equation, to recover the pressure additional solving procedure is inevitable.

Define the total head as:

$$H = \frac{q^2}{2} + \frac{p}{\rho} \tag{11}$$

The gradient of total head is written as

$$\nabla H = -\frac{\partial \boldsymbol{q}}{\partial t} + \boldsymbol{q} \times \boldsymbol{\omega} - \upsilon \nabla \times \boldsymbol{\omega} \tag{12}$$

Hence, the total head is calculated by integrating the RHS of equation (12) along a line in the domain. Further, the relative pressure can be calculated using equation (11).

Alternatively, the pressure can be calculated by making use of the zero velocity on the wall. The momentum equation is simplified as equation (13) because the zero velocity on the wall.

$$\nabla \left(\frac{p}{\rho}\right) = -\upsilon \nabla \times \boldsymbol{\omega} \tag{13}$$

By define the reference pressure at a point on the wall, the pressure elsewhere can be evaluated by integral.

6. General Solving Strategy

The general solving scheme is shown in the Figure 1. Starting from a valid vorticity field $\boldsymbol{\omega}^n$ at time step n, before the marching to the next time step $n + 1$, the solving of the vorticity equation (3) requires velocity of the current step as input. The vorticity induced divergence free velocity field $\boldsymbol{q}^n$ could be recovered by solving the Poisson's equation (8). Then, a vorticity creation scheme is proposed to enforce the non-penetrating boundary condition and no-slip boundary condition of the velocity field.

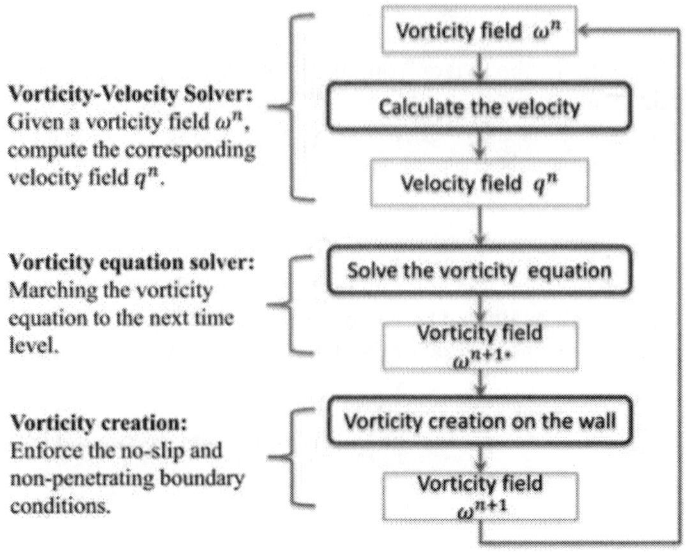

Figure 1: General solving process of VISVE

RESULTS

3-Dimensional hydrofoil with bi-convex sections

In this part, a 3-dimensional rectangular wing with bi-convex sections and no camber is used to run the simulation. The maximum thickness is 8% at the span-wise location y=0. The thickness goes linearly to 0 towards the tip. The velocity field is compared between results of VISVE and that of N-S, since velocity is crucial for the calculation of the pressure in VISVE. For the Navier-Stokes results in terms of the primitive parameters (velocities and pressure) we used the commercial code *ANSYS/Fluent*. While comparing the velocity field, the velocity profile along the line vertical to the wall surface is compared. The grid used in VISVE is shown in Figure 2 and the grid used in N-S is shown in Figure 3.

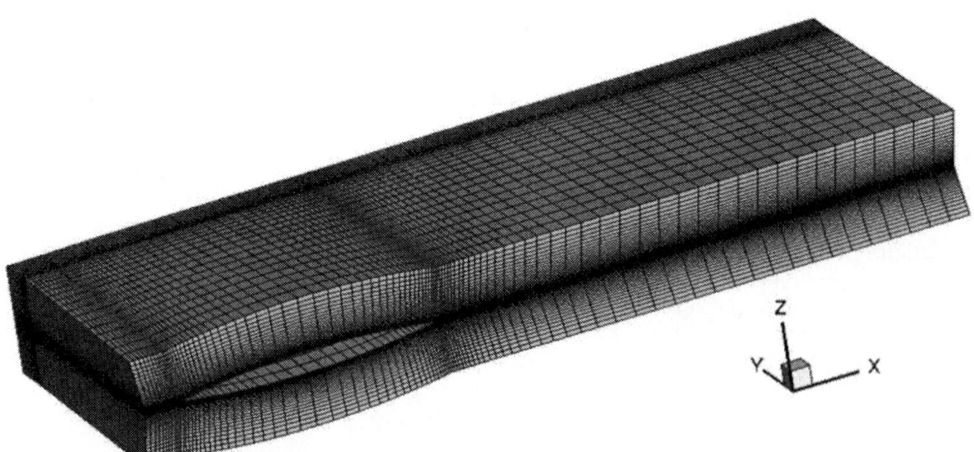

Figure 2: Mesh around the bi-convex wing in VISVE computational domain

In total, a number of 80,100 cells are used in VISVE while a number of 4,447,024 cells are used in N-S even with a symmetric interface. The number of cells in N-S domain with symmetric boundary condition is about 55 times larger than that of VISVE domain as shown in Figure 4. Especially in case of a fully unsteady run for a propeller, the N-S domain is five times of the case with periodic boundary condition. In contrast, the domain of VISVE can be kept unchanged since the influence of other blades on the key blade can be taken into consideration in an iterative way.

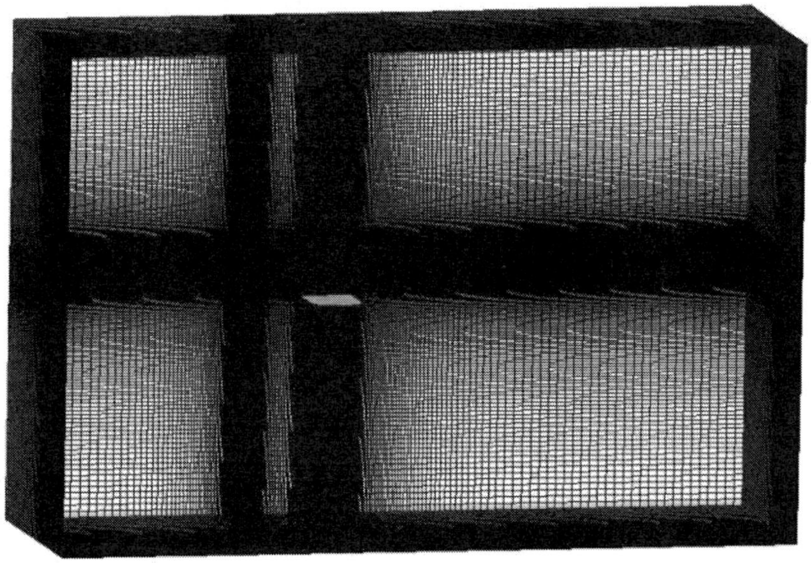

Figure 3: Mesh around the bi-convex wing in N-S computational domain

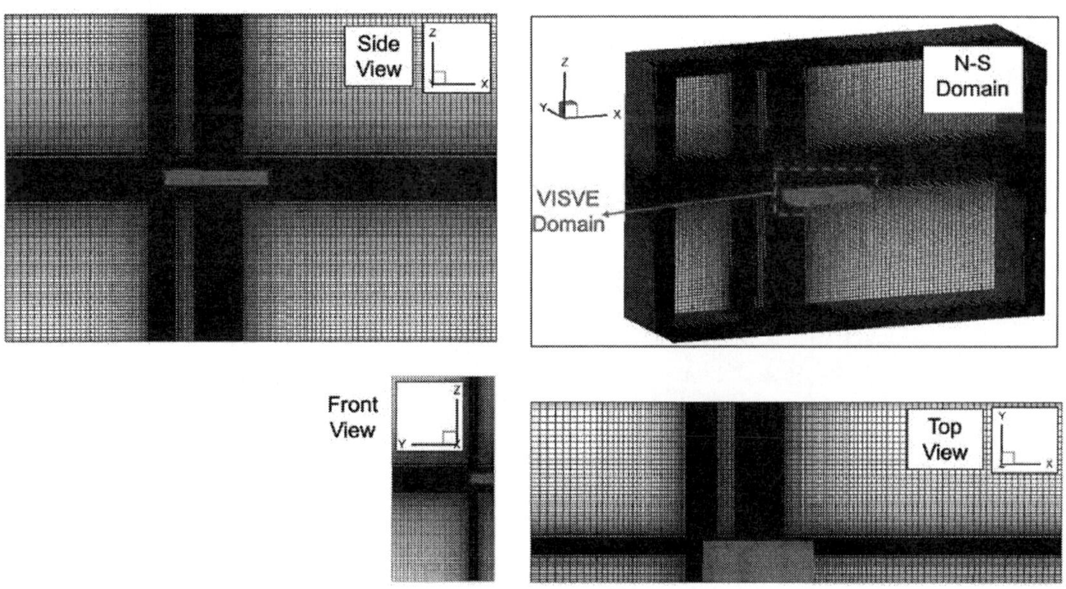

Figure 4: Domain size comparison between VISVE and N-S

The velocity profiles from VISVE are compared with those from N-S in Figure 5 ~ Figure10. Span section 0% is defined at the symmetric interface. z/c is defined as the non-dimensionalized vertical distance from the surface of the hydrofoil. The angle of attack of inflow equals 1 degrees. The Reynolds' number is 5000.

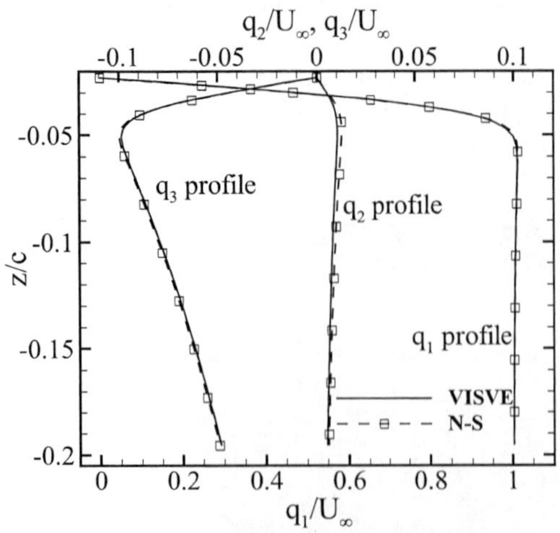

(a) 20% of chord length

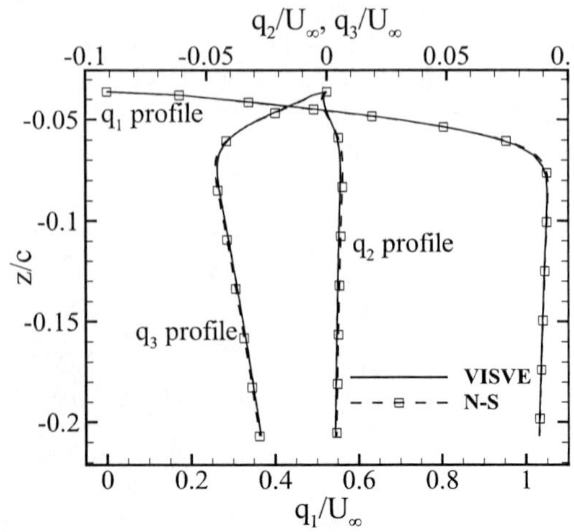

(b) 40% of chord length

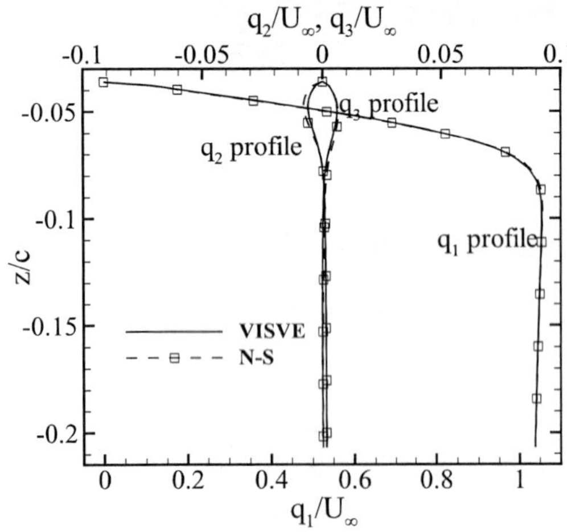

(c) 60% of chord length

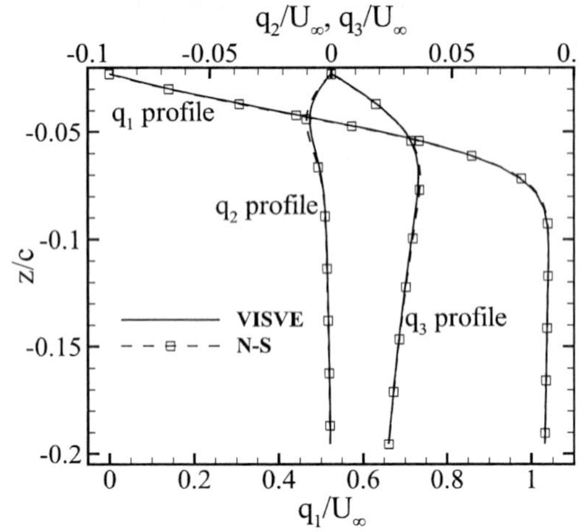

(d) 80% of chord length

Figure 5: Velocity profile comparison on the pressure side at span section 10%, Re = 5000, AOA = 1°

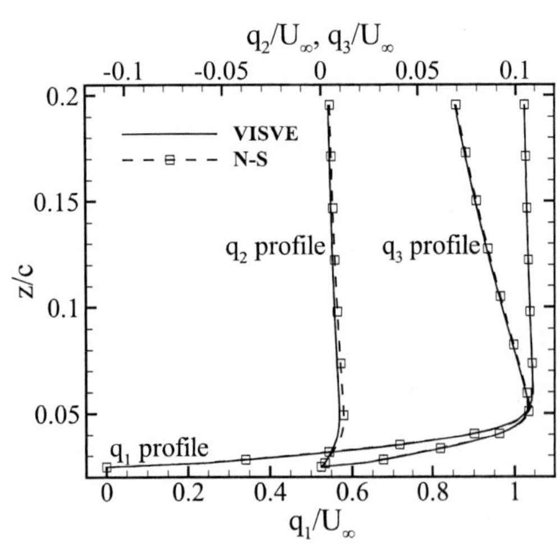

(a) 20% of chord length

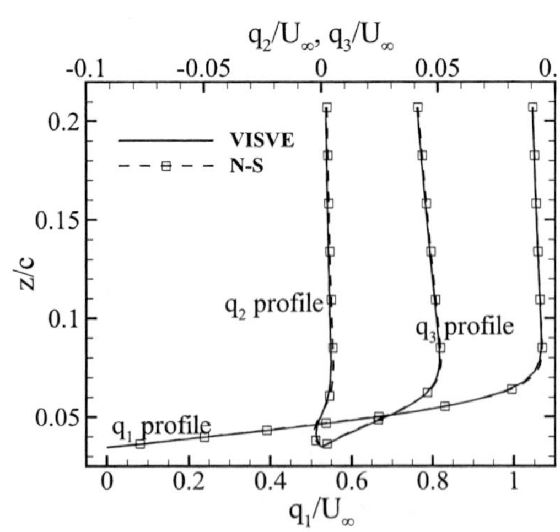

(b) 40% of chord length

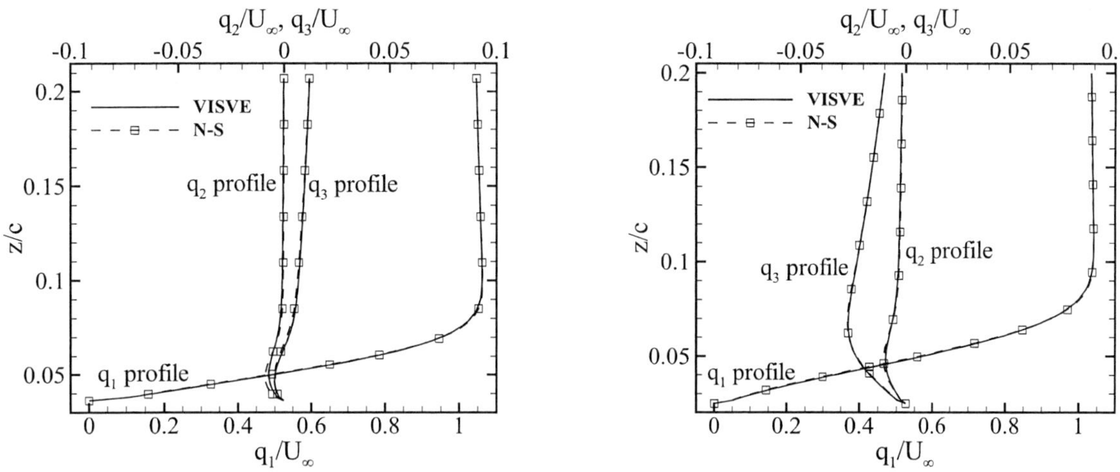

(c) 60% of chord length (d) 80% of chord length

Figure 6: Velocity profile comparison on the suction side at span section 10%, Re = 5000, AOA = 1°

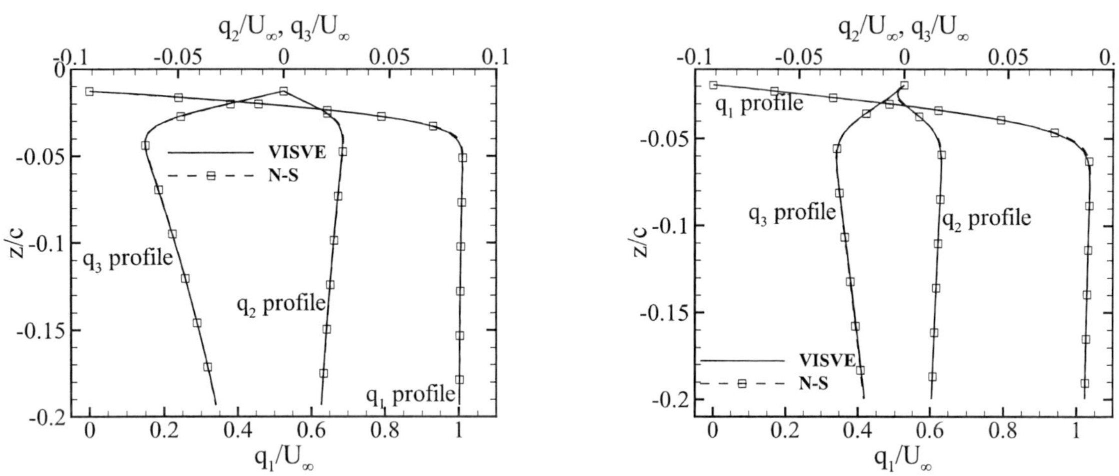

(a) 20% of chord length (b) 40% of chord length

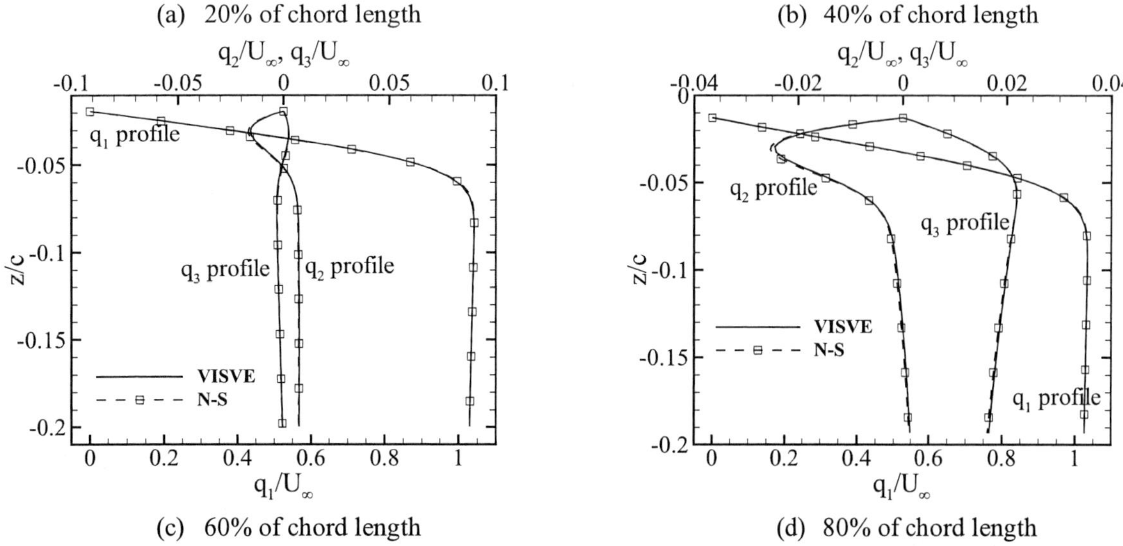

(c) 60% of chord length (d) 80% of chord length

Figure 7: Velocity profile comparison on the pressure side at span section 50%, Re = 5000, AOA = 1°

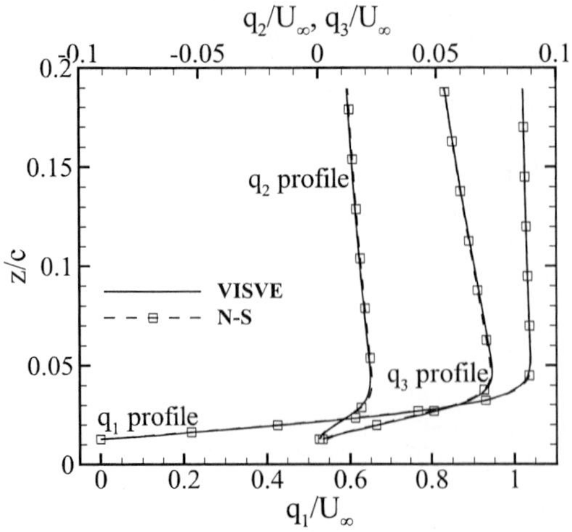

(a) 20% of chord length

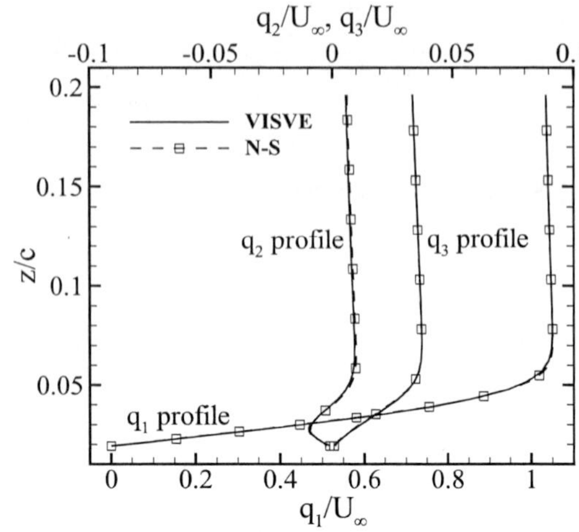

(b) 40% of chord length

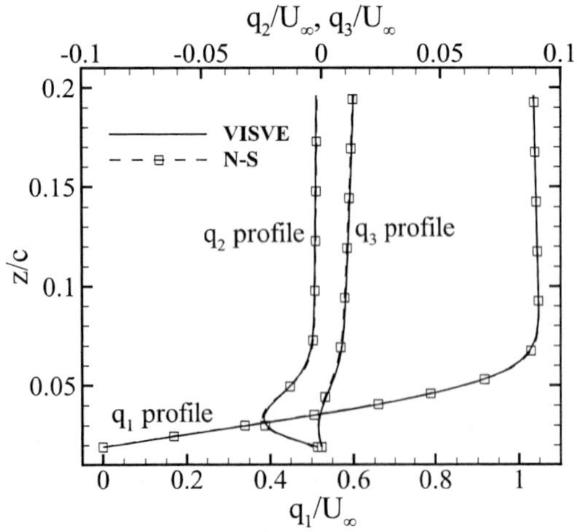

(c) 60% of chord length

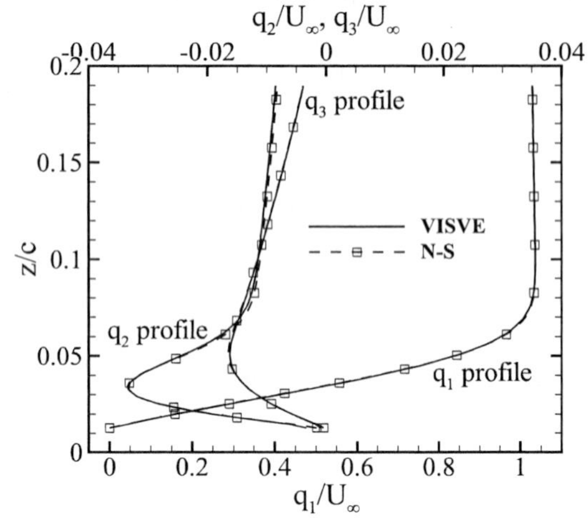

(d) 80% of chord length

Figure 8: Velocity profile comparison on the suction side at span section 50%, Re = 5000, AOA = 1°

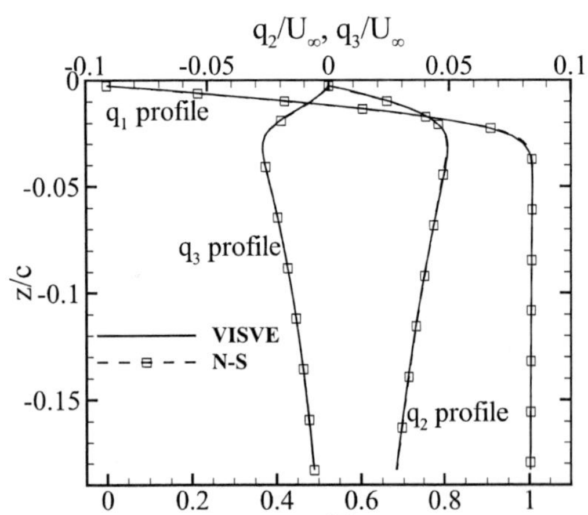

(a) 20% of chord length

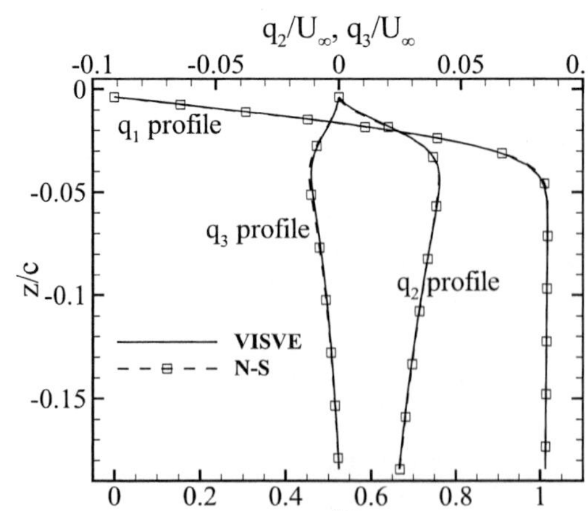

(b) 40% of chord length

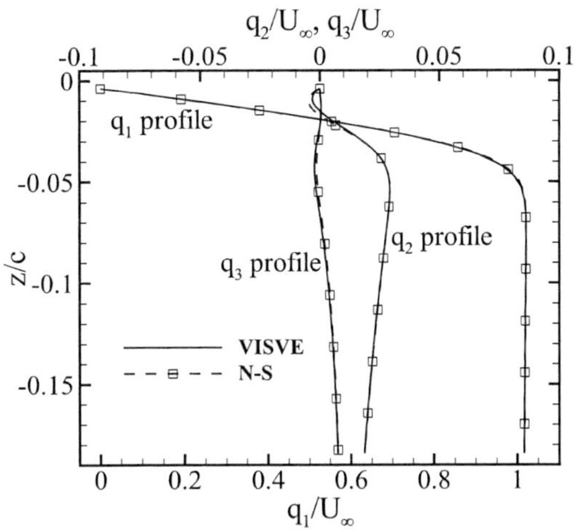

(c) 60% of chord length

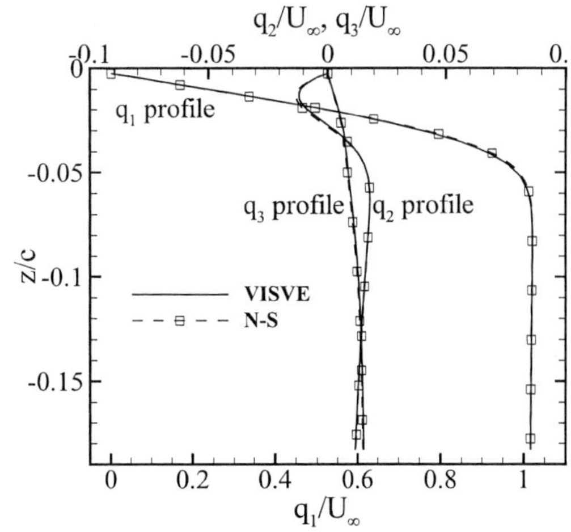

(d) 80% of chord length

Figure 9: Velocity profile comparison on the pressure side at span section 90%, Re = 5000, AOA = 1°

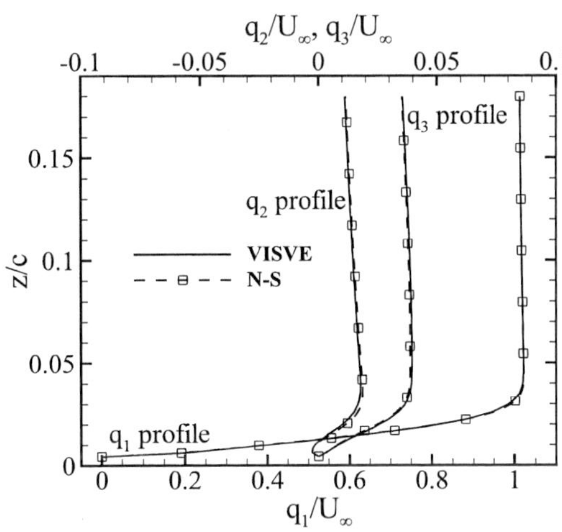

(a) 20% of chord length

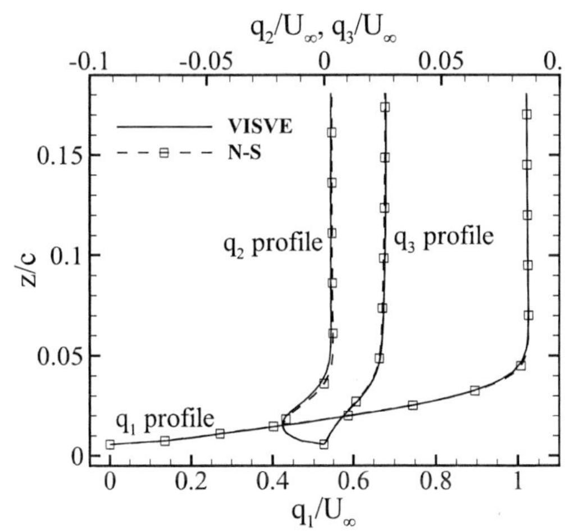

(b) 40% of chord length

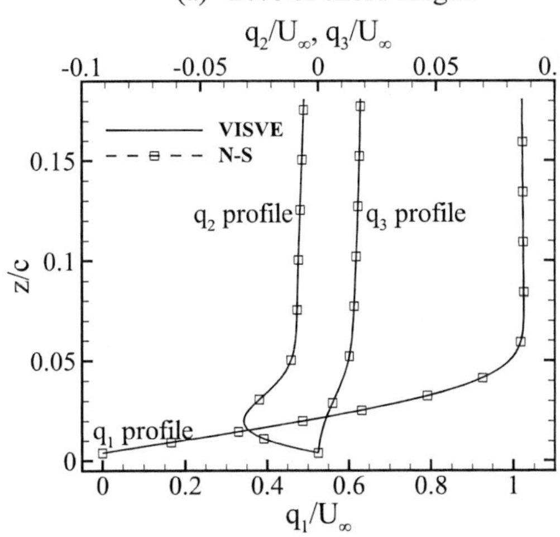

(c) 60% of chord length

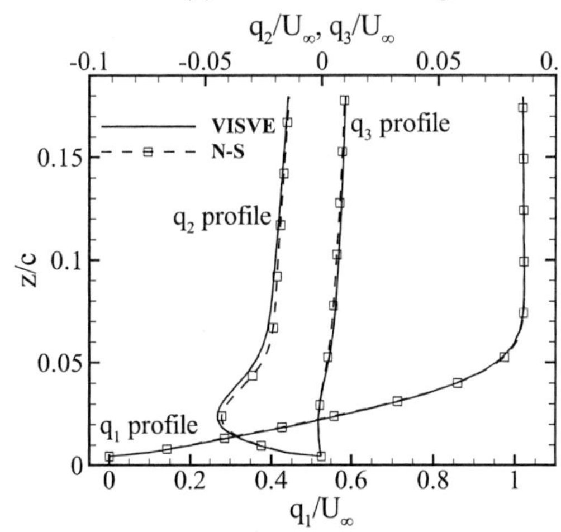

(d) 80% of chord length

Figure 10: Velocity profile comparison on the suction side at span section 90%, Re = 5000, AOA = 1°

The vorticity contours and streamlines of span-wise section 10%, 50% and 90% are shown in Figure 11, 12 and 13.

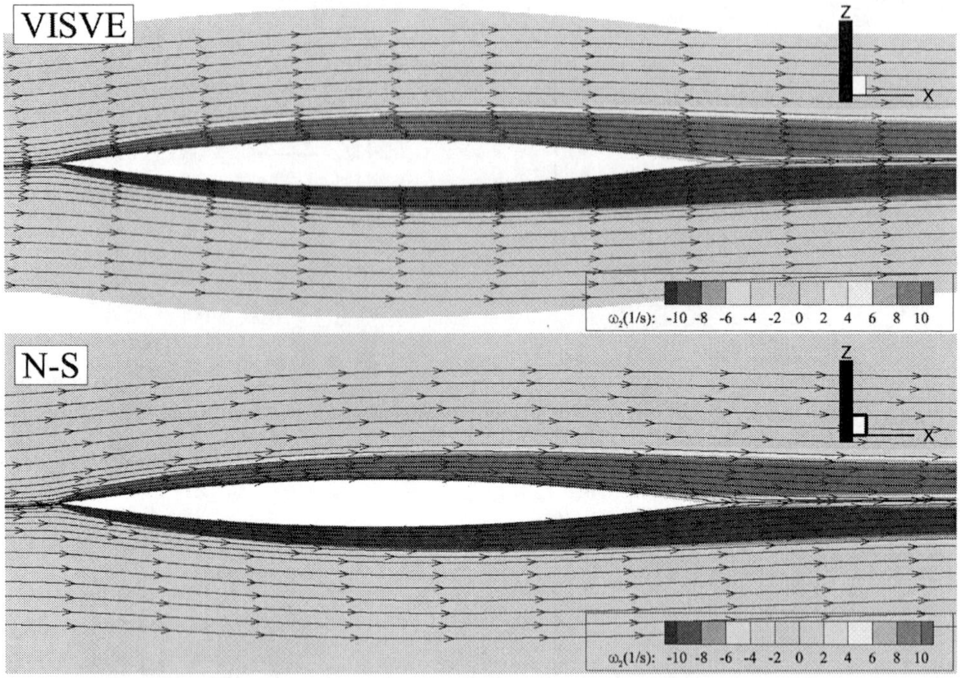

Figure 11: Comparison of vorticity contour and streamlines at span-wise section 10%, Re = 5000, AOA = 1°

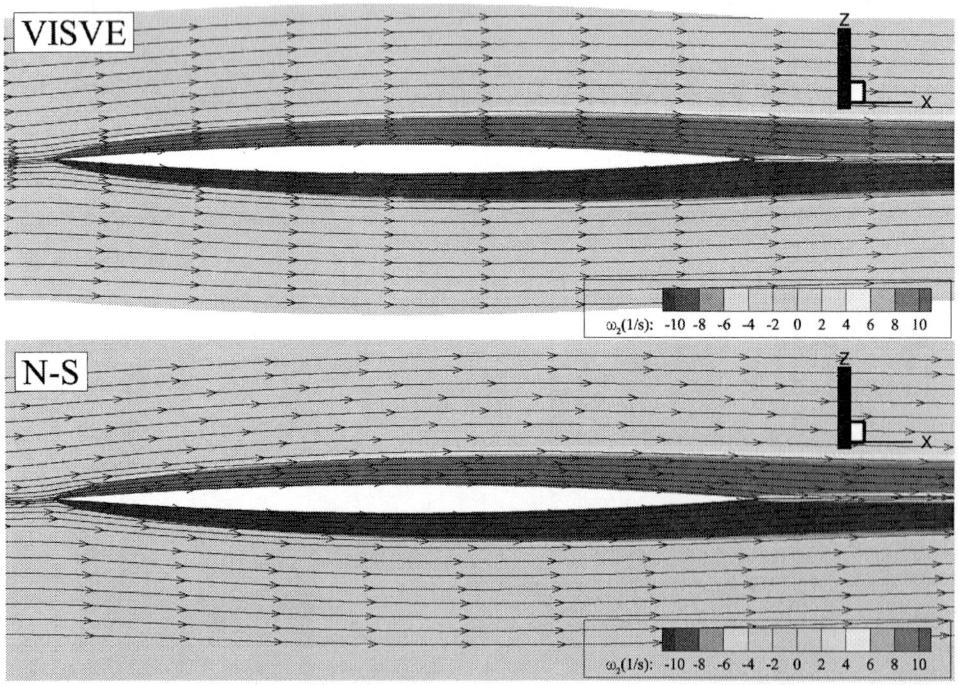

Figure 12: Comparison of vorticity contour and streamlines at span-wise section 50%, Re = 5000, AOA = 1°

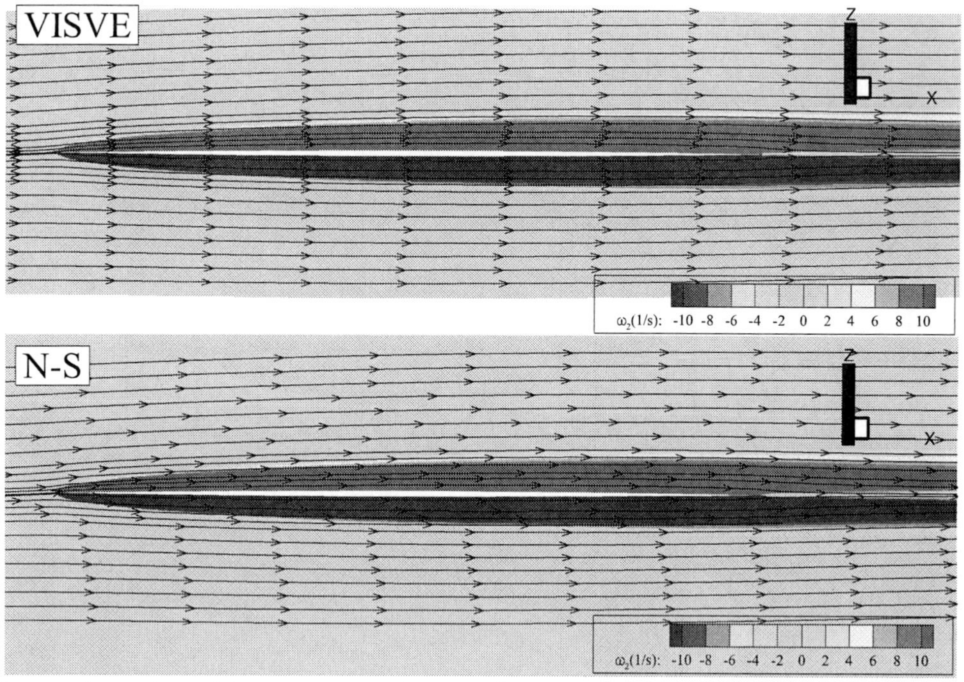

Figure 13: Comparison of vorticity contour and streamlines at span-wise section 90%, Re = 5000, AOA = 1°

From the comparisons of velocity, vorticity and streamlines between results from VISVE and results from N-S, good agreements could be observed, which validated the VISVE method.

3-Dimensional hydrofoil with NACA 4 digits sections

In this part, a 3-dimensional rectangular wing with NACA0008 sections and no camber is used to run the simulation. The maximum thickness is 8% at the span-wise location y=0. The thickness goes linearly to 0 towards the tip. The velocity field is compared between results of VISVE and those of N-S. While comparing the velocity field, the velocity profile along the line vertical to the wall surface is compared. The grid in VISVE is shown in Figure 14.

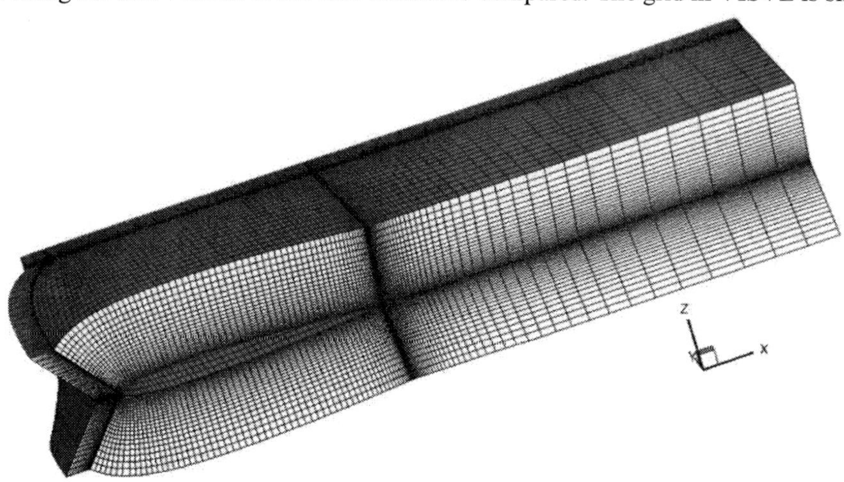

Figure 14: Mesh around the NACA wing in VISVE computational domain

Similarly, the velocity profiles from VISVE with that from N-S are compared in Figure 15 ~ Figure 20. The angle of attack of inflow equals 2 degrees. The Reynolds' number is 500.

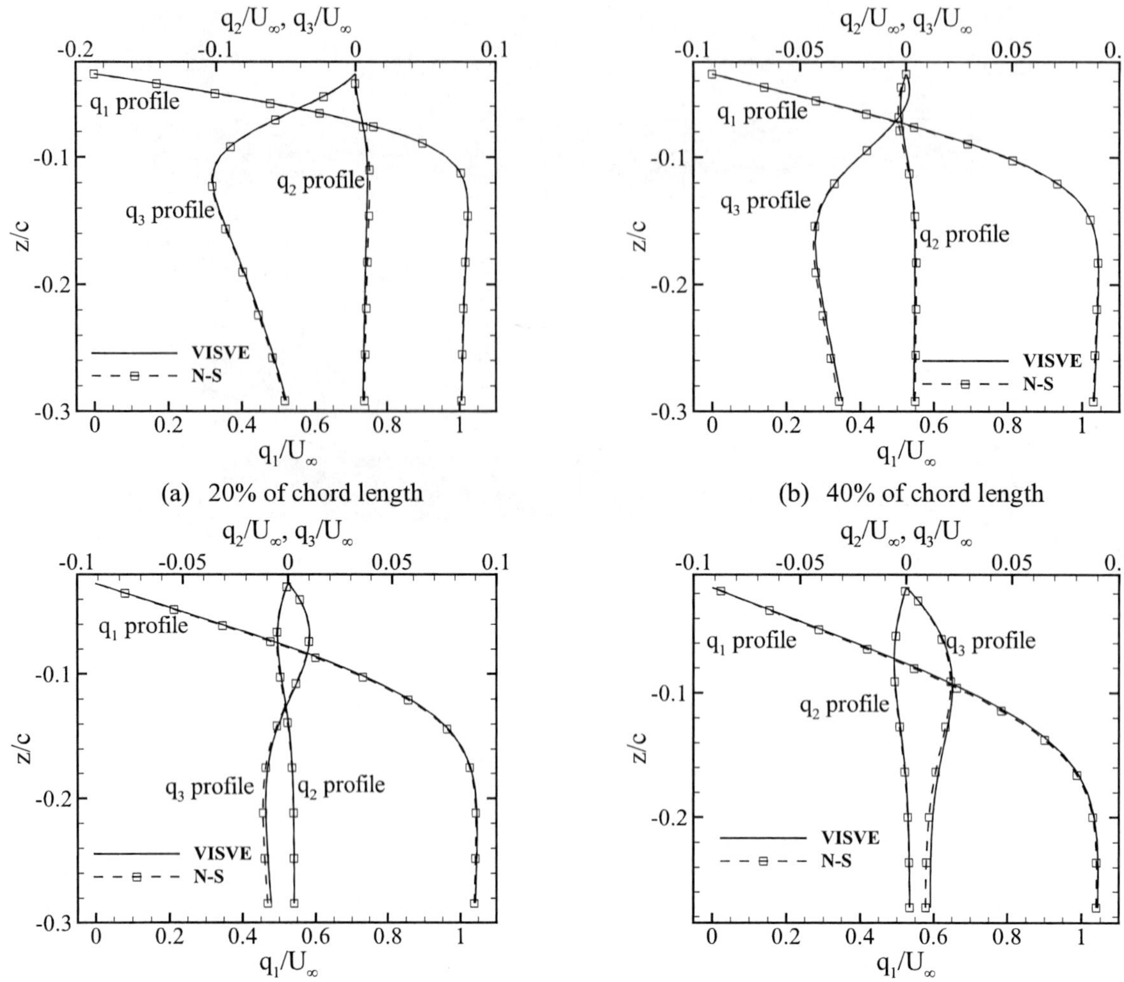

(a) 20% of chord length (b) 40% of chord length

(c) 60% of chord length (d) 80% of chord length

Figure 15: Velocity profile comparison on the pressure side at span section 10%, Re = 500, AOA = 2°

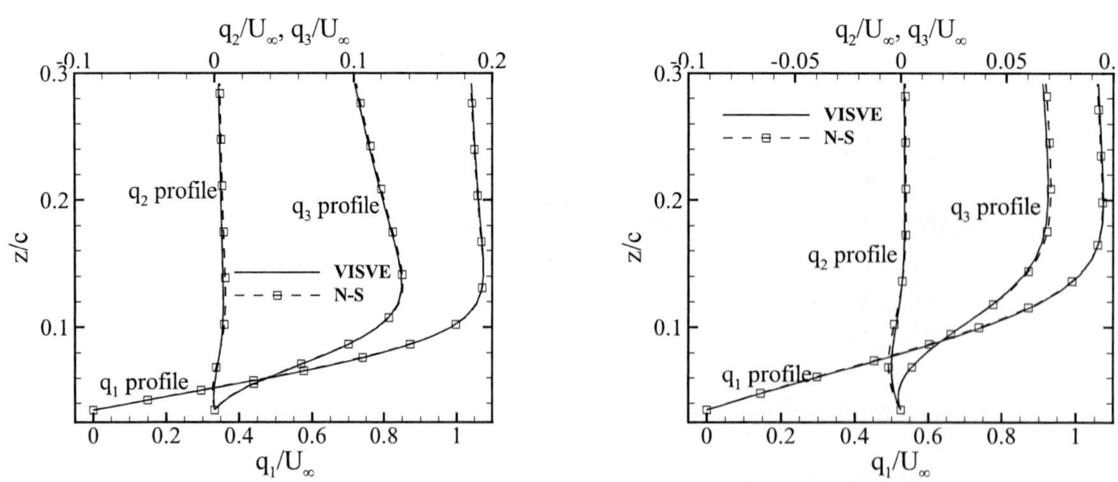

(a) 20% of chord length (b) 40% of chord length

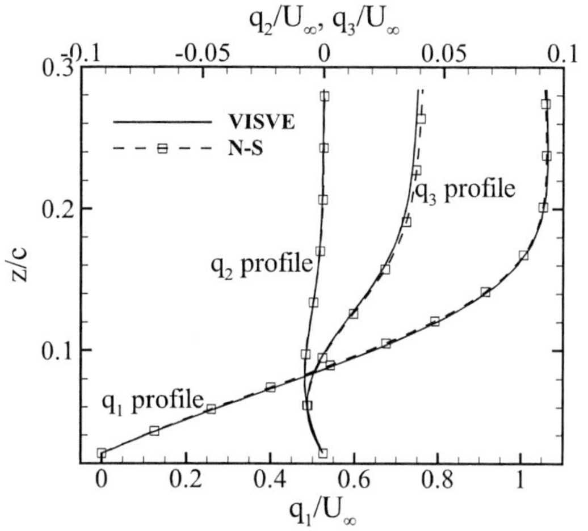

(c) 60% of chord length

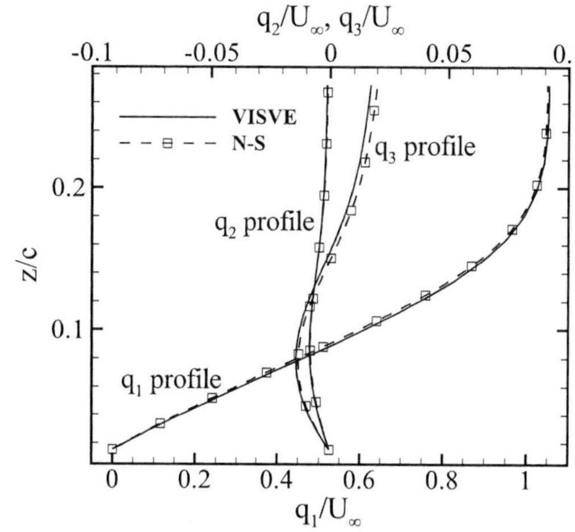

(d) 80% of chord length

Figure 16: Velocity profile comparison on the suction side at span section 10%, Re = 500, AOA = 2°

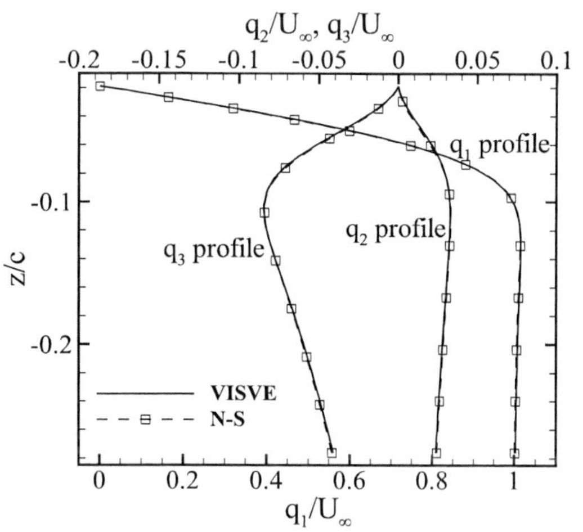

(a) 20% of chord length

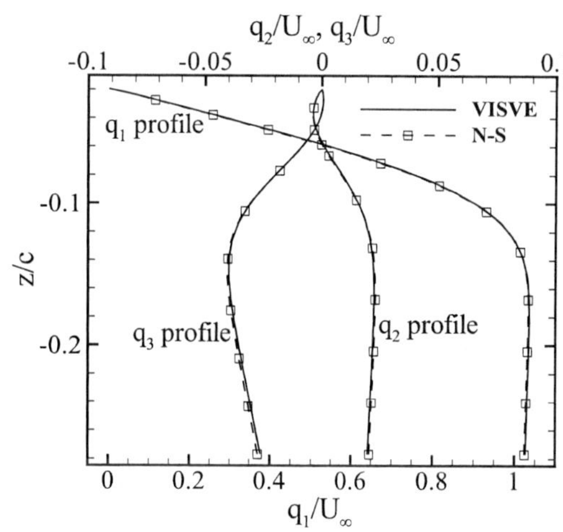

(b) 40% of chord length

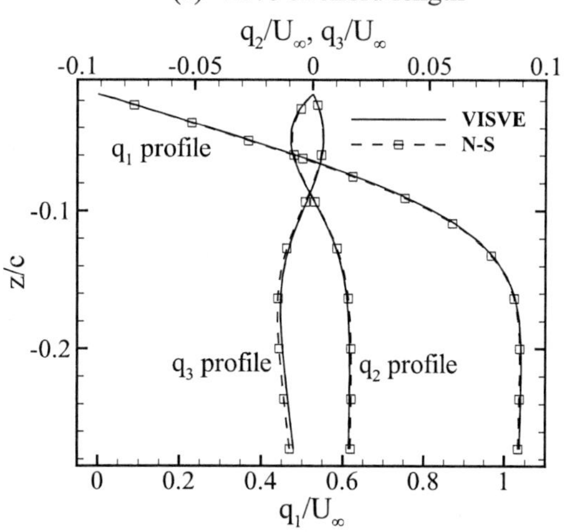

(c) 60% of chord length

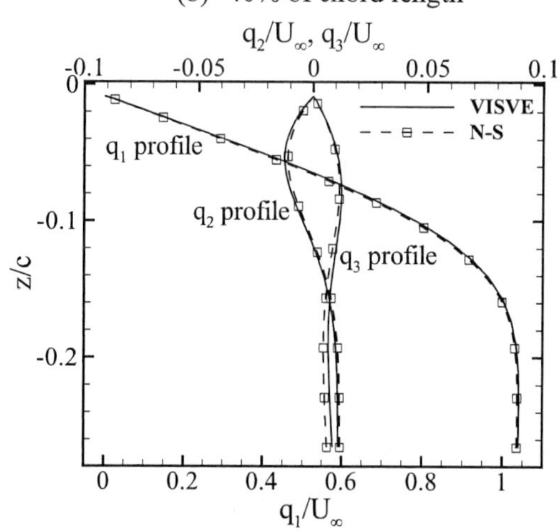

(d) 80% of chord length

Figure 17: Velocity profile comparison on the pressure side at span section 50%, Re = 500, AOA = 2°

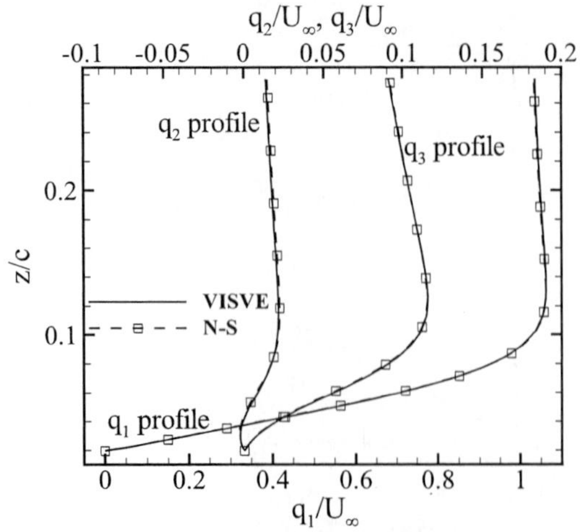

(a) 20% of chord length

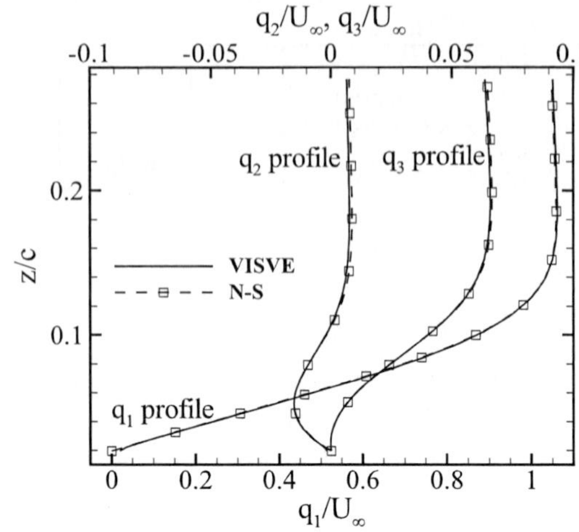

(b) 40% of chord length

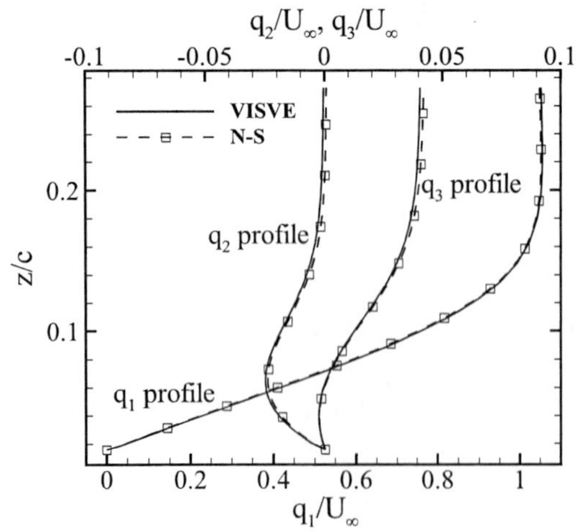

(c) 60% of chord length

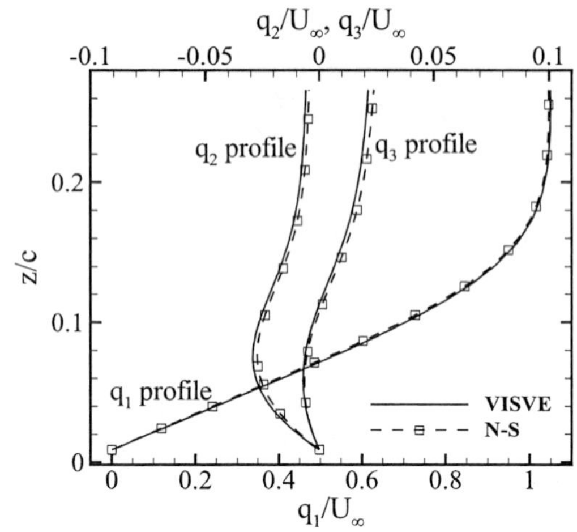

(d) 80% of chord length

Figure 18: Velocity profile comparison on the suction side at span section 50%, Re = 500, AOA = 2°

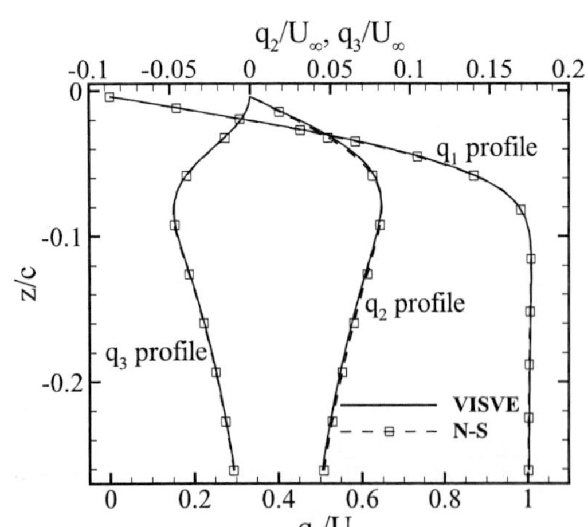

(a) 20% of chord length

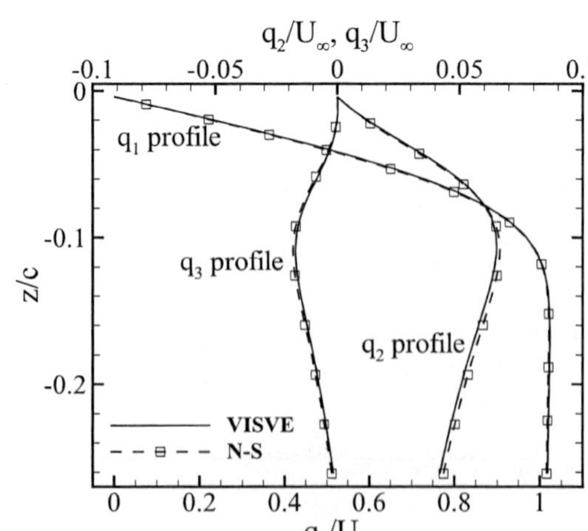

(b) 40% of chord length

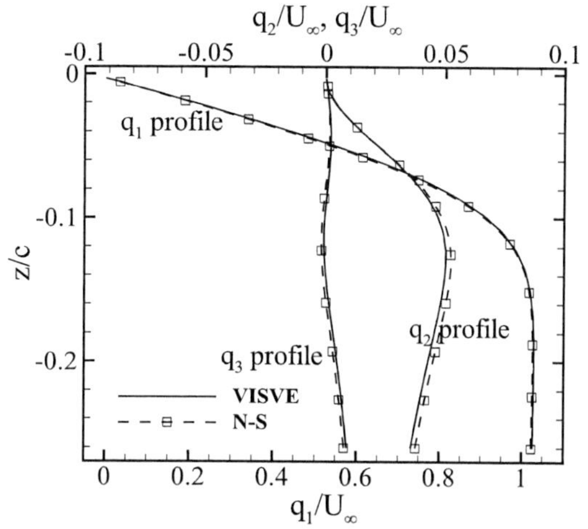

(c) 60% of chord length

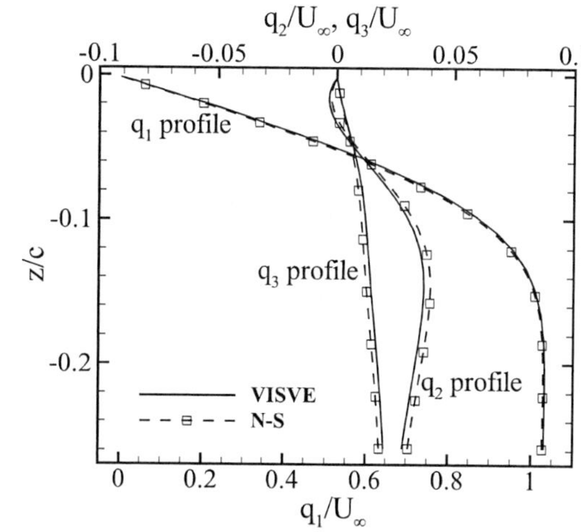

(d) 80% of chord length

Figure 19: Velocity profile comparison on the pressure side at span section 90%, Re = 500, AOA = 2°

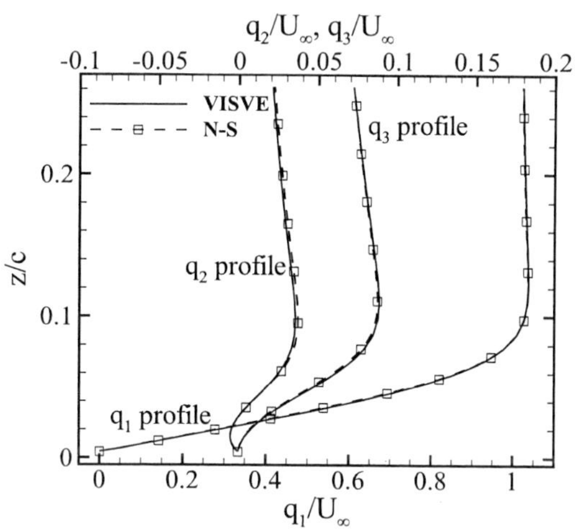

(a) 20% of chord length

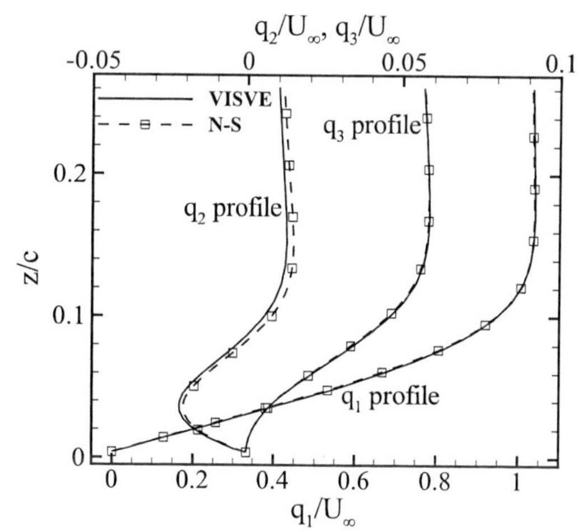

(b) 40% of chord length

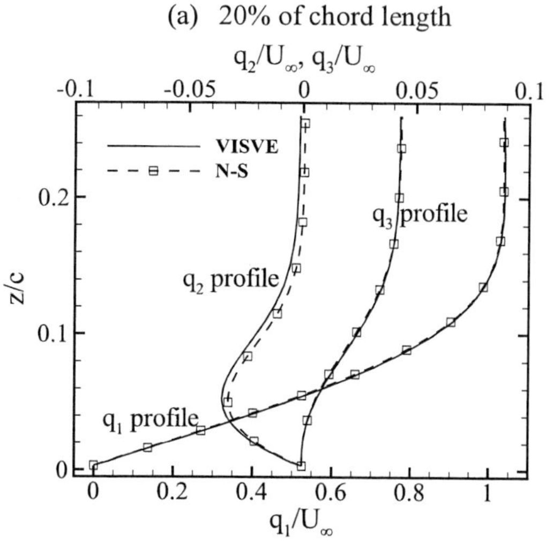

(c) 60% of chord length

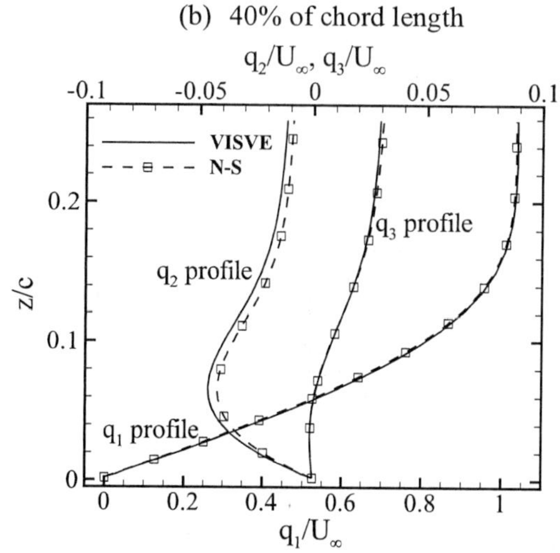

(d) 80% of chord length

Figure 20: Velocity profile comparison on the suction side at span section 90%, Re = 500, AOA = 2°

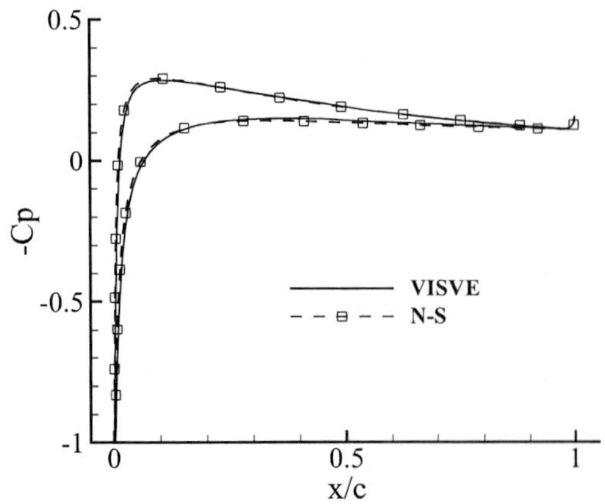

(a) span-wise section 8%

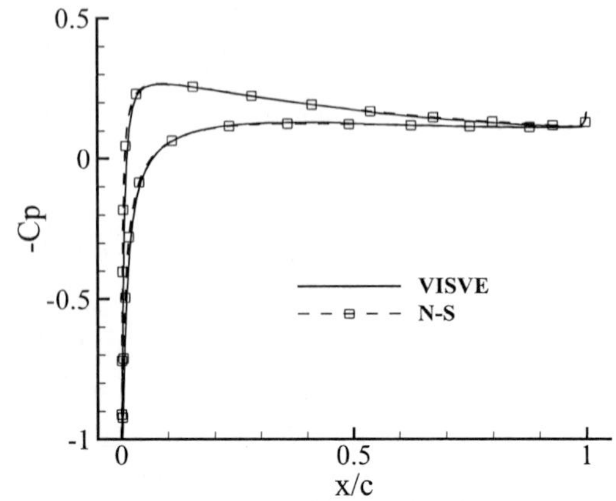

(b) span-wise section 30%

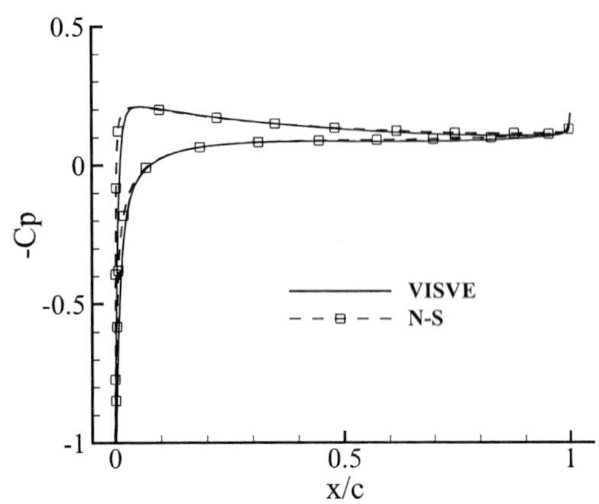

(c) span-wise section 70%

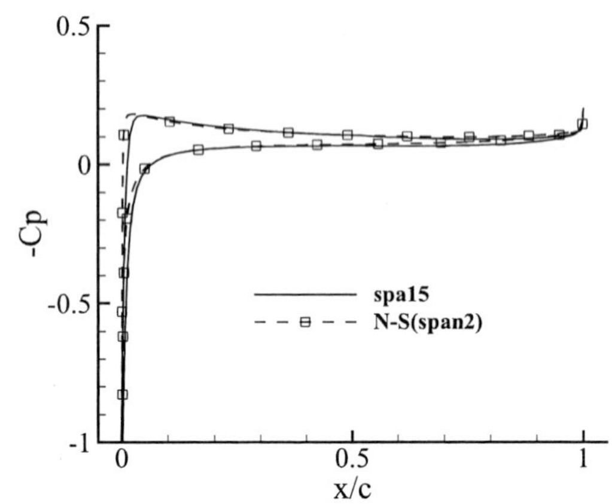

(d) span-wise section 90%

Figure 21: Comparisons of pressure coefficients at different span-wise sections, Re = 500, AOA = 2°

The wall surface pressure coefficients comparisons were shown in Figure 21. The velocity profiles and pressure coefficients distributions of results calculated by VISVE are very much in line with that of N-S. The readers may notice the decreasing accuracy toward the tip of the foil. This is probably caused by the loss of accuracy of BEM results since the thickness of the 3-D hydrofoil also decrease toward the tip. Especially near the tip where the thickness is almost 0, the BEM method will encounter relatively large difficulties. However, this does not affect the results of VISVE significantly.

Finally, the lift coefficient and drag coefficient were obtained. The coefficients were calculated by using Equation 14.

$$\mathrm{C} = \frac{F}{1/2\,\rho(\boldsymbol{U}_{\infty})^2 A} \tag{14}$$

Where, C represents the coefficient, F represents the corresponding force, ρ is the fluid density, $\boldsymbol{U}_{\infty}$ is the unperturbed inflow velocity, and A is the reference area, which equals the chord length times the span width. The comparison of lift and drag coefficients are shown in Table 1.

Table 1 Lift and drag coefficients comparison of the NACA rectangular wing

	Lift		Drag	
	Force (*N*)	Coefficient	Force (*N*)	Coefficient
VISVE	36.833263	7.366457E-2	77.077321	1.541539E-1
N-S	37.048925	7.409785E-2	76.606769	1.532135E-1

CONCLUSIONS

In this paper, a VIScous Vorticity Equation (VISVE) method is proposed to solve the problems of the incompressible laminar flow past 3-dimensional hydrofoils. The proposed method was implemented in steady attached flow past a 3-D hydrofoil in this study. Further validations of separated flow and unsteady flow are necessary. The Results show that this method is efficient and reliable.

Two hydrofoils are tested in this piece of work. Firstly, a hydrofoil with bi-convex sections is used to conduct the simulation. The maximum thickness of the hydrofoil is 8% of the chord length and the thickness goes linearly to 0 to the tip. It shows that the VISVE method is capable of handling the flow past a 3-dimensional hydrofoil with a sharp leading edge. In addition, a 3-dimensional hydrofoil with NACA0008 sections is also tested. Similarly, the section with the maximum thickness is at $y=0$ and the thickness goes linearly to 0 towards the tip. Since the hydrofoil has a blunt leading edge, the grid is not orthogonal any more. This case proves that the VISVE can resolve non-orthogonal grids properly.

The method is currently extended to hydrofoils of elliptical planform, as well as propeller blades, and the results are validated with those from RANS. In addition, the method is extended in order to include the effects of turbulence, via coupling with OpenFOAM.

ACKNOWLEDGEMENTS

Support for this research was provided by the U.S. Office of Naval Research (Grant Nos. N00014-14-1-0303 and N00014-18-1-2276; Dr. Ki-Han Kim) partly by Phases VII and VIII of the "Consortium on Cavitation Performance of High Speed Propulsors". The authors are grateful for the inspiring ideas provided by Dr. Ye Tian to complete this method, as well as the support. The authors also acknowledge the Texas Advanced Computing Center (TACC) at The University of Texas at Austin for providing HPC resources that have contributed to the research results reported within this paper.

REFERENCES

C. Wu, S.A. Kinnas, A VIScous Vorticity Equation (VISVE) method applied to 2-D and 3-D hydrofoil in both forward and backing conditions, 23rd SNAME Offshore Symposium, Houston, Texas, USA, February 14, 2018.

L. Xing, C. Wu, S.A. Kinnas, VISVE, a Vorticity Based Model Applied to Cavitating Flow around a 2-D Hydrofoil, 10th International Symposium on Cavitation, Baltimore, Maryland, USA, May 2018.

P.L. Woodfield, K. Suzuki, and K. Nakabe. "A Simple Strategy for Constructing Bounded Convection Schemes for Unstructured Grids." International Journal for Numerical Methods in Fluids, 46 (2004):1007-1024.

Y.J. Wu. "A Vorticity Based Model and its Application to Flow around an Impulsively Started Cylinder." Master's Thesis, Ocean Engineering Group, The University of Texas at Austin, 2016.

Y. Tian, S.A. Kinnas, A Viscous Vorticity Method for Propeller Tip Flows and Leading Edge Vortex, 4th International

Symposium on Marine Propulsors, Austin, Texas, USA, June 2015.

Z. Li, S.A. Kinnas, VISVE, a Vorticity Based Model Applied to Unidirectional and Alternating Flow around a Cylinder, 22nd SNAME Offshore Symposium, Houston, Texas, USA, February 2017.

SNAME
THE INTERNATIONAL COMMUNITY
FOR MARITIME AND OCEAN PROFESSIONALS

EXPERIMENTAL INVESTIGATION OF A DROPPED MODEL ROCKET IN THE TOWING TANK OF THE UNIVERSITY OF NEW ORLEANS

HALLIE COLLOPY, XIAOCHUAN YU*, YI LI, KEN HAGIMOTO, HAOZHAN MENG, SCOTT ARBEITER**

SCHOOL OF NAVAL ARCHITECT AND MARINE ENGINEERING, UNIVERSITY OF NEW ORLEANS

Proceedings of the 24th Offshore Symposium, February 20th 2019, Houston, Texas
Texas Section of the Society of Naval Architects and Marine Engineers

ABSTRACT

The Space-X company has introduced a reusable first stage rocket that lands on a floating barge. In the event that the first stage rocket falls off the platform, investigation of where it will land on the seafloor in case its sensors are waterlogged is necessary to increase efficiency in the salvage operation. This paper mainly investigates the landing position of a model rocket through a model experiment. The University of New Orleans Towing Tank facility is used to release a model first stage Falcon 9 rocket at initial drop angles from 0° to 90°. The landing location is recorded and compared with the results derived from the 2D module of numerical simulation tool "DROBS" that solves equations of motion for dropped cylinders. The focus of this paper is to experimentally determine the effects of initial conditions on the landing position of this rocket model.

Keywords:* *Dropped Model Rocket, Model Test, Landing Grid, Landing Position, DROBS

INTRODUCTION

Companies such as Airbus and United Launch Alliance are interested in developing methods to recover and reuse rocket engines. The Space-X company is interested in recovering and reusing the entire first stage rocket (Werner, 2015). This is exemplified by recent success of the Space-X company to reuse its first stage of the Falcon 9 rocket. After two failed attempts, the Falcon 9 successfully landed on a floating platform off the coast of Cape Carnival, Florida. Previous attempts to land the Falcon 9 first stage on a barge involved missing the platform entirely in one incident, and tipping over in the second incident. It is probable that in future attempts strong wind or weather induced sea states will cause the rocket to slide off the platform, even after a successful landing. Assuming that the Falcon 9 first stage has fallen off the platform intact, it must be located for recovery of valuable parts, materials, and fairings. Instead of inspecting visually an ever-expanding area in the dark sea floor to find the rocket, it is better to determine the landing location of the fallen rocket using nonlinear dynamics and a statistical data distribution. Since the Falcon 9 rocket seems more like a big cylinder, we intend to investigate how to locate a first stage Falcon 9 rocket using a model test and a numerical simulation tool that solves dynamics with slender body theory, by adopting the technology in the field of offshore engineering.

*- *Corresponding author (xyu5@uno.edu)*; ** - *Former graduate student*

There has been numerous research work on the trajectory and landing distribution of dropped objects in the field of offshore engineering. As one of the main causes of fatalities and serious injuries in the oil and gas industry, dropped objects may accidentally fall down from offshore platforms during lifting or any other offshore operation. In the risk assessment of pipeline protection, it is necessary to estimate the hit probability of freely falling objects on the pipelines on the seabed (DNV, 2010). It is also indicated in ABS's guideline that specialized techniques are needed to predict the trajectory of objects and the subsequent likelihood of striking additional structure and equipment as well as predicting the consequences of such impacts (ABS, 2010). Aanesland experimentally investigated the dynamics of freely falling drill pipes with model tests and a 2D computer program developed to predict the time series of the motion, velocity, and acceleration (Aanesland, 1987). It was found that the process of a rigid cylindrical body falling through water is a highly stochastic process. In fact, the trajectory and landing distribution are sensitive to small changes in longitudinal center of gravity relative to center of buoyancy, initial velocities, initial release angle, etc. More experimental tests about drill pipes have been further performed by Yasseri (2014) and Awotahegn (2015). In order to better simulate the trajectory of dropped cylindrical objects, Xiang et al. (2017a, 2017b, 2016) proposed a modified 3D theory to consider the effect of axial rotation on dropped cylindrical objects with uniform mass distribution by modifying the maneuvering equations for a slender rigid body. A numerical simulation tool called Dropped Objects Simulator (DROBS) has been successfully developed based on this 3D theory to investigate various factors that may affect the trajectories, including drop angle, normal drag coefficient, binormal drag coefficient and rolling frequency. Simulated trajectories have been validated with Aanesland's data from model. Further, Xiang et al. (2019) focuses on considering the influence of water environment (currents and waves) on the dropped cylindrical objects after water entry by modifying the three-dimensional (3D) theory and integrating the effects of currents and waves into DROBS.

The Office of Naval Research (ONR) also supported similar work about the trajectory of mines. Holland observed that the trajectory reaches a stable or quasi stable motion pattern that can be categorized by as spiral, seesaw, straight, straight slant, nose turn, tumbling or combined (Holland, 2001). This steady state behavior is present even in a chaotic sea state environment. Abelev et al. (2003) performed field tests of dropped cylinders in random wind, wave, and current conditions that reached a periodic motion pattern and concluded that precise knowledge of initial conditions may lead to accurate prediction of terminal conditions. Later, freely falling mines were simulated by expanding dynamics of slender body theory to 3D to consider translational and rotational motion in the out of plane direction (Chu and Fan 2006). The 3D simulation was verified experimentally (Chu et al., 2005).

The primary purpose of this study is to experimentally investigate the landing distribution of a first stage rocket free falling through water. Previous experiments on the dynamics of slender body theory have investigated scaled drill pipes and mines. However, a rocket has a much lighter density than these. The lighter density causes different motions than a comparably denser slender body which must be investigated experimentally. The model test is designed to understand the effect of density, water entry angle, and initial velocities on the terminal landing location. The 2D module of an in-house simulation tool "DROBS" is used to compare to the experimental results. The findings are that the experimental data agrees with the numerical simulation tool based on the model's cylindrical properties and initial conditions, and that drop angle greatly influences the landing location.

MODEL TEST

A single model was released entering the water at low initial velocities and at an initial angle. The drop was conducted ten times for angles of $0°, 15°, 30°, 45°, 60°, 75°$ and $90°$ measured from the vertical axis. The experimental design was inspired by similar experiments performed by Aanesland (1987), Gilles (2001), Yasseri (2014) and Awotahegn (2015). The drop apparatus was designed to replicate the device used in the dropped mine experiment by Gilles (2001). A grid to measure landing locations on the tank floor was similar to experimental drops of cylindrical shaped objects in Yasseri (2014).

The model for the experiment was a scaled Falcon 9 v1.2 first stage. The mass of the empty full scale first stage would not sink. The empty fuel tanks are assumed to fill up with just enough salt water to begin to sink. The scaled mass of the model is based on this assumption. The model itself is not free flooding so a constant mass can be analyzed. Froude depth scaling was used and the length to diameter ratio was maintained. The model was constructed with PVC pipe with weights on a threaded rod. The ends were plugged and tightened with a nut to prevent flooding of the interior. The weights were adjusted so that the center of gravity aligned with the center of volume. The appendages representing foldable landing legs were impractical to include at the model size. Initial model iterations were based on the dimensions of fins from preassembled model rocket kits. The first attempt was made with laser cut fins, however, they were prone to breaking. The second attempt was made with 3D printed fins, however, they were permeable. In order to simplify the model, the landing legs were neglected because the full scale legs of the Falcon 9 v1.2 are retractable and are assumed to be in the folded in position. The target scale was 1:75.839. The target scale was chosen based on the available PVC diameter size available. The actual full scale Falcon 9 v1.2 is empty of fuel by the time it lands on the barge. The empty mass is 411 tons. Assuming that it fills with saltwater until it begins to sink, the full scale weight needed to model the physical process is 439 tons. The full scale, target scale, achieved scale model properties and calculated differences are shown in Table 1. Figure 1 shows the clear laser cut appendage prototype, the 3D printed appendage prototype and the final watertight model with no leg appendages.

Table 1: Falcon 9 v1.2 full scale, target model, and achieved model properties

Parameters	Unit	Full Scale	Target Model Scale 1:75.839	Achieved Model Scale	% Difference
Length	'meter'	40.7	0.5366	0.5112	4.74
Mass	'kilogram'	439	1.0072	0.997	1.01
Diameter	'Meter'	3.66	0.04826	0.04826	0

Figure 1: Laser cut leg appendage prototype (clear), the orange 3D printed leg appendage prototype and final watertight model with no leg appendages (from left to right)

The landing grid was placed on the bottom of the tank floor to measure the horizontal excursion. A 3.6 m by 3.6 m grid was constructed of mesh fabric attached to steel pipe frame with string. The squares on the grid are 10 cm by 10 cm. The origin of the grid is marked at the center for reference. The construction and placement of the grid on the tank floor is shown in Figure 2. The experimental set up is shown in Figure 3. The experiment was conducted at the University of New Orleans Towing Tank Facilities. The tank is filled with freshwater to a water depth of 180 cm. The temperature was measured to be 18° C with a corresponding viscosity of 1.0542 $10^{-6}m^2/s$ and density of 998.5986 kg/m^3(ITTC, 2011). The tank is 30.6m long and 4.6 m wide. A platform was placed across the width of the tank .The height of the platform is 0.9 m above the water. Drops were conducted from the platform with a drop apparatus to control the initial angle of release. A digital level was used to determine the initial angle with an accuracy of 0.1

degree. The drop point position above the grid was determined with a plumb bob. The model was dropped ten times for each angle 0°, 15°,30°,45°,60°,75° and 90° as measured from the vertical, which is marked as θ in Figure 3. The final landing position on the grid was recorded manually. Cameras placed outside the tank windows were used to record the trajectory visual and water entry angle.

(a)

(b)

Figure 2: (a) Construction of landing grid (left); (b) Grid on tank floor (right)

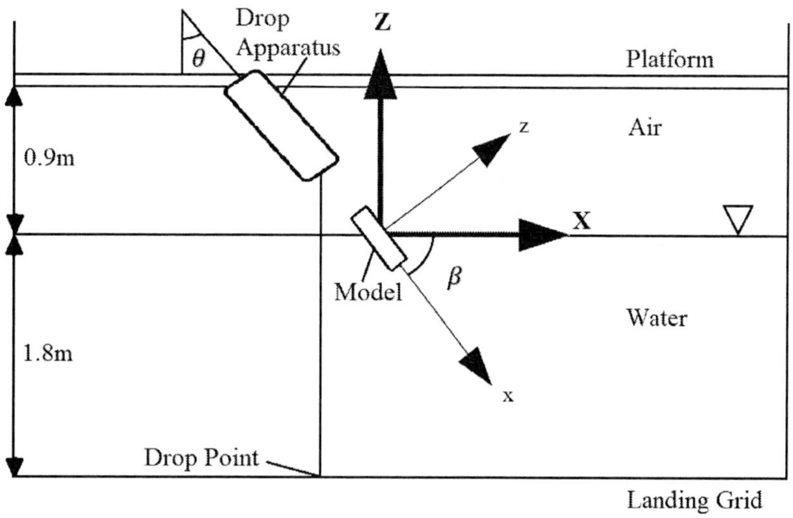

Figure 3: Elevation view of experimental setup

THEORETICAL FORMULAS

The experimental data from model tests will be compared with the numerical tool calculated by DROBS. In order to initially make a simple comparison between numerical and experimental results, it is assumed that there is no rolling motion at initial stage. Therefore, the 2D module of DROBS can be utilized in this study. The equations of motion are

$$(m - \rho\nabla)g\sin(\beta) + F_{dx} = m\,\dot{U}_1 \tag{1}$$

$$-(m - \rho\nabla)g\,cos(\beta) + F_{dz} = \{U_1 m_t U_3 - U_1(x_t m_t)\Omega_2\} + m_{33}\dot{U}_3\} + m(\dot{U}_3 - U_1\Omega_2) \tag{2}$$

$$M_{dy} = \{-U_1(m_{33} + x_t m_t)U_3 + U_1\,x_t^2 m_t \Omega_2\} + m_{55}\dot{\Omega}_2 + M_{55}\dot{\Omega}_2 \tag{3}$$

where, the velocity components are U_1 (surge), U_3 (heave), and Ω_2 (pitch); m is the mass; ρ is the water density; ∇ is the displaced volume of cylinder; g is the gravitational acceleration; m_{55} and m_{33} are the added mass in the pitch and heave motions, respectively; M_{55} is the pitch moment of inertia; x_t represents the longitudinal position of effect trailing edge (Newman, 1977); m_t is the added mass coefficient at the trailing edge. In addition, viscous forces and moment, F_{dx}, F_{dz} and M_{dy}, are calculated by

$$F_{dx} = -0.664\pi\sqrt{\nu\rho^2 L}\,U_1\sqrt{|U_1|} - \frac{1}{8}\rho\pi C_{dx} D^2 U_1 |U_1| \tag{4}$$

$$F_{dz} = \frac{1}{2}\int_{-0.5L}^{0.5L} \rho C_{dz} D U_z(x)|U_z(x)|dx \tag{5}$$

$$M_{dy} = -\frac{1}{2}\int_{-0.5L}^{0.5L} \rho C_{dz} D x U_z(x)|U_z(x)|dx \tag{6}$$

where, $U_z(x)$ is the local relative velocity between water and cylinder; ν is the kinematic viscosity of the water; C_{dz} is the drag coefficient in the z- direction; C_{dx} is the drag coefficient in the x- direction; L is the length of cylinder; and D is the diameter of the cylinder. The velocity components are solved at each time step. The motions in the local body coordinate system are transformed to the global coordinates at each time step. The velocity and time values give a trajectory path and predicted landing distribution for various angles, mass properties, and initial velocities. For more discussions about the 2D theory, please see Aanesland (1987) and Xiang et al. (2017b).

RESULTS AND DISCUSSIONS

The drop location and angle were recorded for a total of 70 drops (10 at each angle). The model is released from one end of the drop apparatus. The other end where the model exits is measured as the "drop point". The platform above the tank, the travel of the model in the water and the landing location scenario is demonstrated in Figure 4. The experimental results were recorded and converted from Cartesian coordinates to Polar coordinates. The experimental polar excursion from the drop point was averaged for ten drops at each angle. Because the drop point above the water was 90 cm, the test article traveled horizontally and gained velocity before entering the water. The initial velocities and excursion from the drop point to the water were calculated using simple kinetic equations. The velocities are important initial conditions to input into DROBS. The air excursion away from the drop point is simply added to the simulation results to obtain the theoretical prediction for landing distribution

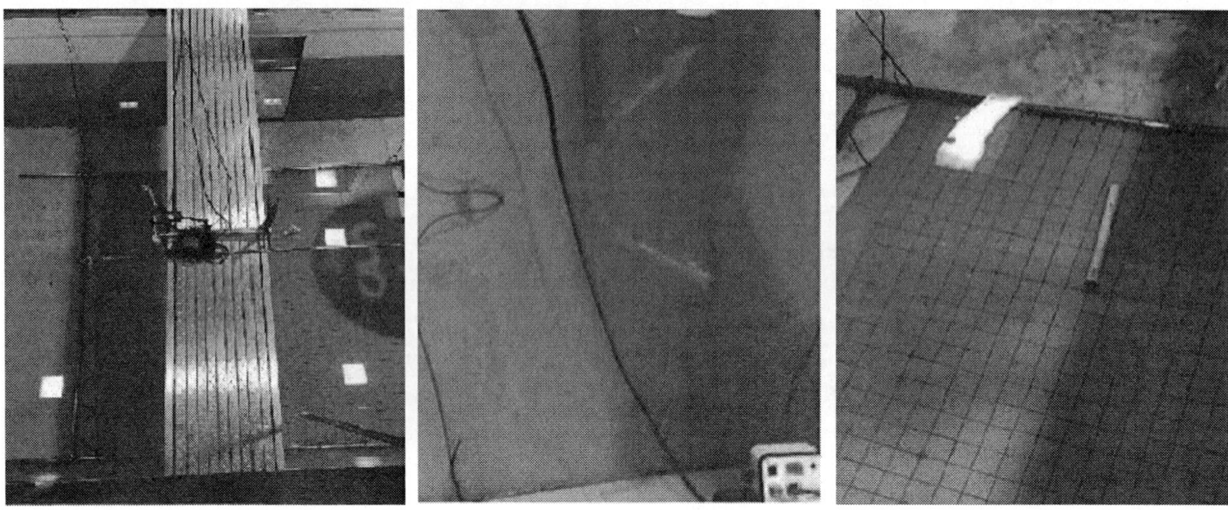

(a) (b) (c)

Figure 4: Experiment scenario: from left to right (a) platform above water (b) recorded path (c) recorded landing points.

DROBS was run for each angle with inputs of the model rocket's mass properties including length, mass, and diameter. The initial velocities can be obtained by directly solving the kinematic equation. The simulation was adjusted for viscosity and density of freshwater at the temperature of 18°C.

For the 0° drop angle case, the predicted mean radius of landing distribution from DROBS is 0 cm, as shown by the blue dot at the origin in Figure 5. For the 90° drop angle case, the predicted mean radius is 0 cm with a straight trajectory as well. In fact, the simulated 90° and 0° cases predict that the cylinder will follow a straight path ideally in the numerical simulation, because there is no any other disturbance considered in the calculation in the 2D module of DROBS. Experimentally, however, this is not the case. There must be various types of disturbances in the experimental tests of model rocket. Therefore, you can see the scattering distribution of rocket model on the tank floor for the 0° drop angle case and 90° drop angle case, as indicated by Figure 5. In a similar way, for the 30° drop angle case, the predicted mean radius of landing distribution from DROBS is 186.4 cm from the point of water surface entry, as shown in Table 2. It is added to the horizontal travel in the air to yield a theoretical prediction of 204.4 cm. It should be noted that the friction in the drop apparatus is neglected, and the angle of release is assumed to be maintained until water entry in the calculation. Since this paper mainly introduces the experimental study of dropped model rocket, the trajectories form DROBS are not shown here. It focuses on the landing distribution of the dropped rocket model.

Table 2 shows the theoretical excursion values for each drop angle scenario. The 60° case has the furthest displacement for the experimental landing distribution. This is consistent with theoretical predicted landing distribution from DROBS. Figures 5-8 show the comparison of landing distribution of experimental results to the theoretical prediction for the drop angle cases.

Table 2: Theoretical predicted landing distribution (mean radius) on tank floor

Drop Angle (deg)	Travel in Air (cm)	Travel in Water from DROBS (cm)	Total (cm)
0	0	0	0
15	7.4	65.0	72.4
30	18.0	186.4	204.4
45	33.7	178.8	212.5
60	52.0	244.4	296.4
75	43.7	130.0	173.7
90	0	0	0

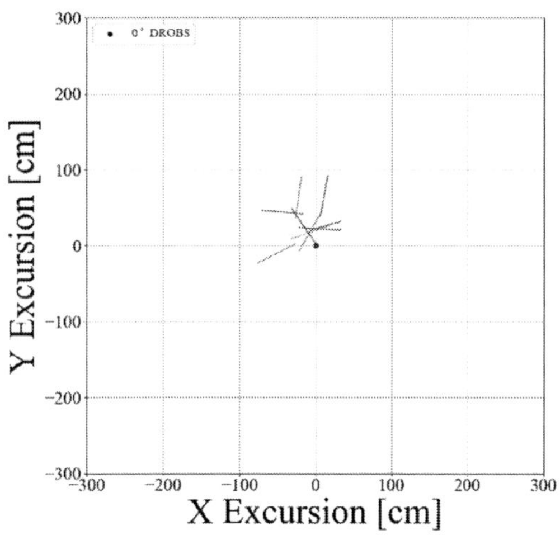

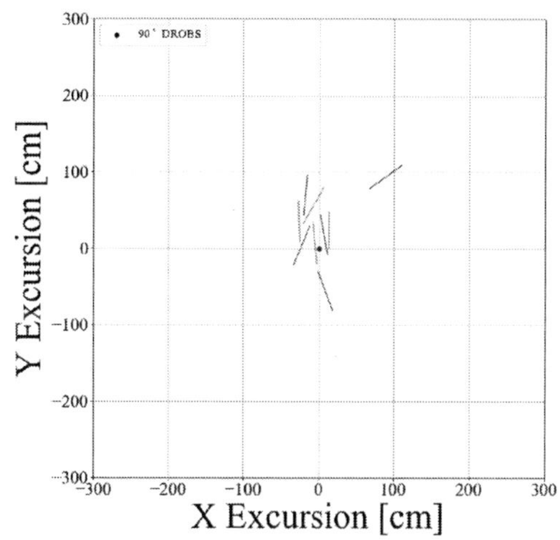

Figure 5: Comparison of experimental results with the mean radius calculated from DROBS – drop angle 0° case (left) and drop angle 90° case (right)

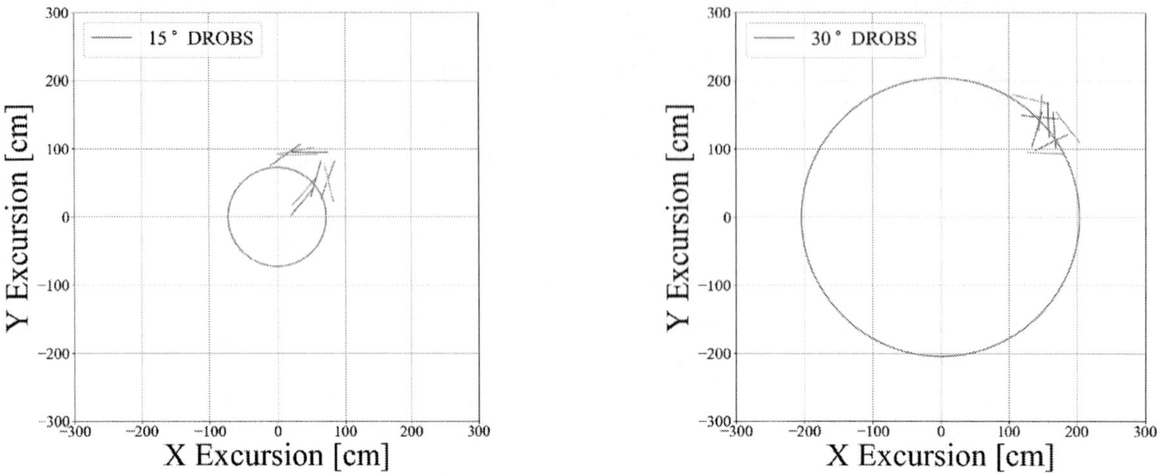

Figure 6: Comparison of experimental results with the mean radius calculated from DROBS – drop angle 15° case (left) and drop angle 30° case (right)

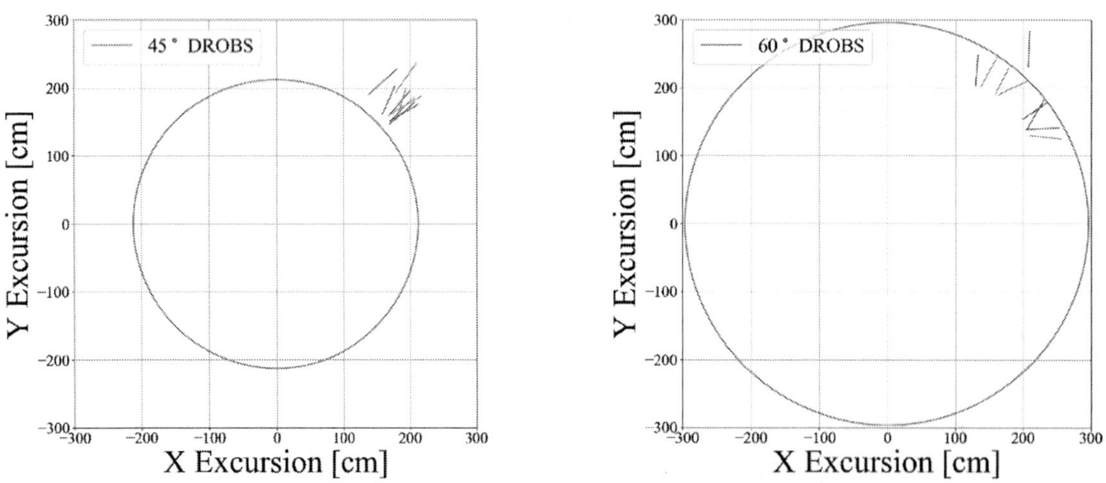

Figure 7: Comparison of experimental results with the mean radius calculated from DROBS – drop angle 45° case (left) and drop angle 60° case (right)

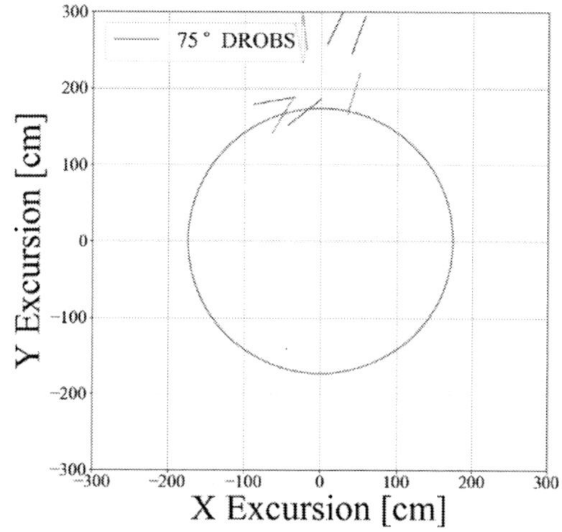

Figure 8: Comparison of experimental results with the mean radius calculated from DROBS – drop angle 75° case

Further, the theoretical predicted total displacement is compared to a scatter plot of experimental results for each angle. The experimental drop locations are measured from the drop point and compared with the total theoretical predicted landing distribution on the tank floor from Table 3.

The percent difference between the mean experimental excursion and total theoretical excursion is calculated as the difference between experimental and theoretical divided by experimental. The difference is shown in Table 3. For both theoretical and average experimental landing distribution the 60° case had the largest excursion. DROBS overpredicted the excursion hence the negative percent difference. The largest difference was for the 75° case. A possible reason for this is that the angle may have changed more than the other cases during water entry.

Table 3: Comparison of Theoretical to Mean Experimental Landing Distribution

Drop Angle (deg)	Theoretical (cm)	Mean Experimental (cm)	Percent Difference
0	0	39.0	N/A
15	72.4	84.2	14.0%
30	204.4	207.7	1.6%
45	212.5	258.9	17.9%
60	296.4	275.1	-7.7%
75	173.7	235.6	26.3%
90	0	48.0	N/A

The percent difference is not calculated for the vertical and horizontal case. There exists strong directional stability which is indicated by the trend of experimental landing points biased in the direction of initial velocity. Possible sources of experimental error may result from human error in releasing the model and observing landing locations to a resolution of the size of the grid. The test plugs on the ends of the model may have caused an additional appendage drag. The simple kinetic equations of initial velocity are not detailed enough to consider air drag or the impact of the model on the water surface. However, DROBS is capable of providing a good estimate of the landing points of the model Falcon 9 first stage rocket with inputs of initial velocity and initial angle.

Rigid bodies freely falling through water is indeed a highly chaotic process sensitive to very small perturbations. This is evident as seen in the experiment, where no drop trial landed in exactly the same place even with the same angle. Computational fluid dynamics may be more accurate. However, it is computationally demanding since the problem requires a large set of analysis to cover all statistical parameters. The simple method approach in DROBS may be better if used with Monte Carlo method to further predict a range of drop locations based on variation in these initial conditions.

CONCLUSIONS AND FUTURE WORK

In this paper, the experimental drop tests of a rocket model – first stage Falcon 9 rocket at the towing tank of University of New Orleans has been introduced. In order to simplify the model, the landing legs are neglected because they were small relative to the length. The drop location, angle, and water depth were recorded for a total of 70 drops (10 at each angle). The model is released from one end of the drop apparatus. The other end where the model exits is measured as the "drop point". The platform above the tank, the travel of the model in the water and the landing location scenario is demonstrated in Figure 4. The experimental results were recorded and converted from Cartesian coordinates to Polar coordinates.

Overall, the landing position matches well with the numerical results derived by the 2D module of simulation tool "DROBS". It is validated again that DROBS is a reliable numerical tool which can be utilized into the estimation of trajectory and landing points for cylinders.

Recommendations for future work would be to use 3D module of DROBS and conduct more model tests. The numerical simulation could be also improved by considering water impacts on the cylinder, air entrapment and cavitation, etc.

ACKNOWLEDGEMENTS

The authors wish to express our appreciation to Louisiana Space Grant Consortium for their generous support of this study. The authors also thank Ryan Thiel and George Morrisey for their assistance with the model test.

REFERENCES

Aanesland, V(1987). "Numerical and Experimental Investigation of Accidentally Falling Drilling Pipes," Proceedings of the19th Annual Offshore Technology Conference, Houston, Texas, USA, pp.117-125.

Abelev, AV, Valent, PJ, Plant.NG and Holland, KT (2003). "Evaluation and quantification of randomness in free-fall trajectories of instrumented cylinders," Celebrating the Past ... Teaming Toward the Future (IEEE Cat. No.03CH37492), San Diego, CA, USA, Vol.(4), pp. 2355-2365.

ABS (2010). Guidance Notes on Accidental Load Analysis and Design for Offshore Structures, American Bureau of Shipping.

Awotahegn, MB (2015). Experimental Investigation of Accidental Drops of Drill Pipes and Containers, Master's Thesis, University of Stavanger, Stavanger, Norway

Chu, PC and Fan, CW (2006). "Prediction of Falling Cylinder through Air-water-sediment Columns," ASME Journal of Applied Mechanics Review, 73, 300–314.

Chu, PC, Gilles, A and Fan, CW (2005). "Experiment of Falling Cylinder through the Water Column," Experimental Thermal and Fluid Science, 29, pp. 555–568.

DNV (2010). Risk Assessment of Pipeline Protection, RP-F107, Norway.

Gilles, MB (2001). Mine Drop Experiment (MIDEX), Master's Thesis, Naval Postgraduate School, Monterey, California.

ITTC (2011). Fresh Water and Seawater Properties. RP-7.5-02-01-03.

Newman, JN (1977). Marine Hydrodynamics. The MIT Press .

Werner, D (2015). "Reusable Rocket Renaissance." Aerospace America, Vol.(53), No. 8, Sept. 2015, pp. 38-43.

Xiang, G, Birk, L, Yu, X and Li, X. (2017a). "Study on the Trajectory and Landing Points of Dropped Cylindrical Object with Different Longitudinal Center of Gravity". International Journal of Offshore and Polar Engineering, Vol. (27), No. 3, pp. 274-282

Xiang, G, Birk, L, Yu, X and Lu, H. (2017b) "Numerical Study on The Trajectory of Dropped Cylindrical Objects". Ocean Engineering, Vol. (130), pp. 1-9.

Xiang, G, Birk, L, Li, L, Yu, X and Luo, Y. (2016). "Risk Free Zone Study for Cylindrical Objects Dropped into Water". Ocean Systems Engineering, Vol. (6), No. 4, pp. 377-400.

Xiang, G., Li, X., Yu, X., Luo, Y. and Cao, Y. (2019). Motion Dynamics of Dropped Cylindrical Objects in Flows after Water Entry. Ocean Engineering, Vol.(173), pp. 659 - 671.

Yasseri, S. (2014). "Experiment of free-falling cylinders in water". Underwater Technology. 32(3), pp.177-191

SNAME
THE INTERNATIONAL COMMUNITY
FOR MARITIME AND OCEAN PROFESSIONALS

APPLYING 3D VISCOUS/INVISCID INTERACTION METHOD TO PREDICT FLOWS AROUND WINGS AND PROPELLERS

WEIKANG DU
PH.D. CANDIDATE, THE UNIVERSITY OF TEXAS AT AUSTIN

SPYROS A. KINNAS
PROFESSOR, THE UNIVERSITY OF TEXAS AT AUSTIN

Proceedings of the 24th Offshore Symposium, February 20th 2019, Houston, Texas
Texas Section of the Society of Naval Architects and Marine Engineers

ABATRACT

In the past, the boundary element method (BEM) and vortex lattice method (VLM) have been widely used to predict the potential flow around hydrofoils and propellers. The viscosity, which not only causes friction and reduces the efficiency but also affects the pressure distributions, is considered via an empirical skin friction coefficient. In this paper, the three-dimensional VII method, which couples a lower order BEM or VLM with a two-dimensional integral boundary layer solver (the viscous/inviscid interaction method, or VII) is used to solve the flow around three-dimensional wings and propellers. It is assumed that the boundary layer growth is mainly in the chord-wise direction and simplifies the three-dimensional boundary layer equations into two-dimensional. The three-dimensional velocities, which are from either the BEM or the VLM, are used as the edge velocities. The pressure distributions and open water characteristics are compared with either full-blown RANS simulations or with existing experimental data, and it is shown that the 3D VII method can improve the results significantly.

Keyword: Panel method, Vortex lattice method, Boundary layer, Viscosity, Wings, Propellers

INTRODUCTION

In high Reynolds number external flows, the viscous effect is confined to a boundary layer which forms along the body. Outside of these regions, the flow is essentially inviscid and irrotational. The boundary layer is assumed to be thin compared with the characteristic body length, and the pressure is assumed to be constant in the direction normal to the wall. When these simplifications are applied to the Navier-Stokes equations, the boundary layer equations are obtained (Schlichting, 1979).

In the outer regions, the flow can be solved by using methods based on the potential theory, which has been proved to be a great success in designing and analyzing various propulsor devices, including propellers, water jets, ducts, etc. Among these methods, the most popular ones are the Boundary Element Method (BEM, more commonly known as the panel method) and the vortex lattice method (VLM).

In the classic BEM and VLM, the flow is assumed to be inviscid. However, the viscosity is essential in predicting the performances (thrust, torque and efficiency) accurately, and it also affects the pressure distributions on propeller blades, resulting in a change in cavitation inception, development and decay. For conventional designs of marine propellers, the effect of viscosity was often considered through an empirical drag to lift ratio (Kerwin and Lee, 1978; Kerwin et al., 1987). However, as the types of modern propellers become increasingly complicated and the flow around propellers becomes more complex, the effects of viscosity cannot be accurately predicted by empirical viscous corrections.

Another way to consider the viscous effect is by solving the boundary layer equations after the inviscid solution is obtained, known as the viscous/inviscid interaction method (VII). Drela (1989) coupled a linear-vorticity panel method with a two-dimensional (2D) boundary layer integral equation solver and developed an iterative design and analysis tool XFOIL. Du and Kinnas (2018) coupled XFOIL and the panel method in a flow separation model to predict flows around hydrofoils and 2D axisymmetric ducts with blunt trailing edges. Nishida and Drela (1995) and Milewski (1997) developed numerical schemes which couple the potential flow methods with three-dimensional (3D) boundary layer solvers. However, these methods are too computationally expensive and can only be applied to simple 3D geometries like wings, not for propulsor devices.

In propeller flows, it was shown by Jessup (1989) in an experiment that the boundary layer development in the chord-wise direction is much more significant than that in the span-wise direction. Based on this, Hufford (1992) developed the boundary layer strip theory, which assumes that the viscous flow around propellers can be simplified into two-dimensional and solved strip by strip by using the 2D boundary layer equations, which greatly reduces the required computational efforts. Sun (2008) coupled a lower order panel method with the boundary solver XFOIL and predicted the viscous flows around cavitating propellers. Sun (2008)'s work was further improved by Kinnas et al. (2012), Yu (2012) , Kinnas et al. (2013) and Purohit (2013) by using 3D influence coefficients instead of 2D in the boundary equations, and considered the influence of the boundary layer at other strips.

In this study, a 3D viscous/inviscid interaction method is applied to predict the flows around 3D wings and propellers, which coupled either a low order panel method or a vortex lattice method with a 2D boundary layer solver. This method is based on the work of Sun (2008), Yu (2012) and Purohit (2013), validated in a systematic way in 2D hydrofoils and 3D wings before it is applied to propellers. Results are compared with those from the Reynolds-averaged NavierStokes (RANS) simulations and with the experimental measurement.

METHODOLOGY

The panel method

In the panel method, the total flow velocity $\overrightarrow{q}$ is decomposed into an incoming flow $\overrightarrow{q_{in}}$ and a propeller-induced flow $\overrightarrow{U_{ind}}$ (the perturbation velocity),

$$\overrightarrow{q} = \overrightarrow{q_{in}} + \overrightarrow{U_{ind}}. \tag{1}$$

$\overrightarrow{U_{ind}}$ can be treated as the potential flow which is governed by the Laplace's equation

$$\overrightarrow{U_{ind}} = \nabla\phi, \quad \nabla^2\phi = 0, \tag{2}$$

where ϕ is the perturbation potential.

The Kutta condition is required which means the velocity must remain finite

$$|\nabla\phi| < \infty \quad \text{at the trailing edge.} \tag{3}$$

By using the Green's identity, equation (2) can be written in the boundary integrated form:

$$2\pi\phi_p = \iint_{S_H} \left[\phi_p \frac{\partial G(p,p')}{\partial \mathbf{n}_{p'}} - G(p,p')\frac{\partial \phi_p}{\partial \mathbf{n}_{p'}}\right] dS + \iint_{S_W} \Delta\phi_W \frac{\partial G(p,p')}{\partial \mathbf{n}_{p'}} dS, \tag{4}$$

where S_H is the surface of the hydrofoil or propeller blade, duct, and hub, S_W the surface of the trailing wake, G the Green's function, which is defined as $1/R(p,p')$ in 3D and $2lnR(p,p')$ in 2D, and R the distance between the two points p and p'.

In the panel method, how to predict the trailing wake is of great importance. The simplest way is to use a helical wake, which means that the wake is not aligned with the inflow and the wake angle does not change as it goes downstream. Greeley and Kerwin (1982) invented a fast wake alignment scheme (PSF-2 type alignment), in which the wake is aligned on a coarse grid first and then interpolated onto a fine grid. Tian and Kinnas (2012) showed that proper wake alignment model can improve the accuracy of the predicted propeller performances in the panel method, especially at low advanced ratios and, therefore, introduced a pseudo-unsteady wake alignment scheme (full wake alignment scheme, or FWA). This scheme was extended by Kinnas et al. (2015) and Kim et al. (2018) for ducted propellers by including the duct-induced velocities in the FWA. Similar approaches were adopted by Baltazar et al. (2012), who used a rudimentary way to take into account the boundary layer over the duct inner surface by assuming a power law velocity profile and reduced the pitch of the blade wake near the tip.

The vortex lattice method

In the vortex lattice method, line vortices and sources are placed on the mean camber surface of the blade and the trailing wake, and the strengths of those singularities are determined by solving the potential flow which satisfies the kinematic boundary condition and dynamic boundary condition. The resultant velocity $\vec{V}$ can be written as the summation of the individual contributions as:

$$\vec{n}\cdot\vec{V} = \vec{n}\cdot\left(\vec{V_\Gamma}+\vec{V_Q}+\vec{V_I}+\vec{V_B}+\vec{V_O}\right) = 0, \tag{5}$$

where $\vec{V_\Gamma}$ is the velocity induced by the line vortices on the key blade and wake, $\vec{V_Q}$ the velocity induced by the cavity sources on the key blade and wake, $\vec{V_I}$ the inflow velocity, $\vec{V_B}$ the velocity induced by the line sources, which represents the thickness of the blade, and $\vec{V_O}$ the induced velocity by the other blades and wakes.

The numerical code using the vortex lattice method is MPUF-3A. Some major improvements in MPUF-3A include: the thickness loading coupling scheme by Kinnas (1992), the unsteady wake alignment scheme and the non-linear terms in pressure evaluation by He (2010). All of these improvements are included in this study to predict the edge velocities in the viscous/inviscid interaction method.

Wall Transpiration Model in 2D

In a propeller system, using x_1 as arc-length in the chord-wise direction, x_2 perpendicular to that on the blade surface, and x_3 normal to x_1 and x_2, the three-dimensional boundary equations are given as (Schlichting, 1979)

$$\frac{1}{V^2h_1}\frac{\partial}{\partial x_1}(V^2\theta_{11})+\frac{\delta_1^*}{Vh_1}\frac{\partial U_1}{\partial x_1}+\frac{1}{V^2h_2}\frac{\partial}{\partial x_2}(V^2\theta_{12})+ \frac{\delta_2^*}{Vh_2}\frac{\partial U_1}{\partial x_2}-K_2\left(\theta_{12}+\theta_{21}+\frac{U_2\delta_1^*}{V}\right)+K_1\left(\theta_{22}-\theta_{11}+\frac{U_2\delta_2^*}{V}\right)=\frac{\tau_{w1}}{\rho V^2} \tag{6}$$

and

$$\frac{1}{V^2h_1}\frac{\partial}{\partial x_1}(V^2\theta_{21})+\frac{\delta_1^*}{Vh_1}\frac{\partial U_2}{\partial x_1}+\frac{1}{V^2h_2}\frac{\partial}{\partial x_2}(V^2\theta_{22})+ \frac{\delta_2^*}{Vh_2}\frac{\partial U_2}{\partial x_2}-K_1\left(\theta_{12}+\theta_{21}+\frac{U_2\delta_1^*}{V}\right)+K_2\left(\theta_{11}-\theta_{22}+\frac{U_1\delta_1^*}{V}\right)=\frac{\tau_{w2}}{\rho V^2}, \tag{7}$$

where V denotes the resultant velocity at the edge of the boundary layer ($V^2 = U_1^2+U_2^2$). h_1 and h_2 are the metric coefficients along x_1 and x_2, and K_1 and K_2 are the geodesic curvatures, defined as

$$K_1 = -\frac{1}{h_1h_2}\frac{\partial h_2}{\partial x_1}; K_2 = -\frac{1}{h_1h_2}\frac{\partial h_1}{\partial x_2}. \tag{8}$$

The two equations above contain six distinct integral thicknesses to be determined δ_1^*, δ_2^*, θ_{11}, θ_{22}, θ_{12} and θ_{21}.

By assuming that the dominating effect of boundary layer in propeller blade is in the chord-wise direction, equation (6) and equation (7) can be reduced in the two-dimensional form along the streamline

$$\frac{d\theta}{ds} + (2+H)\frac{\theta}{U_e}\frac{dU_e}{ds} = \frac{C_f}{2}, \tag{9}$$

where U_e is the edge velocity, C_f the skin friction coefficient and H the shape factor, defined as

$$H = \frac{\delta^*}{\theta}, \tag{10}$$

and

$$C_f = \frac{\tau_w}{\frac{1}{2}\rho U_e^2}. \tag{11}$$

Equation (9) is the Karman momentum integral equation for low velocity incompressible flows, where the March number M_e equals 0. The kinetic energy equation for two-dimensional boundary layer is

$$\theta\frac{dH^*}{ds} + H^*(1-H)\frac{\theta}{U_e}\frac{dU_e}{ds} = 2C_D - H^*\frac{C_f}{2}, \tag{12}$$

where H^* is the kinetic energy shape parameter, defined as

$$H^* = \frac{\theta^*}{\theta}, \tag{13}$$

where θ is the momentum thickness

$$\theta = \int_0^\delta \frac{u}{U_e}\left(1-\frac{u}{U_e}\right)dy, \tag{14}$$

and θ^* is the kinetic energy thickness

$$\theta^* = \int_0^\delta \frac{u}{U_e}\left(1-\frac{u^2}{U_e^2}\right)dy. \tag{15}$$

The following equations are used for laminar and turbulent flows respectively:

Laminar closure (laminar flow):

$$\frac{d\tilde{n}}{ds} = \frac{d\tilde{n}}{dRe_\theta}(H_k)\frac{dRe_\theta(H_k,\theta)}{ds} \tag{16}$$

Turbulent closure (turbulent flow):

$$\frac{\delta}{C_\tau}\frac{dC_\tau}{ds} = 5.6[C_{\tau EQ}^{1/2} - C_\tau^{1/2}] + 2\delta \times \left\{\frac{4}{3\delta^*}\left[\frac{C_f}{2} - \left(\frac{H_k-1}{6.7H_k}\right)^2\right] - \frac{1}{U_e}\frac{dU_e}{ds}\right\} \tag{17}$$

where δ is the boundary layer thickness, defined as

$$\delta = y\,|_{\frac{u}{Ue}=0.99} \tag{18}$$

δ^* the displacement thickness

$$\delta^* = \int_0^\delta \left(1-\frac{u}{U_e}\right)dy \tag{19}$$

H_k the kinematic shape factor

$$H_k = \frac{H - 0.290M_e^2}{1+0.113M_e^2} \tag{20}$$

M_e the boundary layer edge March number (equals 0 in propeller flows) and $\tilde{n}$ is the most-amplified of the Tollmien-Schlichting wave. The momentum equation (9), kinetic energy equation (12) and the closure equations are combined together and can be solved iteratively by the 2D boundary layer solver XFOIL (Drela, 1989).

In the viscous/inviscid interaction method, the effect of viscosity can be taken into account by placing boundary layer sources on the surface S_H and the S_W in equation (4). The edge velocity of the boundary layer is from solving the potential flow by using the panel method. The strength of the boundary layer sources can be obtained from a boundary layer solver (Drela, 1989) and the viscous velocity on the ith panel is obtained by

$$\overrightarrow{U_{i,vis}} = \overrightarrow{U_{i,ind}} + \sum_{j=1}^{N+N_W} C_{ij}\sigma_j, \tag{21}$$

where $\overrightarrow{U_{i,ind}}$ is the edge velocity on the ith panel, N and N_W the number of panels on S_H and S_W, C_{ij} the source influence coefficient matrix, and σ_j the boundary layer source strength on the jth panel.

Drela (1989)'s model can predict the transition points from the laminar flow to turbulent flow on both the pressure side and the suction side. However, in the presented method, the flow is assumed to be fully turbulent. Forced transitions at 1% chord length from the leading edge are applied in the VII method throughout this study, and results are compared with fully turbulent RANS simulations.

The 3D VII method

Derivation of the equations for the 3D VII method can be seen in Purohit (2013). The viscous velocity on the Kth strip is

$$\begin{aligned} U_K^{vis} &= U_K^{inv} + D_K A_{KK}^{-1} B_{KK}\sigma_K \\ &+ \sum_{J=1,J\neq K}^{M} D_K A_{KK}^{-1} B_{KJ}\sigma_J \\ &+ \sum_{J=1,J\neq K}^{M} D_K A_{KK}^{-1} A_{KJ}\left[\phi_J^{inv} - \phi_J^{vis}\right], \end{aligned} \tag{22}$$

where M is the number of strips in the span-wise direction, D_K the differentiation matrix with respect to the arc-length on the Kth strip, A the dipole induced influence coefficient matrix and B the source induced influence coefficient matrix. The first line of equation (22) represents the edge velocity plus the influence of the boundary layer sources of the Kth strip, which is similar to the 2D form in equation (21). The second line of equation (22) represents the influence of the boundary layer sources from other strips and the last line represents the dipoles from other strips, which is missing in the work of Yu (2012).

It should be noted that ϕ^{vis} on the right hand side of equation (22) is from the previous iteration. A flow chart of the 3D VII method is shown in figure 1.

The Reynolds-Averaged Navier-Stokes (RANS) method

In the VII method, the unknown sources and dipoles are placed on the surface of the body and the wake. In the RANS method, the computational domain extends from the body to far upstream and downstream. The velocity field u_i is decomposed into its mean U_i and the fluctuation u_i' (Reynolds, 1894), as

$$u_i = U_i + u_i'. \tag{23}$$

The governing equations in the RANS method are the continuity equation and the momentum equations, as

$$\nabla \cdot \vec{u} = 0, \tag{24}$$

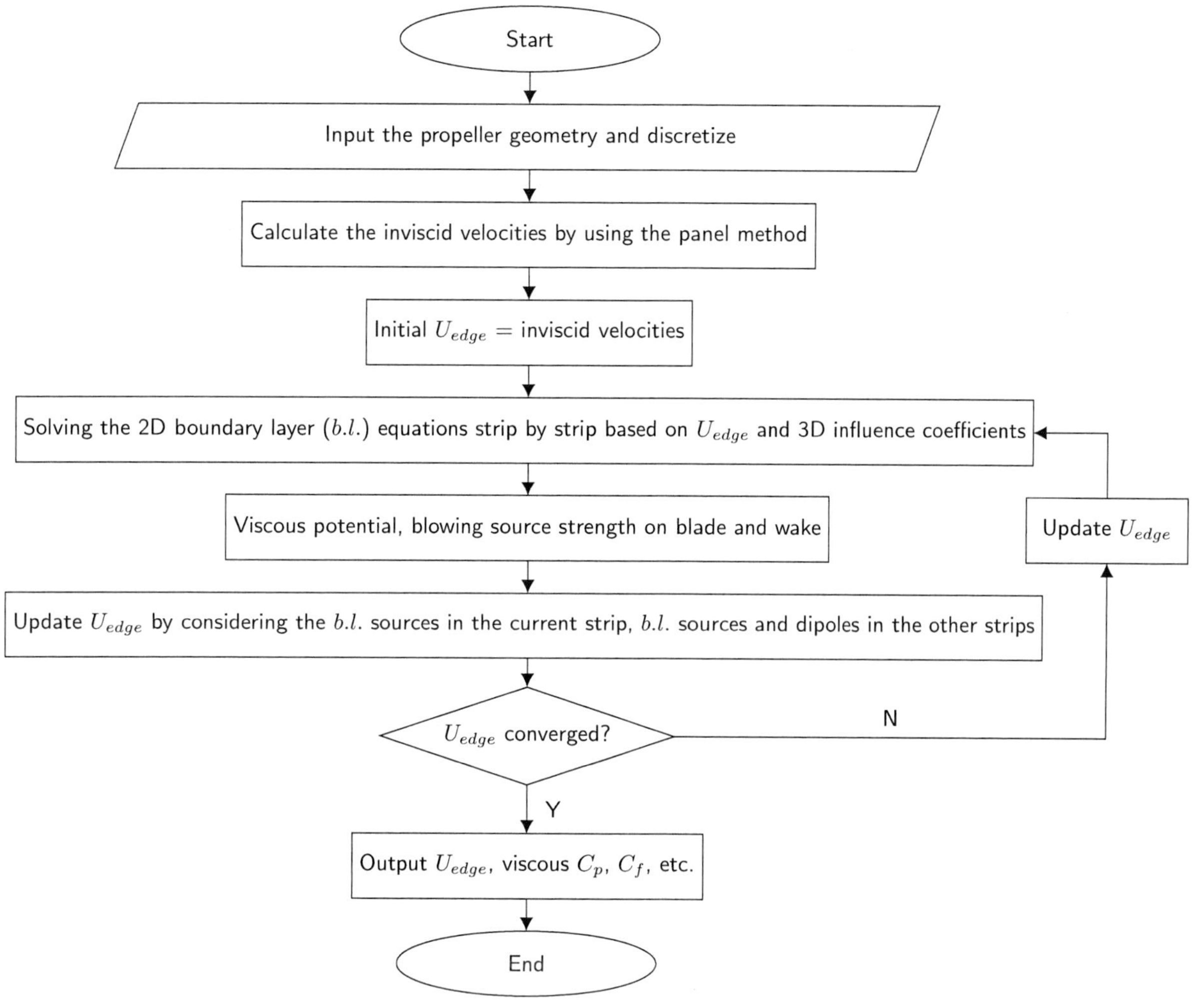

Figure 1 Flowchart of applying the three-dimensional viscous/inviscid interaction method to predict viscous flows around wings and propellers

and

$$\frac{\partial U_i}{\partial t} + U_j \frac{\partial U_i}{\partial x_j} = -\frac{\partial P}{\partial x_i} + \frac{\partial}{\partial x_j}\left(\nu \frac{\partial U_i}{\partial x_j} - \overline{u_i' u_j'}\right), \tag{25}$$

where P is the pressure, ν the kinematic viscosity and $\overline{u_i' u_j'}$ the Reynolds stress.

In this study, the pressure distribution from the VII method will be compared with the RANS simulation by using the commercial CFD software Ansys-Fluent. The flow is assumed to be incompressible, and the $k-\omega$ SST model (Pope, 2000) at fully turbulent condition is used. Detailed numerical settings are listed in table 1.

Table 1 Numerical settings in ANSYS-Fluent simulations

Flow	2D, steady
Turbulence modeling	$k-\omega\ SST$ (fully turbulent)
Pressure-velocity coupling scheme	Coupled
Gradient spatial discretization	Least squares cell based
Pressure spatial discretization	Second order
Discretization of momentum, turbulent kinetic energy, specific dissipation rate	Second order upwind
Residuals	All below 1.0×10^{-6}

RESULTS AND DISCUSSIONS

2D hydrofoil

The viscous/inviscid interaction method is firstly tested in a 2D hydrofoil case before it is applied to 3D geometries. The hydrofoil, as shown in figure 2, has a NACA 0005 thickness distribution and a NACA a=0.8 mean line with the maximum camber equal to 2% of the chord length (Abbott and Von Doenhoff, 1959). The angle of attack is 1.0 degree. In the RANS simulation, the total number of quadrilateral cells is 207.6k and the $y+$ is smaller than 2.5. The boundary conditions are shown in figure 3, and other settings in the RANS simulation are shown in table 1. To make sure that perturbations caused by the foil are not affected by the boundary, the velocity inlet boundaries are 50 times chord length away from the foil, and the pressure outlet boundaries are 100 times chord length away.

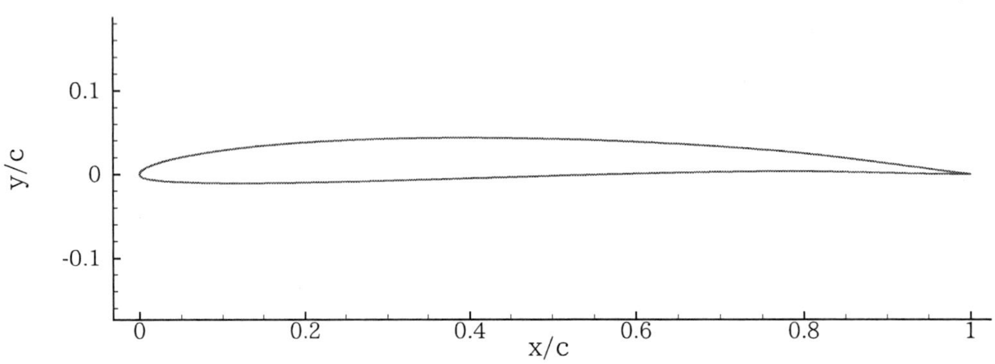

Figure 2 A 2D hydrofoil to test the VII method (NACA 0005 thickness distribution and a NACA a=0.8 mean line with the maximum camber equal to 2% of the chord length)

In XFOIL, a linear-vorticity panel method, which is based on the stream function approach, is used to obtain the inviscid solution for the flow around hydrofoil. In this section, this method is replaced by a nonlinear perturbation boundary element method, and the pressure distributions from different methods are shown in figure 4 and compared with the results from the RANS method. Without coupling with the boundary layer solver, there are some differences between the pressure distributions from the BEM and the RANS simulations, both near the leading edge and trailing edge. However, by coupling with XFOIL, the correlation with the RANS method improves significantly. It is also found that both methods in VII work successfully with the boundary layer solver.

It should be noted that compared with the RANS method, the VII methods is very computational efficient. By using 1 CPU, it took less than 0.2 seconds for VII methods (both BEM/XFOIL or XFOIL) while by using 32 CPU, it took 20 minutes for the RANS method. The time for preprocessing and post precessing in RANS is not included.

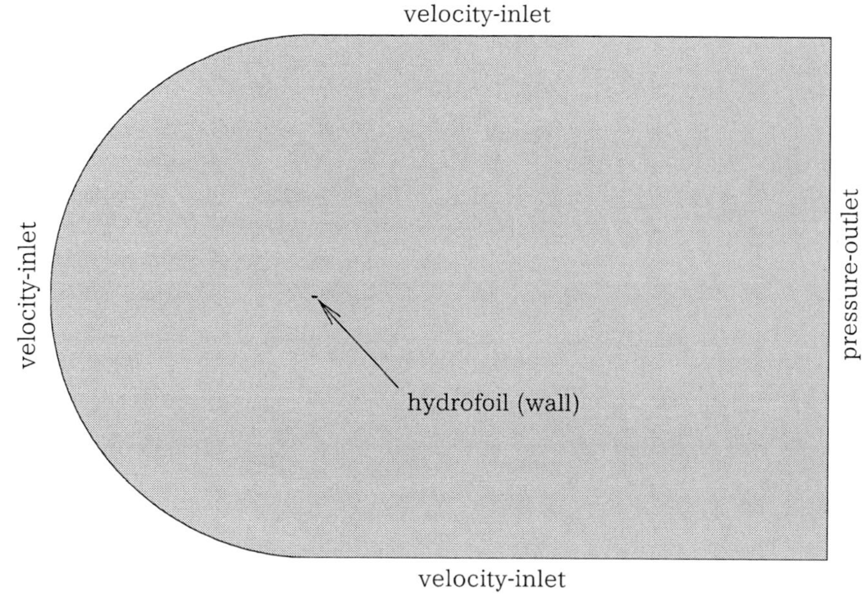

Figure 3 The computational domain and boundary conditions in RANS simulations

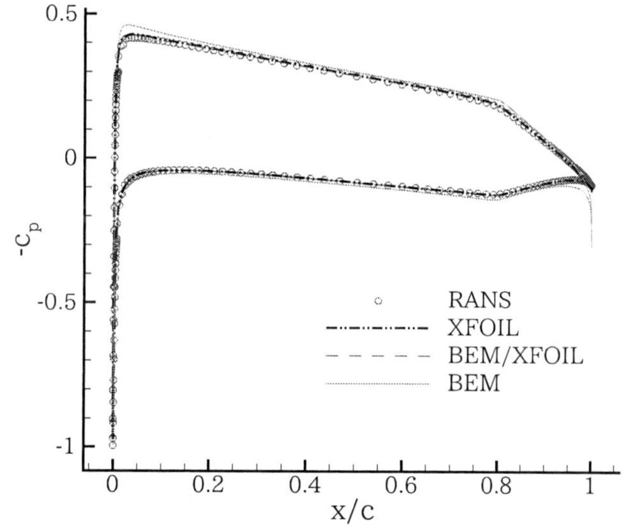

Figure 4 The pressure predicted from VII agrees well with those from RANS in 2D hydrofoils ($Re = 1.0 \times 10^6$, which is defined by the chord length c as $Re = \frac{cV_\infty}{\nu}$)

3D wing

The VII method is applied to a 3D wing and the pressure distributions at different sections are compared with those from the RANS simulations. The cross section of this wing is shown in figure 5, and it has a NACA 66 thickness distribution with the maximum thickness equal 10% of the chord length at every section.

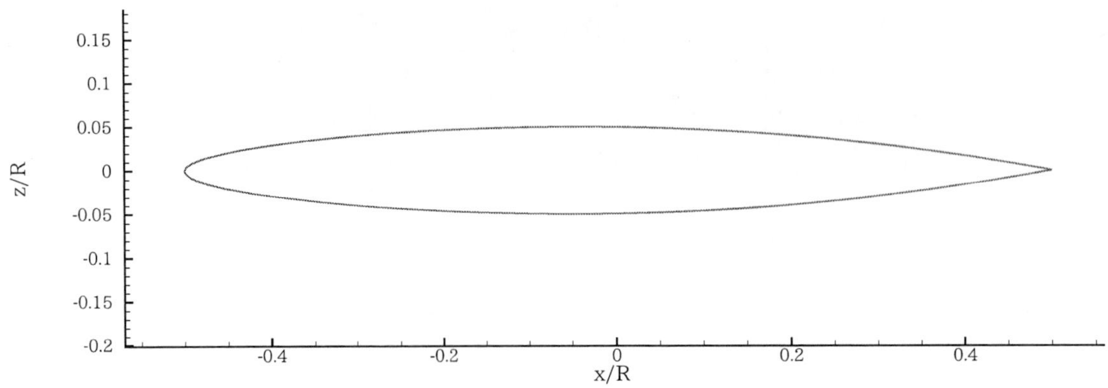

Figure 5 Cross section of the wing at $y/R = 0$, where R is the length of the wing.

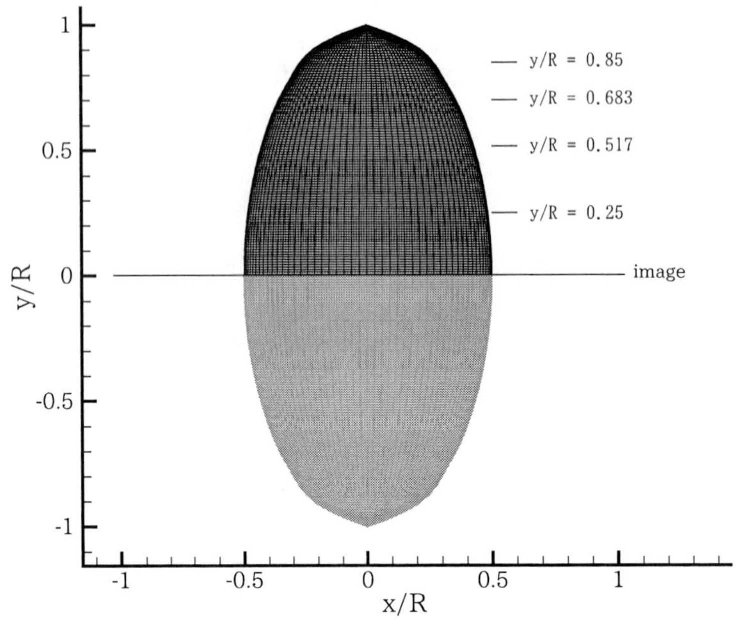

Figure 6 Image model is used in the VII method for the elliptic wing and pressure distributions of the shown sections will be compared between VII method and the RANS simulations

As shown in figure 6, the length of the wing is R, so is the chord length at $y/R = 0$. The chord length of the wing follows an elliptic distribution and it goes to zero at $y/R = 1.0$. An image model is used in the VII method so only half of the wing is solved. Similarly, in the RANS simulation, symmetry boundary condition is used at the bottom of the domain to reduce the number of grids by half. As shown in figure 7, the size of the computational domain is large enough so that the perturbations caused by the wing are not affected by the boundary. The number of hexahedral cells is 4.25 million in the RANS method and all the other settings are the same as in table 1. As shown in figure 8, denser grids are used near the surface of the wing and in the wake region.

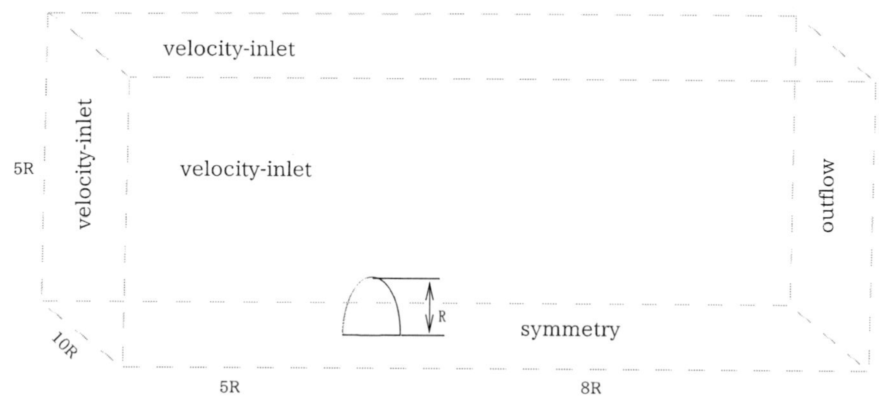

Figure 7 Computational domain size and boundary conditions in the RANS simulation for the elliptic wing, note the symmetry boundary condition at the bottom to reduce the number of grids by half

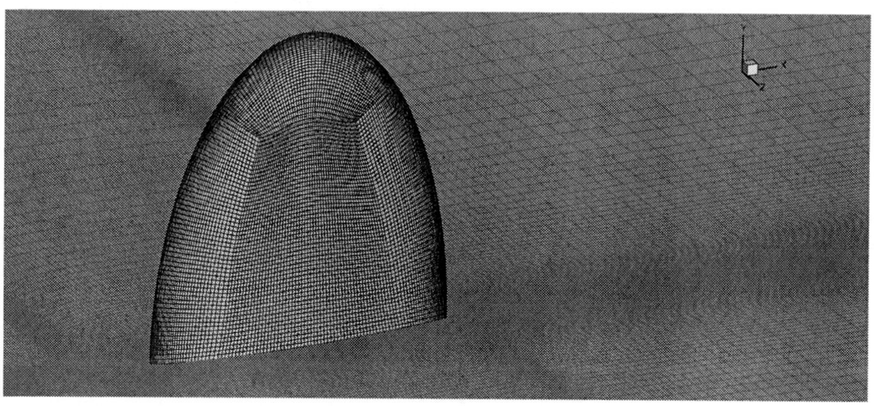

Figure 8 The mesh on the surface of the wing and at the bottom of the domain, note that denser grids are used closer to the wing and in the wake region

The pressure coefficient is defined as

$$c_{p_i} = \frac{p_i - p_\infty}{\frac{1}{2}\rho v_\infty^2}, \tag{26}$$

with p_∞ and v_∞ being the pressure and velocity at far upstream respectively, and ρ being the fluid density. Correlation for the pressure distributions predicted by the 3D VII method and the RANS method is shown in figure 9 and 10. As shown in figure 6, four sections are considered, which are at $y/R = 0.25$, 0.517, 0.683 and 0.85. The angle of attack, which is the angle between the inflow direction and the X-Y plan is 0 degree and 2 degree. Since this wing has no camber, when the angle of attack equals 0 degree, the pressures on both sides of the wing overlap, as shown in figure 9.

Pressures predicted by the panel method are close to those in the RANS simulations of the inviscid flow analysis. In the viscous flow analysis, the transition from laminar to turbulent flow is forced at 1% chord length from the leading edge in the VII method. Results are compared with RANS simulation with fully turbulent condition. It is found that if the panel method does not couple with the boundary layer solver, the predicted pressure deviate from those from the

RANS simulations near the trailing edge. However, the 3D VII method will improve the correlation with the RANS method significantly near the trailing edge.

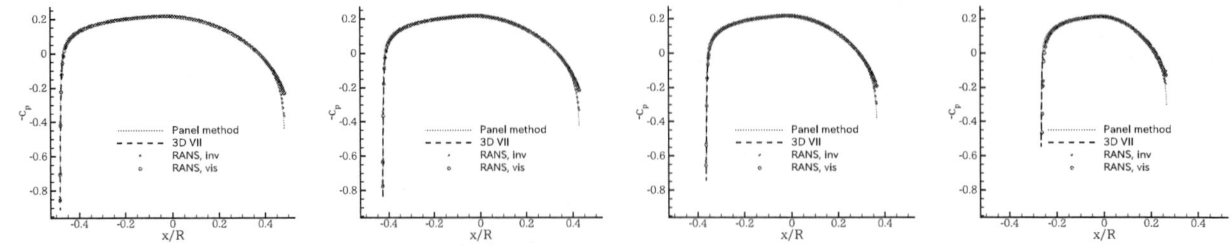

(a) $y/R = 0.25$ (b) $y/R = 0.517$ (c) $y/R = 0.683$ (d) $y/R = 0.85$

Figure 9 Correlation of the pressure distribution by the present method and RANS simulations at different sections of the elliptic wing (the angle of attack is 0 degree)

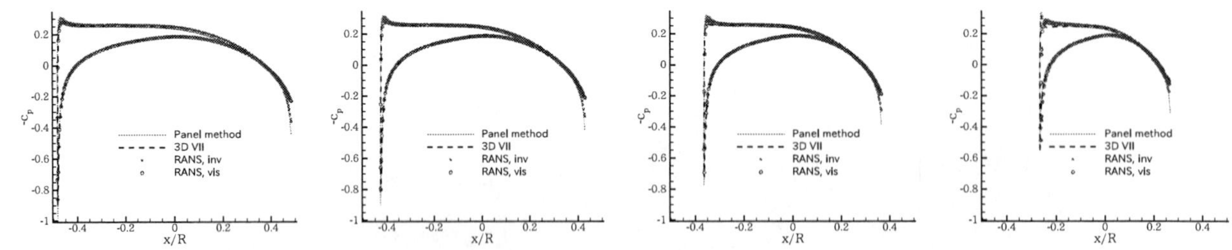

(a) $y/R = 0.25$ (b) $y/R = 0.517$ (c) $y/R = 0.683$ (d) $y/R = 0.85$

Figure 10 Correlation of the pressure distribution by the present method and RANS simulations at different sections of the elliptic wing (the angle of attack is 2 degree)

Pressure prediction in a 5-blade propeller

The three-dimensional viscous/inviscid interaction method is applied to a five-blade propeller (Propeller 4381) subject to uniform inflow, as shown in figure 11. The advance ratio is 0.889, which is defined as

$$J = \frac{V_s}{nD}, \tag{27}$$

where V_s is the ship speed, n the propeller rotational frequency and D the propeller diameter.

Results from the three-dimensional VII are compared with those from the RANS method. In the VII method, the edge velocities are from the panel methpd. The RANS domain is shown in figure 12. The periodic boundary condition is used so that only one blade is included in the domain, which greatly reduced the total number of grids to be 4.4 million. The mesh on the propeller blade and hub is shown in figure 13. All the other settings are the same as in table 1.

The predicted pressure distributions at strip $r/R = 0.7$ are compared between the present method and the RANS method, as shown in figure 14. In propeller case, the pressure coefficient c_p is defined as

$$c_p = \frac{p - p_\infty}{\frac{\rho}{2} n^2 D^2}, \tag{28}$$

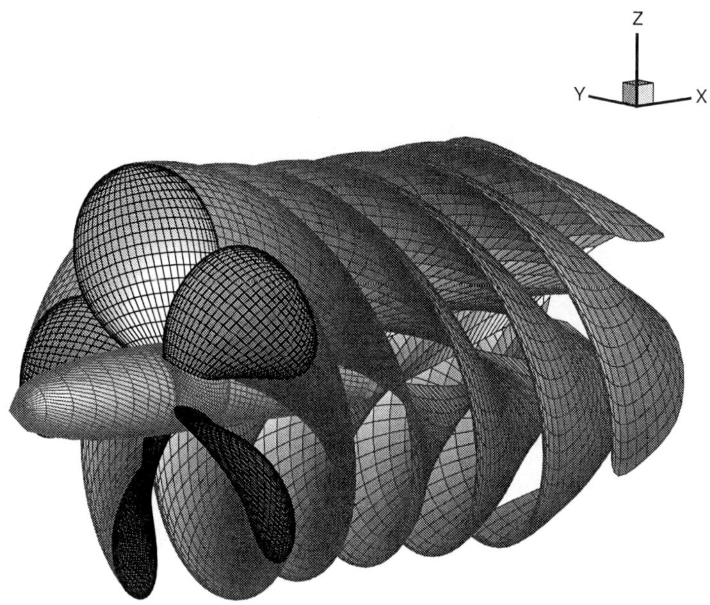

Figure 11 The mesh and wake in the panel method for Propeller 4381, predicted by the full wake alignment scheme (Tian and Kinnas, 2012) with $J = 0.889$

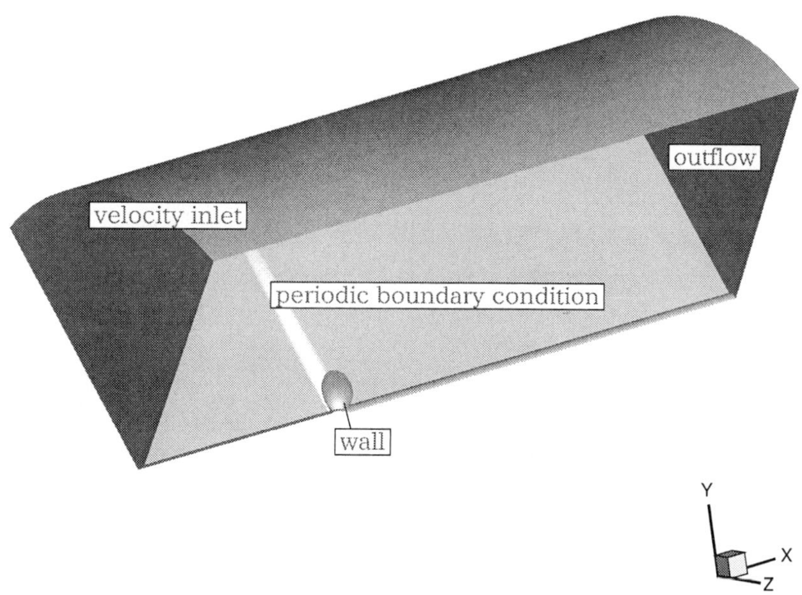

Figure 12 The RANS domain and the boundary conditions used in the simulation

By coupling with the boundary layer solver, the pressure near the trailing edge correlates much better with the results from the RANS method, compared with the results from using only the panel method.

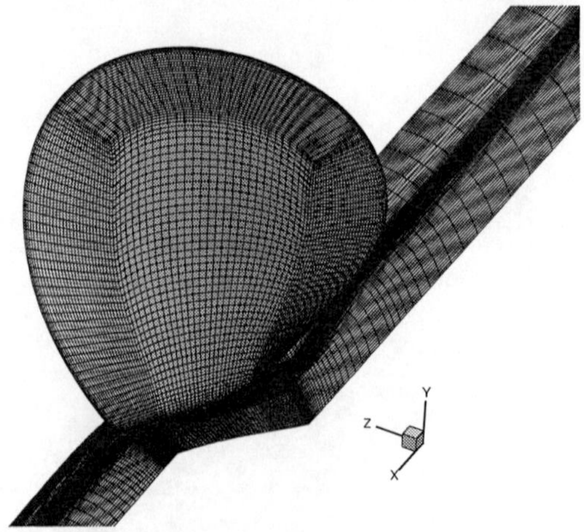

Figure 13 The grid on the propeller blade and hub in the RANS simulation

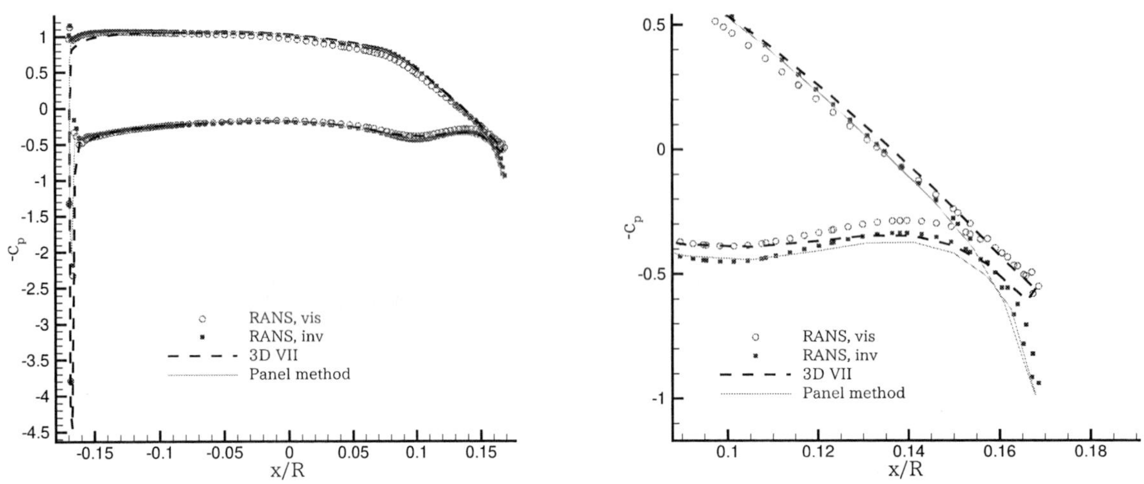

(a) full view

(b) detailed view close to leading edge

Figure 14 Comparison of pressure coefficients on the strip, $r/R = 0.7$, of Propeller 4381, predicted by RANS (ANSYS/Fluent), panel method and the present method.

In the viscous/inviscid interaction method, the edge velocity can also be evaluated by the vortex lattice method, in which the unknown line vortices and sources are placed on the mean camber surface of the blade and the trailing wake. The thickness loading coupling (Kinnas, 1992), full wake alignment scheme and the non-linear terms in pressure (He, 2010) are included to improve the accuracy of this method. The inviscid pressure distributions from the vortex lattice method are compared with those from the panel method. It should be noted that in VLM, the hub is represented by an image model while in the panel method, hub panels are used. Consequently, there are some differences in the predicted pressure near the hub. However, as the sections are closer to the blade tip, the correlation between these two methods is improved, as shown in figure 15.

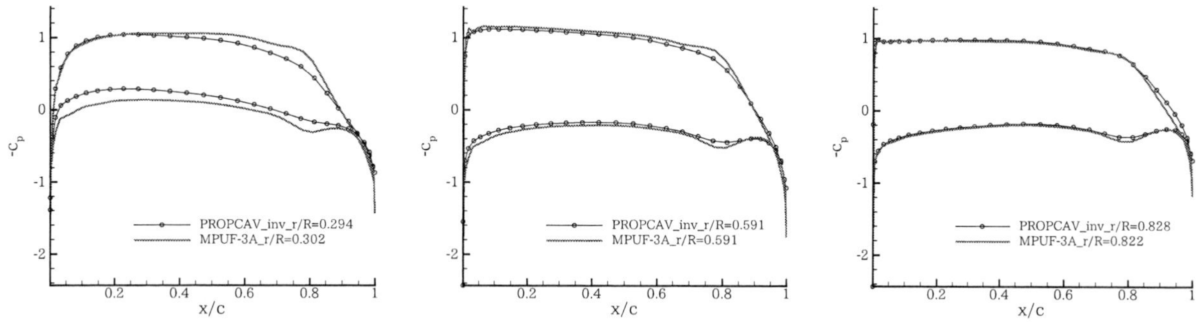

(a) $y/R = 0.3$ (b) $y/R = 0.6$ (c) $y/R = 0.82$

Figure 15 The inviscid pressure distribution from the vortex lattice method (MPUF-3A) and the panel method (PROPCAV) (note that sections in these two methods are not at exactly the same radius)

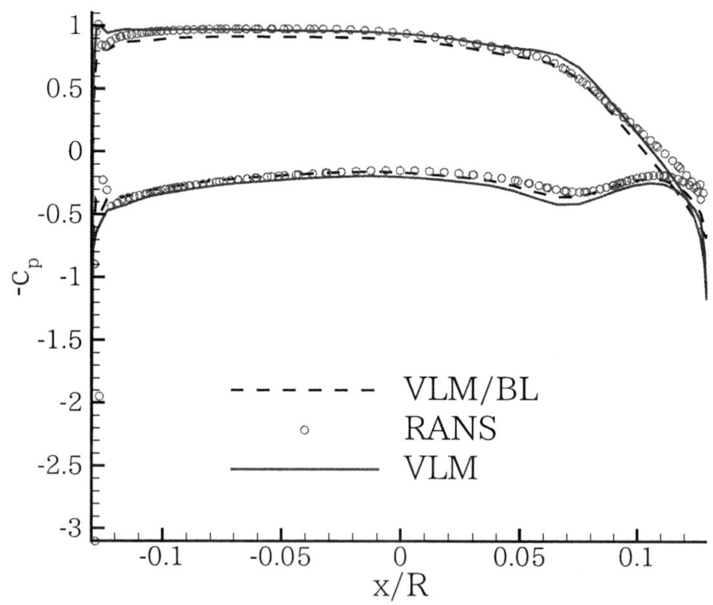

Figure 16 The inviscid and viscous pressure distribution, predicted by coupling the vortex lattice method with the boundary layer solver, compared with results from the RANS method

The viscous pressure from coupling the vortex lattice method and the boundary layer solver is shown figure 16. Similarly, the correlation with the RANS method near the trailing edge is improves. The 3D edge velocities are only in the chord-wise direction, which is consistent with the assumption that the dominating effect of the boundary layer in propeller blades is in the chord-wise direction. The Reynolds number for each blade section is based on the chord length of the blade section and the relative velocity (including the axial and tangential components). It should also be noted that in MPUF-3A/XFOIL, 2D influence coefficients matrices are used in the boundary layer equations while in PROPCAV/XFOIL, 3D influence coefficient matrices are used. This is determined by the character of the vortex lattice method, and in general, the panel method is a more accurate method. However, the vortex lattice method is still widely used by the industry and this study gives an efficient way of including the effect of the boundary layer.

The open water characteristics of a 5-blade propeller

The three-dimensional viscous/inviscid interaction method (coupled with the panel method) is applied to a skewed five-blade propeller (Propeller 5168) subject to uniform inflow, as shown in figure 17, for a range of advance ratios. The predicted forces (thrust and torque) are compared with existing experimental measurements.

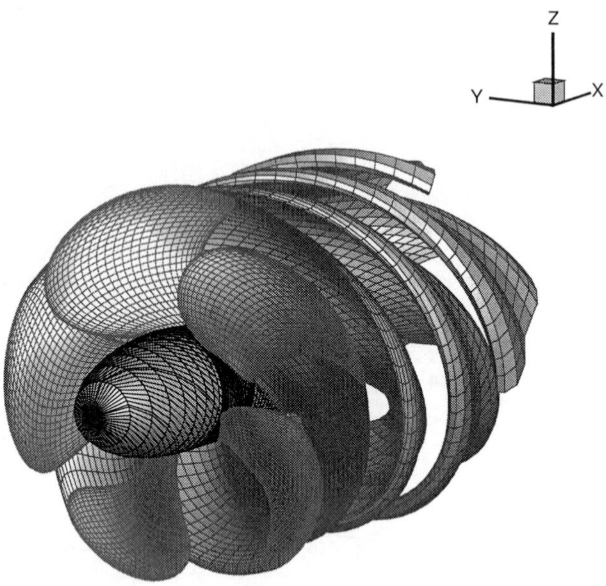

Figure 17 The mesh and wake in the panel method for Propeller 5168, predicted by the full wake alignment scheme with $J = 0.9$

The thrust coefficient on the propeller is defined as

$$KT = \frac{T}{\rho n^2 D^4}, \tag{29}$$

where T is the thrust generated by the propeller. The torque coefficient is defined as

$$KQ = \frac{Q}{\rho n^2 D^5}, \tag{30}$$

where Q is the torque on the propeller.

As shown in figure 18, the viscous effect has a higher influence on the torque than on the thrust. By only applying the panel method without any corrections, the torque will be under-predicted. The viscous effect can be considered by applying an empirical constant frictional coefficient over the blades and empirically correct the pitch angles, known as the viscous pitch correction (VPC) (Kerwin and Lee, 1978). The effect of viscosity can also be considered by the VII method. Both VPC and VII will significantly improve the correlation of the predicted torque with the experiment measurements, as shown in figure 18.

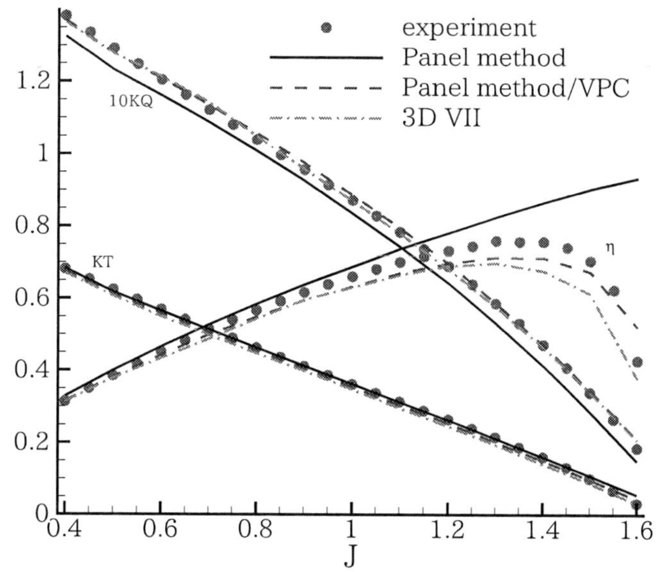

Figure 18 The open water characteristics predicted from the VII method and the experimental measurements

Conclusions

In this study, the three-dimensional viscous/inviscid interaction method, which couples a low order panel method or a vortex lattice method with a 2D integral boundary layer equation solver, is used to predict flows around 2D hydrofoil, 3D wings and propellers. The pressure distributions are compared with results from full-blown RANS simulations and the open water characteristics of a 5-blade propeller are compared with existing experimental measurements. It is found that both the 3D VII method and the empirical skin friction can improve the predicted torque with the experiment, but the 3D VII method significantly improved the pressure distribution near the trailing edge in hydrofoils, wings and propellers.

Future work

In the present work, the 3D VII method is applied to propellers with sharp trailing edges due to lack of models to predict the flow separation downstream of the blunt trailing edges. In the future, the 3D VII method will be coupled with a flow separation model, as an extension of the work shown in Du and Kinnas (2018). In addition, the span-wise terms in the boundary layer calculation will also be added properly.

Acknowledgements

Support for this research was provided by the US Office of Naval Research (Grant Number N00014-14-1-0303 and N00014-18-1-2276; Dr. Ki-Han Kim) partly by Phases VII and VIII of the Consortium on Cavitation Performance of High Speed Propulsors.

References

Ira Herbert Abbott and Albert Edward Von Doenhoff. *Theory of wing sections, including a summary of airfoil data.* Courier Corporation, 1959.

J. Baltazar, J. A. C. Falcão de Campos, and J. Bosschers. Open-water thrust and torque predictions of a ducted propeller system with a panel method. *Int. J. Rotating Machinery*, 2012, 2012.

M. Drela. XFOIL: An analysis and design system for low Reynolds number airfoils. In *Low Reynolds number aerodynamics*, pages 1–12. Springer, 1989.

W. Du and S.A. Kinnas. A flow separation model for hydrofoil, propeller and duct sections with blunt trailing edges. *Journal of Fluid Mechanics*, 861:180199, 2018. doi: 10.1017/jfm.2018.859.

D. S. Greeley and J. E. Kerwin. Numerical methods for propeller design and analysis in steady flow. *Transactions of Society of Naval Architects and Marine Engineers*, 90:415–453, 1982.

L. He. *Numerical simulation of unsteady rotor/stator interaction and application to propeller/rudder combination (also UT-OE Report 10-05)*. PhD thesis, Ocean Engineering Group, The University of Texas at Austin, Austin, TX, 2010.

G.S. Hufford. Viscous flow around marine propellers using boundary layer strip theory. Master's thesis, Massachusetts Institute of Technology, 1992.

S.D. Jessup. An experimental investigation of viscour aspects of propeller blade flow. *The Catholic University of America, Dissertation for Ph. D.*, 1989.

J.E. Kerwin and C.-S. Lee. Prediction of steady and unsteady marine propeller performance by numerical lifting-surface theory. In *Transactions of Society of Naval Architects and Marine Engineers*, 1978.

J.E. Kerwin, S.A. Kinnas, J.-T. Lee, and W.-Z. Shih. A surface panel method for the hydrodynamic analysis of ducted propellers. In *Transactions of Society of Naval Architects and Marine Engineers*, 1987.

S. Kim, S.A. Kinnas, and W. Du. Panel method for ducted propellers with sharp trailing edge duct with fully aligned wake on blade and duct. *Journal of Marine Science and Engineering*, 6(3), 2018. ISSN 2077-1312. doi: 10.3390/jmse6030089. URL `http://www.mdpi.com/2077-1312/6/3/89`.

S.A. Kinnas. A general theory for the coupling between thickness and loading for wings and propellers. *Journal of Ship Research*, 36:59–68, 1992.

S.A. Kinnas, X. Yu, and Y. Tian. Prediction of propeller performance under high loading conditions with viscous/inviscid interaction and a new wake alignment model. In *Proceedings of the 29th Symposium on Naval Hydrodynamics, Gothenburg, Sweden*, pages 26–31, 2012.

S.A. Kinnas, H. Fan, and Y. Tian. A panel method with a full wake alignment model for the prediction of the performance of ducted propellers. *J. Ship Research*, 59(4):246–257, 2015.

Spyros A Kinnas, Chan-Hoo Jeon, Jay Purohit, and Ye Tian. Prediction of the unsteady cavitating performance of ducted propellers subject to an inclined inflow. In *International symposium on marine propulsors SMP*, volume 13, 2013.

W.M. Milewski. *Three-dimensional viscous flow computations using the integral boundary layer equations simultaneously coupled with a low order panel method.* PhD thesis, Massachusetts Institute of Technology, 1997.

B. Nishida and M. Drela. Fully simultaneous coupling for three-dimensional viscous/inviscid flows. In *13th Applied Aerodynamics Conference*, page 1806, 1995.

S. B. Pope. *Turbulent flows*. Cambridge University Press, 2000.

J.B. Purohit. An improved viscous-inviscid interactive method and its application to ducted propellers. Master's thesis, Ocean Engineering Group, The University of Texas at Austin, Austin, TX, 2013.

O. Reynolds. On the dynamical theory of incompressible viscous fluids and the determination of the criterion. *Philos. Trans. R. Soc. London Ser. A*, 186:123–164, 1894. ISSN 02643820. URL `http://www.jstor.org/stable/90643`.

H. Schlichting. *Boundary layer theory*. McGraw-Hill, Inc., 1979.

H. Sun. *Performance prediction of cavitating propulsors using a viscous/inviscid interaction method.* PhD thesis, Ocean Engineering Group, The University of Texas at Austin, Austin, TX, 2008.

Y. Tian and S.A. Kinnas. A wake model for the prediction of propeller performance at low advance ratios. *Int. J. Rotating Machinery*, 2012, 2012.

X. Yu. Three dimensional viscous/inviscid interactive method and its application to propeller blades. Master's thesis, Ocean Engineering Group, The University of Texas at Austin, Austin, TX, 2012.

DEMONSTRATING TARGET SAFETY LEVEL FOR LIFE EXTENSION OF OFFSHORE STRUCTURES

PIOTR SZALEWSKI
PRINCIPAL ENGINEER, DNV GL HOUSTON

GRZEGORZ MALINOWSKI
PRINCIPAL ENGINEER, DNV GL HOUSTON

JENNY YAN LU
SENIOR PRINCIPAL ENGINEER, DNV GL HOUSTON

Proceedings of the 24th Offshore Symposium, February 20th 2019, Houston, Texas
Texas Section of the Society of Naval Architects and Marine Engineers

ABSTRACT

This paper presents a case study on how a target safety level can be demonstrated by using Structural Reliability Analysis (SRA), which is especially beneficial for structural life extension assessment.

Life extension requires operator to document that the structure will continue performing safely past its original design life.

The most frequently asked question by the industry in the process of life extension is:
1) Whether Design Fatigue Factor (DFF) can be reduced during life extension assessment
While the most frequently asked question by the regulator in the process of life extension is:
2) What is the current condition? When a particular component is non-inspectable, how do we know the condition and is it still safe to operate?

Both questions are related to structural reliability. Reliability analysis is to quantify uncertainties. The key criteria for life extension is to remain at the same level of safety as original design. When fatigue failure mode is of concern, the objective is not to maintain the same DFF, rather to maintain the same target safety.

This paper first discusses the background of DFF's and how it accounts for various uncertainties. Then the structural reliability methodology is introduced to explain the relationship between target safety and DFF. Probability of failure in terms of fatigue cracks grows over time. A few numerical examples are used to demonstrate how in-service inspection helps to reduce uncertainties and no findings of fatigue cracks reduces documented probability of failure and therefore increase target safety.

Further the paper addresses how to demonstrate that target safety is maintained for non-inspectable areas, such as TLP tendons, spar soft tank to truss, mooring piles, etc. A case study is used to illustrate a critical TLP tendon connection representing non-inspectable detail with high failure consequences. By collecting data such as fabrication and maintenance records, metocean and load measurements, more sophisticated fatigue analysis and structural reliability analysis can be used to demonstrate that same target safety level can be achieved with longer fatigue life. Historic data can be used to more accurately quantify the past fatigue damage, and introduction of monitoring at life extension will reduce uncertainties during continued operation. Effects on documented target safety level practically achievable with these data sets are quantified in the paper.

Overall, for both inspectable and non-inspectable structures, it is possible with proper structural reliability analysis to demonstrate that target safety level is maintained for many years past its original design life. Quality operational records and monitoring data can help, and careful selection of statistical parameters for SRA are critical in such assessments.

Keywords:* *SNAME, Offshore, Symposium, Structural Reliability Analysis, Fatigue, Risk Based Inspection

INTRODUCTION

Offshore structures are designed and maintained to the recognized industry standard and must comply with relevant shelf state requirements. Compliance with the design code requirement ensures that the required safety level has been met. For fatigue strength, that is achieved through a combination of design and in-service requirements. It shall be documented that fatigue life is not less than planned design life multiplied by a design fatigue factor (DFF). Depending on consequences of failure, and DFF achieved in design, in-service inspection may be required.

Life extension requires the operator to document that the target safety level will continue to be maintained past original design life. In most cases it is required to document remaining fatigue life to the most recent design codes, incorporating updated operating and metocean conditions. This often leads to more stringent fatigue requirements. When multiple details with no remaining calculated fatigue are identified, but detailed inspections do not find any cracks the analysis gets questioned as too conservative. For non-inspectable structures, questions are asked whether lower DFFs can be justified based on findings from inspectable details. This is where structural reliability analysis (SRA) may be required to provide confidence required by the regulators and other stakeholders.

DFFS IN DESIGN CODES - TECHNICAL BACKGROUND

The recommended DFF value depends on:

- consequences of failure (CoF), which can be a threat to human life, loss of property or the environmental impact, but also costs associated with repairs of fatigue cracks,
- ability to detect in-service cracks, including crack growth characteristics, reliability of inspection methods, and access for inspection,
- uncertainties in fatigue life prediction, including uncertainties in load and stress prediction, S-N curve, Palmgren-Miner rule.

Design codes specify DFFs required to meet target safety level for a given type of structure. An example from NORSOK N-001 is presented in Table 1; other design codes provide very similar tables and DFFs.

Table 1. Typical recommended DFF values [1].

Classification of structural components based on damage consequence	Not accessible for inspection and repair or in the splash zone	Accessible for inspection, maintenance and repair, and where inspections or maintenance is planned	
		Below splash zone	Above splash zone or internal
Substantial consequences	10	3	2
Without substantial consequences	3	2	1

Relationship between DFF and target safety level can be demonstrated by SRA for specific structural details. A time dependent limit state function for fatigue capacity is:

$$M(t) = R(t) - S(t) \tag{1}$$

where:

$M(t)$ - time dependent limit state function
$R(t)$ - fatigue capacity in function of time
$S(t)$ - load effect in function of time

Fatigue failure is indicated with:

$$M(t) < 0 \tag{2}$$

R(t) and S(t) hence M(t) can be modelled as stochastic variables, and probability of fatigue failure can be calculated as:

$$P_F(t) = P[\, M(t) < 0 \,] \tag{3}$$

where:

$P_F(t)$ - calculated probability of fatigue failure in function of time (PoF)

All inputs to R(t) and S(t) may be modelled with associated uncertainties, through for practical reasons it is convenient to model some parameters as deterministic and assign all uncertainties to limited number of input, for example:

- uncertainty in fatigue stress prediction, including metocean data, load prediction, modelling accuracy, stress concentration factor (SCF) are represented by a normal distribution with a coefficient of variance (CoV). For a simplified fatigue analysis the CoV is typically about 0.3-0.4, for full SFA about 0.12-0.15.
- Uncertainty in Palmgren-Miner damage summation rule, typically log-normal distribution with mean 1.0 and CoV of 0.3.
- Uncertainty in S-N curve can be found in fatigue standards, logC standard deviation around 0.2. Slope m is usually assumed as deterministic.

PoF can be defined as through thickness crack in a single detail. In order to assess target safety level, the fatigue failure needs to be related to the entire structural system, i.e. system PoF. In most cases the offshore structures are redundant, hence even with fatigue crack in the local detail, the load will be redistributed without significant consequences. In some cases, such as TLP tendons, a presence of a single fatigue crack in any of the girth welds may develop into system collapse. Consequences of fatigue crack can be expressed as P_{SYS}, with typical values taken as:

$P_{SYS} = \sim 10^0 - 10^1$ for a high consequence crack, e.g. crack in a TLP tendon pipe girth weld
$P_{SYS} = \sim 10^{-2}$ for a medium consequence crack, e.g. crack in a redundant brace on a semi-submersible
$P_{SYS} = \sim 10^{-3} - 10^{-4}$ for a low consequence crack, e.g. crack in a longitudinal end connection

Accumulated probability of failure target may be expressed as:

$$P_{F\ accumulated\ Target} = \frac{10^{-4}}{P_{SYS}} \tag{4}$$

what leads to:

$P_{F\ \text{accumulated target}} = \sim 10^{-4} - 10^{-3}$ for high consequence crack
$P_{F\ \text{accumulated target}} = \sim 10^{-2}$ for medium consequence crack
$P_{F\ \text{accumulated target}} = \sim 10^{-1} - 10^{0}$ for low consequence crack

Structure meets required target safety level when:

$$P_F < P_{F\ accumulated\ Target} \quad (5)$$

The PoF target is based on evaluation of crack CoF in terms of impact on system PoF, which can be challenging to determine accurately. Hence, design codes and regulators favor determination of target PoF based on calibration towards known case which can be supported by industry experience. Calibration approach is also less dependent on underlying methodologies and inputs, as the same analysis is performed for both the known case and the actual detail.

PoF as function of time can be presented on the chart as on Fig.2, where linear x-axis represent time (years in service), logarithmic y-axis represents PoF. Target PoF is a horizontal line. PoF curve below target PoF line indicates meeting required target safety level.

As structure deteriorates, if no action is taken the PoF will eventually exceed PoF target. When new information is introduced, e.g. in-service inspection is performed not revealing a crack, the PoF can be updated using Bayesian statistics. Effect of in-service inspection on PoF can be calculated as conditional probability:

$$P(F|I) = \frac{P(F \cap I)}{P(I)} \quad (6)$$

where:

$P(F|I)$ - calculated probability of failure F given known inspection result I

Inspections are modelled as stochastic variables and have associated probability of detection (PoD) which is a function of a crack size, hence SRA requires modelling of crack growth by fracture mechanics calculation. Crack growth model is calibrated to S-N fatigue results by aligning respective PoFs. Inspection with no crack detection reduces calculated PoF after inspection, which can be instantly seen on the graph as drop in PoF at the time of inspection, meaning the confidence in prediction fatigue performance increases.

A more thorough discussion on above concepts is available in the literature, e.g. [3] and [4].

To document relationship between PoF and DFFs recommended by design codes, 5 representative structural FSPO details, and TLP tendon pipe girth weld details have been evaluated, including effect of in-service inspections. Design life of 25 years is assumed. Deterministic S-N fatigue analysis was first performed using DNVGL curve D in-air with Weibull distributed hot-spot stress history, shape factor h=1.0 for FPSO details, h=0.7 for tendon pipes. Weibull scale factor q was calibrated to result in factored FL equal to design life, e.g. for a detail with DFF=3.0, unfactored FL is 75 years, factored FL is 25 years.

PoF was then calculated for each detail accounting for uncertainties in load and stress prediction, damage summation and S-N curve as discussed earlier in the paper. Probabilistic crack growth model was calibrated to result in PoF corresponding to S-N fatigue; this is necessary step to account for inspection and PoD curve. PoD curves are per [3], see Figure 1 for comparison of PoD for CWI and NDT.

PoF calculations were performed with probabilistic risk based inspection (RBI) software PROFAST [8]; input parameters are summarized in Table 2. Inspection plan was generated with prescribed 5 year cycle corresponding to standard class renewal survey (RS) cycle at 5, 10, 15 and 20 years of life. Resulting PoF curves are presented in Figure 2.

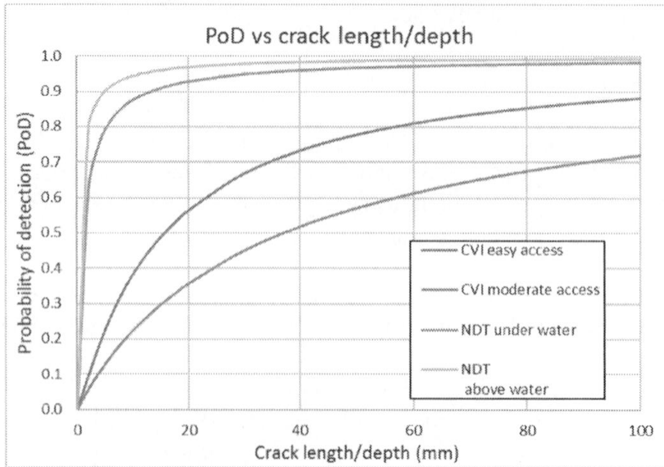

Figure 1. Probability of detection for CVI and NDT

Table 2. Case study No 1 - DFF background, structural details and analysis parameters.

Case ID	Detail	Inspection	DFF	FL in years unfactored	Consequence of failure	P_F accumulated target	Fatigue design analysis method	Weibull scale factor CoV
Base 01	FPSO side longitudinal end connection detail	CVI every 5y	1	25	Low	$\sim 10^{-1}$	Simplified	0.3
Base 02	FPSO main deck to topside stool support	CVI every 5y	1	25	Medium	$\sim 10^{-2}$	Component SFA with FE based SCF	0.2
Base 03	FPSO side shell plating	CVI every 5y	2	50	Medium	$\sim 10^{-2}$	Component SFA	0.25
Base 04	FPSO Moonpool under waterline	UW-NDT every 5y	2	50	High	$\sim 10^{-3}$	Component SFA with FE based SCF	0.2
Base 05	FPSO deck connection under PFP	No inspection	3	75	Medium	$\sim 10^{-2}$	Component SFA with FE based SCF	0.2
Base 06	TLP tendon girth weld	No inspection	10	250	Very High	$\sim 10^{-4}$	Tendon fatigue analysis	0.25

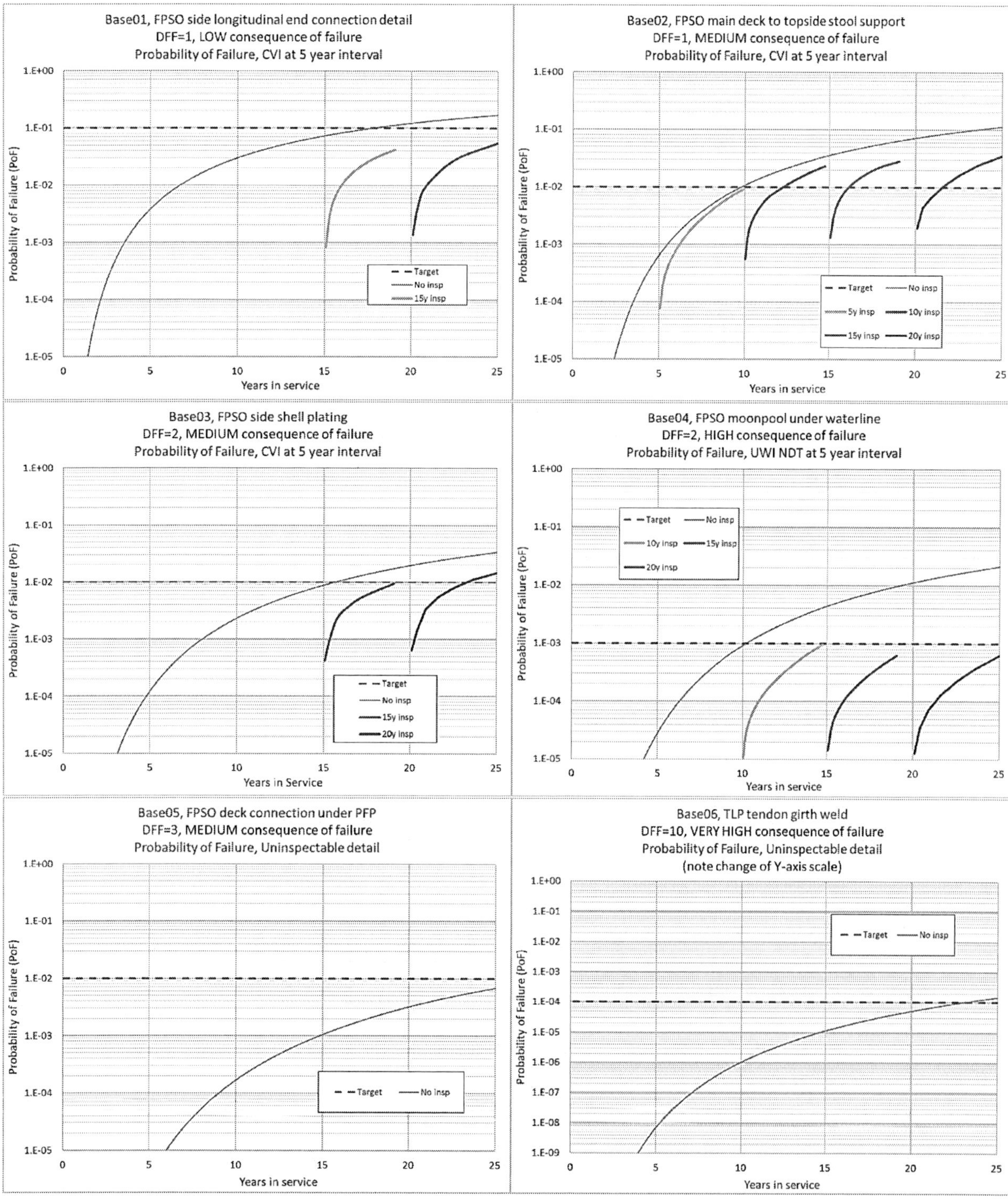

Figure 2. Case study No 1, PoF against target PoF for selected structural details with factored FL equal to design life of 25 years, DFF and CoV corresponding to minimum design code or class requirements [1], [2].

The calculation shows that the considered structural details, when designed in accordance with code recommended DFFs using required fatigue methodologies and inspected in service as required, are within the target PoF evaluated by equation (4) throughout the entire design life (note the target PoF based on (4) is specified as a range). The analysis can be used as calibration towards known case to establish target PoF specific to this analysis approach.

For side longitudinals and side shell plating (Base01 and Base03), PoF is below target PoF for first 15 years of service even without inspection. CVI at 15 at 20 years is required to maintain the PoF below equation (4) target. In practice these details will still be subject to inspection during e.g. general visual inspection, and inspection requirement would increase if cracks were found.

PoF for FPSO topside stool support at main deck detail (Base02) is approximately $3.5\text{x}10^{-2}$, which is higher than 10^{-2} but considered adequate given additional consideration explained further, which are not modelled in SRA in this case study, but would normally be taken into account during actual life extension SRA. Topside stool supports details are subject to spot check NDT every 5 years where locations with lowest design FL will usually be inspected. NDT scope may be increased in case spot checks reveals any cracks. The details are also very easily accessible and larger cracks would be promptly noticed by vessel crew between formal inspections, and adequate mitigations implemented involving relevant stakeholders.

For critical details with high consequences of failure such as FPSO moonpool details (Base04), the calculation shows that NDT every 5 years is necessary to maintain target safety level.

For non-inspectable details with up to medium CoF (Base05), DFF=3 is sufficient to maintain the target throughout design life. For non-inspectable details with very high CoF (Base06), DFF=10 is required.

DESIGN STAGE POF IMPROVEMENTS

Longer design FL obviously results in lower PoF during in-service life. A more stringent FL design basis can be expressed as higher recommended DFF value. Together with higher DFF, a more accurate fatigue analysis method, with corresponding smaller CoV on fatigue stress prediction, can also be used for design fatigue analysis, what further reduces PoF (provided the result of direct analysis shows the same FL). For details with longer design FL inspection requirement can be reduced already during original design life, and the structure will be more likely to meet future life extension requirements. The concept is reflected as additional class notations. FMS (DNVGL) recommends both increased DFFs and full spectral fatigue analysis. SFA (ABS) requires full spectral analysis though not necessarily an increased DFF (longer fatigue life may be specified). The comparison between the main class and FMS requirements are summarized in Table 3.

Table 3. Main Class vs FMS notation requirements based on [2] (condensed table is presented)

Structure	Examples	Main Class DFF	FMS DFF
Accessible structures accessible for regular inspection and repair	Main deck plate, topside supports, tranvsverse frames, longitudinal end connections	1	2
Accessible structures with limited access for inspection and/or repair	Shell and bottom plate, bilge keel, structure welded to submerged part.	2	3
Non-inspectable structures	Topside supports with passive fire protection	3	5
In addition to increased DFF, FMS notation also requires use of direct stochastic fatigue analysis, reducing uncertainty in design fatigue life prediction.			

The effect of increased DFFs and use of more sophisticated fatigue design analysis is demonstrated in Case study No 2. The same structural details as in Case Study No1 are used. See Table 4 for inputs and Figure 3 for PoF curves.

Table 4. Case study No 2, structural details and analysis parameters, Increased DFF and direct fatigue analysis

Case ID	Detail	Inspection	DFF	Unfactored FL	Consequence of failure	P_F accumulated target	Fatigue design analysis method	Weibull scale factor CoV
Ext01	FPSO side longitudinal end connection detail	CVI every 5y	2	50	Low	$\sim 10^{-1}$	Component SFA with FE based SCF	0.25
Ext02	FPSO main deck to topside stool support	CVI every 5y	2	50	Medium	$\sim 10^{-2}$	Full SFA	0.15
Ext03	FPSO side shell plating	CVI every 5y	3	75	Medium	$\sim 10^{-2}$	Component SFA with FE based SCF	0.25
Ext04	FPSO Moonpool under waterline	UWI NDT every 5y	3	75	High	$\sim 10^{-3}$	Full SFA	0.15
Ext05	FPSO deck connection under PFP	No inspection	5	125	Medium	$\sim 10^{-2}$	Full SFA	0.15
Ext06	TLP tendon girth weld	No inspection	20	500	Very High	$\sim 10^{-4}$	Tendon fatigue analysis	0.25

The calculation shows that use of higher DFFs and more sophisticated fatigue methodologies allows to achieve significantly lower PoF throughout the design life. Additional fatigue design margins will help documenting required remaining fatigue life at life extension.

For side longitudinals, topside support details at main deck and side shell plating and (Ext01, Ext02, Ext 02), PoF is below target PoF for entire 25 years design life even without inspection. This is reflected in class requirements [7] which allows extension of inspection intervals based on increased design DFF.

For critical details with high consequences of failure such as FPSO moonpool details (Ext04), the calculation shows that increase of DFF=3 instead of DFF=2 and use of direct SFA instead of component SFA, results in target safety level being maintained until 17 years of service without NDT requirement.

For non-inspectable details with up to medium CoF (Base05), DFF=5, and non-inspectable details with very high CoF (Base06), DFF=10, the target safety level is maintained well passed original design life, which is as expected given longer calculated fatigue life.

Note that surplus of design fatigue life should not automatically translate to remaining fatigue life past design life, and may need to be re-evaluated at life extension using available historical records and actual condition of the structure. This is discussed later in this paper.

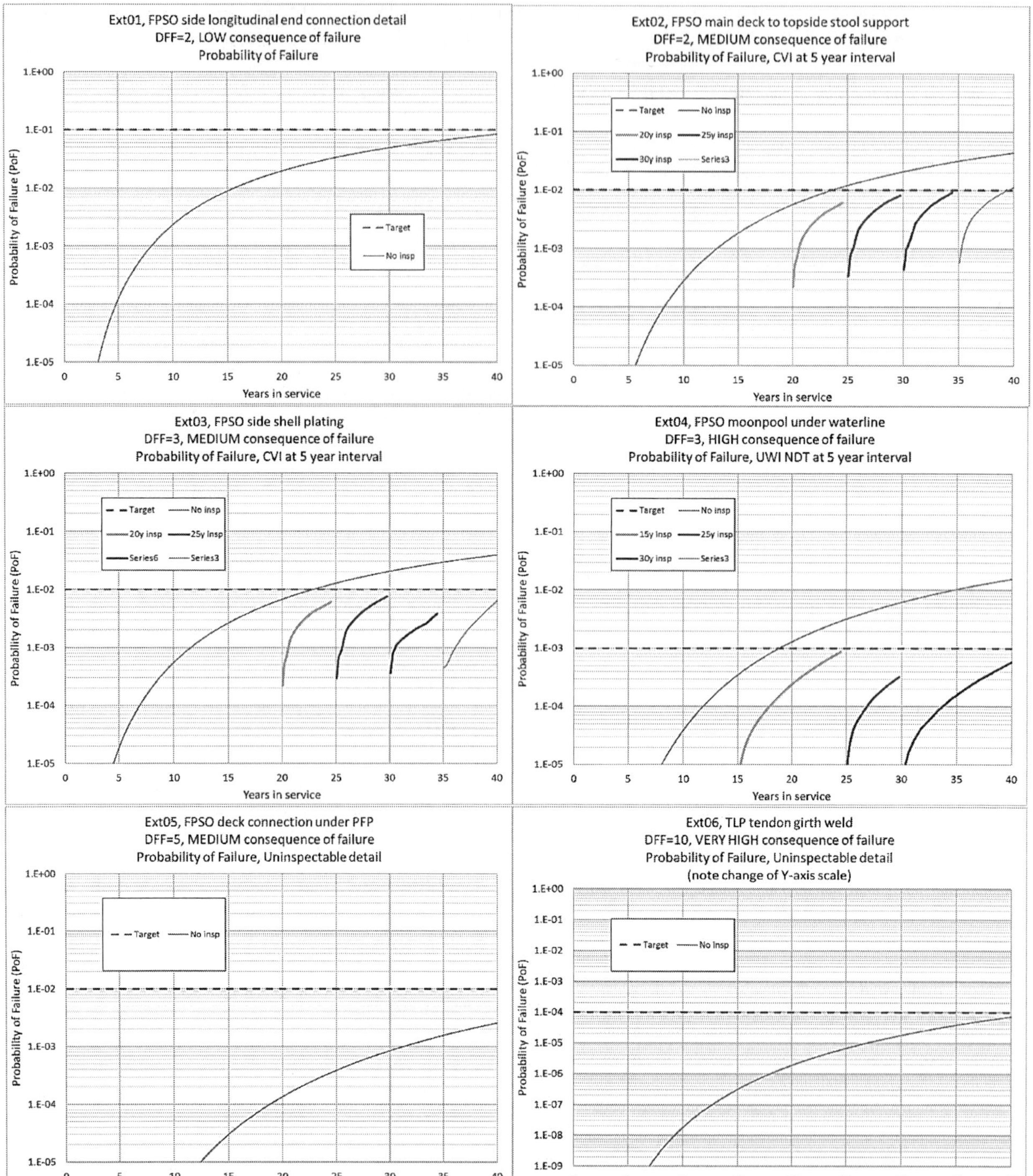

Figure 3. Case study No 2, PoF against target PoF for selected structural details with factored FL equal to design life of 25 years, increased DFF and reduced CoV corresponding to recommendation of FMS class notation [2]

LIFE EXTENSION STAGE – DOCUMENTATION OF TARGET POF WITH SRA

It is common that remaining FL documented by deterministic analysis is not sufficient for planned life extension. Possible reasons include insufficient design margins, relocation to a harsher operational site, higher fatigue damage rate based on monitoring, higher corrosion rates, coating deterioration, load increases due to unit modifications, updates in metocean data, changes in operational profile and other deviations from the original design assumptions. Sometimes shorter FL may be a result of using more sophisticated fatigue analysis methods compared to design method, or even errors in design FL prediction discovered during life extension planning.

Unlike in deterministic fatigue analysis, uncertainties in FL prediction can be directly accounted for in SRA, which can be used to evaluate whether the structure continues to maintain target safety level. For inspectable details, SRA can provide sufficient evidence to limit or even avoid extensive and costly structural repairs, what can be a difference between a "go" or "no go" for life extension. For non-inspectable details, including details with very high CoF, SRA may be capable to document higher safety level even when deterministic analysis indicated shorter FL.

Accuracy of inspection method, expressed as PoD, can be directly accounted for in SRA. Full credit can be taken for use of a more reliable inspection method. Inspection results have direct impact on calculated PoF, hence SRA can be used to derive required inspection interval to maintain required target PoF.

A life extension plan will typically include a combination of repairs and in-service inspections. With SRA, lifetime cost can be estimated for each acceptable scenario with associated confidence limits. A solution with lower cost uncertainty may be preferred over marginally less expensive plan but highly uncertain cost; such information is not available through deterministic analysis.

Use of SRA will typically involve a combination of multiple steps including:

- Fatigue re-analysis using advanced methods.
- Accounting for collected operational history data.
- Fatigue life monitoring during life extension.
- Implementation of more frequent inspections or use of more accurate inspection methods.
- Accounting for correlation effects to reduce inspection scope.

Introduction of stress monitoring or additional NDT inspection will only be practical for limited number of details with lowest calculated FL identified through fatigue analysis, e.g. topside stool support with lowest fatigue life, and the most critical TLP tendon. For the approach to work, it is essential that correct detail locations are identified as critical.

It is also useful to reduce uncertainties in FL prediction as much as possible, which can be achieved through fatigue re-analysis using advanced analytical methods, e.g. full SFA preferably introducing available operational data. During life extension planning phase the unit has typically been in service for minimum 15 years, and some information had been collected at least for some time periods. The operational data may include:

- For mobile units, time spent at actual location and associated weather data.
- Actual loading history, e.g. loading computer data or information from onboard daily loading reports.
- Unit heading history including relative to-wave, to-wind and to-current headings.
- Actual metocean measurements collected during operational life of the unit, e.g. site-specific weather buoy data.
- Motion data
- Load monitoring data
- Stress measurements
- Equipment live load history, e.g. crane loading data

It is common that the data is incomplete, and the operator may not be fully aware of the available data sets. Still, even an incomplete additional information is better than sole design analysis which is normally based on assumptions. E.g. correlation studies of time stamped data often provide useful validations and increase confidence in PoF calculations.

Fatigue reanalysis will also be required for structure which was modified, added or removed during service life. Historical UTM records may be useful, as design analysis was likely based either on gross or half-corroded state, which may be different in real life. Coating and corrosion protection system condition can provide basis for selection of S-N curves appropriate to corrosion state in particular service life periods. Fabrication records including achieved manufacturing tolerances, material certificates, NDT inspection records or even fatigue testing results can be used to reduce uncertainties in fatigue life prediction for life extension, hence document higher safety level.

Case study No3 is presented to illustrate the process on how SRA can be used to document required safety level. Two details from Case study No1 are presented, which is also used as calibration of target PoF. See inputs in Table 5 and resulting PoF curves in Figure 4.

Detail 1 (Life1): Topside support stool on the FPSO main deck (same detail as Base02) – inspectable structure with DFF=1, Medium consequences of failure.

3 subcases are evaluated and PoF curves presented on Figure 4-LEFT:

- Life1A: is original detail with fatigue life prediction based on design analysis equal to 25 years, i.e fatigue life being equal to design life with DFF=1.0 The detail has been subject to CVI during first 20 years of service per class requirement. Based on Medium criticality, target PoF was assessed at around $\sim 10^{-2}$. CoV=0.20 on fatigue stress prediction corresponds to component stochastic method with FE based SCF. The results are presented on Figure 4-LEFT showing that based on SRA no inspection is necessary until first 10 years of service. At the end of 25 years of service life, with CVI the PoF is 3.5×10^{-2} which can be used as calibrated target PoF for cases Life 1B and 1C. Past 25 years design life, continuing the 5 year CVI inspection cycle will maintain about the same PoF, i.e. within the calibrated target.
- Life1B: is the same details as 1A, however in this case fatigue re-analysis during life extension planning revealed that calculated fatigue life is about 50% of the original design assumption. Compared to 1A, SRA was performed with lower CoV=0.15 corresponding to lower uncertainty in stress prediction when direct SFA is used. As detail now has nearly 50% shorter FL, calculated PoF is obviously higher than 1A, and lower uncertainty is not enough to bring it down to the calibrated target PoF of 3.5×10^{-2} – see Figure 4-LEFT
- Life1C: Following lower than design FL in 1B, NDT is performed at 25 years (at life extension) and again at 30 years (5 years past original design life), with no cracks discovered. With superior PoD, NDT has significant effect on calculated PoF, see Figure 4-LEFT It can be demonstrated that introducing NDT inspection the calculated PoF is clearly below calibrated target PoF of 3.5×10^{-2}.

Detail 2 (Life2): Girth weld on TLP tendon pipe (same details as Base06) – non-inspectable structure with DFF=10, Very High consequences of failure.

3 subcases are evaluated and PoF curves presented on Figure 4-RIGHT:

- Life2A: is original detail with fatigue life prediction based on design analysis equal to 25 years, i.e fatigue life being equal to 25 years design life with DFF=10, or unfactored calculated fatigue life of 250 years. No in-service inspection is planned for the welds. Uncertainty in fatigue life prediction at design stage was represented with CoV=0.25 on fatigue stress suggested by [6]. Calculated PoF reaches 1.4×10^{-4} at 25 years, which very close to PoF $\sim 10^{-4}$. As 1.4×10^{-4} corresponds to code requirement of DFF=10, it is used as target PoF calibrated to know case for cases Life2B and Life2C.
- Life2B: is the same girth weld if TLP tendon pipe, but in this case PoF was calculated using fatigue stress based on calibration with stress measurements. Reduced uncertainty is represented with CoV=0.15 was suggested by (ref [6]) based on about 8 months top tendon tension measurement campaign. Resulting PoF is compared against 1.4×10^{-4} calibrated target PoF established in Life2A giving 43 service life. Comparing against PoF=10^{-4} gives 40 years service life.

- Life2C: is the same inputs as Life2B, but with top tendon tension measurements showing higher loads than predicted in design, leading to the calculated fatigue life of about 66% of design requirement (165 years unfactored FL vs 250 in cases 2A and 2B). PoF curve (Figure 4-RIGHT) reaches 1.4×10^{-4} calibrated target PoF at 30 years of service, and 10^{-4} PoF at 28 years, indicating somehow extended service life even with increased measured tendon loads.

Table 5. Case study No 3, structural details and analysis parameters

Case ID	Detail	Inspection	Target DFF	FL in years unfactored	Achieved DFF	Consequence of failure	P_F accumulated target based on equation (4)	P_F accumulated target calibrated	Fatigue design analysis method	Weibull scale factor CoV
Life1A	FPSO main deck to topside stool support	CWI every 5y	1	25	1	Medium	$\sim10^{-2}$	$3.5\text{x}10^{-2}$	Component SFA with FE based SCF	0.25
Life1B		CWI every 5y	1	13	0.52				Full SFA	0.15
Life1C		CWI every 5y, NDT every 5y from year 25	1	13	0.52				Full SFA	0.15
Life2A	Girth weld on TLP tendon pipe	No inspection	10	250	10	Very High	$\sim10^{-4}$	$1.4\text{x}10^{-4}$	Time domain design analysis	0.25
Life2B		No inspection	10	250	10				Measurements	0.15
Life2C		No inspection	10	165	6.6				Measurements	0.15

CONCLUSIONS

Technical background of DFFs was illustrated using 5 typical FPSO details with Low and Medium CoF, and TLP tendon girth weld with Very High CoF. Target PoF for each detail was established based on CoF and system PoF. For all considered details it can be demonstrated that the DFFs recommended by design codes, in combination with in-service inspection requirements, lead to a consistent safety level. It is also identified that some reduction of inspection scope, especially at the beginning of service life, may be possible using SRA.

Meeting increased DFFs in combination with more direct design fatigue analysis allows for SRA justification of even less in-service inspections. Chances of meeting remaining FL in life extension are also greater, especially that analysis input in terms of loading and structural capacity might have changed.

When planning for life extension, fatigue re-analysis may be required based on actual operational history. Fatigue re-analysis may also be necessary to select correct fatigue critical details for additional inspection or monitoring. All available operational data should be collected to support the re-assessment. Identified critical details can be analyzed using SRA to identify RBI program required to maintain target safety level.

Required target safety level may be identified from calibration against know case, which in this case may be DFF recommended by the code.

For inspectable details with low DFF it was demonstrated that even significantly lower FL is identified during fatigue re-analysis at life extension, both increase of CVI inspection frequency and introduction of NDT can be implemented to maintain required target safety level. As NDT has greatly superior PoD, it leads to significant PoF reduction if no cracks are found.

For non-inspectable details with high DFF increase confidence in fatigue loading is very effective to achieve lower PoF hence document longer fatigue life. In presented example introduction of monitoring data for limited time period allowed to document 70% longer design life. Even if the measured loads are approximately 10% higher than design predictions, 20% additional design life can be documented.

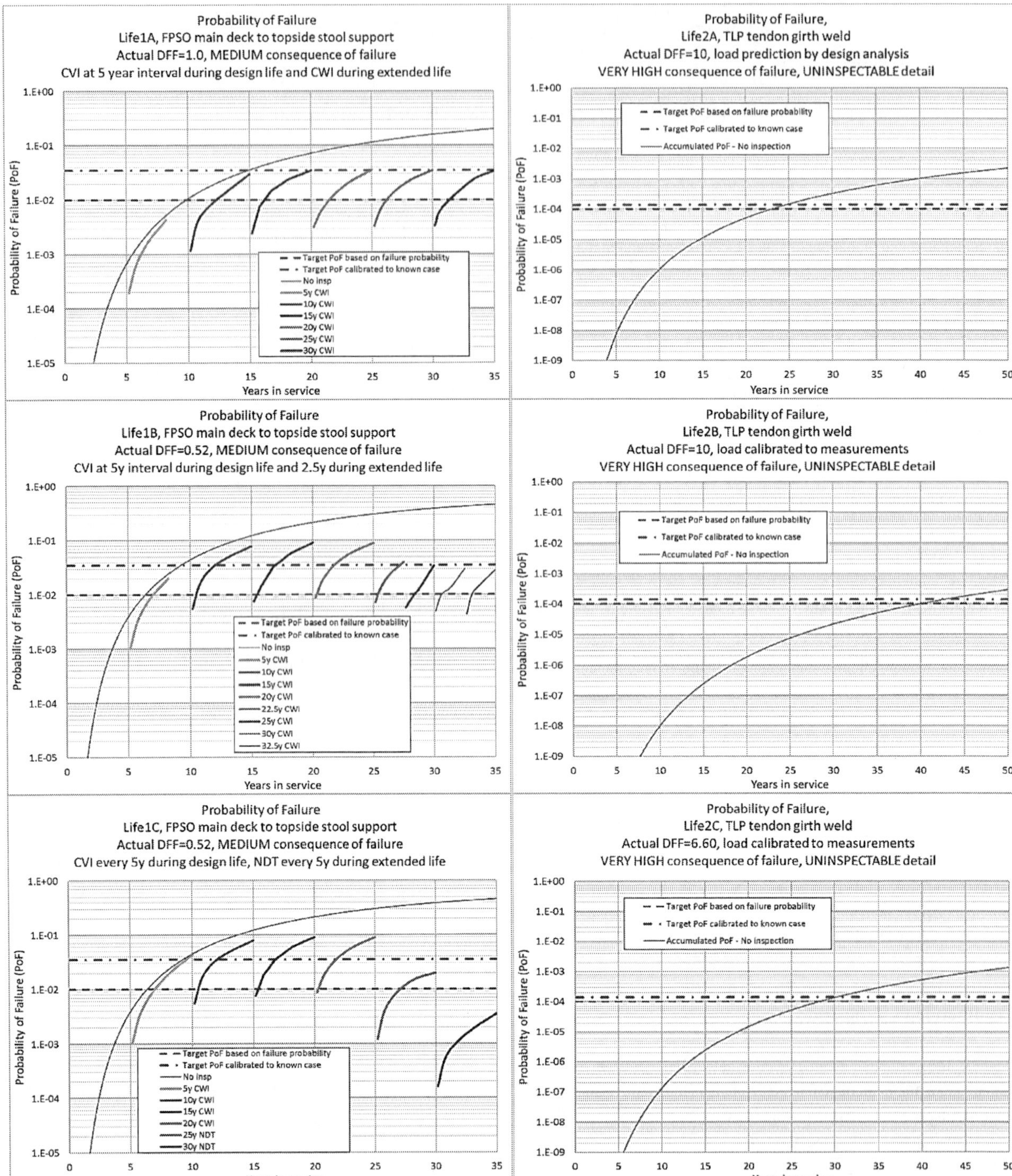

Figure 4. Case study No 3, PoF and corresponding target values for cases Life1A/B/C FPSO topside stool support detail (LEFT) and Life2A/B/C TLP tendon pipe girth weld (RIGHT)

ACKNOWLEDGEMENTS

The opinions expressed herein are those of the authors and shall not be construed as necessarily reflecting the views of DNV GL.

GLOSSARY

Terms and abbreviations used in the text are defined below.

SRA – Structural Reliability Analysis.
FL – fatigue life.
D – fatigue damage calculated from S-N curve using Palmgren-Miner summation. D=1.0 means fatigue failure as defined by design S-N curve.
DFF – Design Fatigue Factor.
Factored FL – fatigue life including DFF used as basis of design, factored FL = design life / (D x DFF), unless otherwise noted FL is reported in years.
Unfactored FL – fatigue life disregarding DFF, unfactored FL = design life / D, unless otherwise noted FL is reported in years.
PoF – Probability of Failure, unless otherwise noted, text refers to calculated accumulated PoF in fatigue failure mode, where fatigue failure is defined as through thickness crack.
Accumulated PoF – PoF accumulated from beginning of service life.
CoF – Consequence of Failure, unless otherwise noted it means CoF of the local fatigue failure, i.e. single through thickness crack of the individual structural detail.
System PoF – PoF of the entire structure, e.g. loss of unit. Unless otherwise noted, text refers to accumulated PoF.
PoD – Probability of Detection associated with inspection method, PoD is expressed as function of crack length or depth.
CoV – coefficient of variance.
SCF – Stress Concentration factor.
RBI – Risk Based Inspection.
CVI – close visual inspection.
NDT – non-destructing testing.
UW-NDT – under water NDT.
UTM – ultrasonic thickness measurements.
Simplified fatigue analysis – deterministic S-N fatigue life prediction with simplified representation of fatigue loading and normally using tabulated SCF. Fatigue damage is typically obtained from closed form formulas assuming long term Weibull distribution of stress history with distribution parameters determined from simplified analysis or using prescriptive formulas in design codes. For ship-shaped units, classification ship rules may be used.
Stochastic fatigue analysis (SFA) – deterministic S-N fatigue life prediction with fatigue stress prediction using frequency domain direct hydrodynamic analysis. S-N fatigue damage is first determined for individual seastates in the scatter diagram from closed form formulas assuming short-term Rayleigh distributed stress history, and total fatigue damage is calculated as sum of damages from individual seastates considering seastate occurrence probabilities.
Component SFA – stochastic fatigue analysis where fatigue stress is established as superposition of stress components factored to hydrodynamic load responses from frequency domain direct hydrodynamic analysis, e.g. local plate bending stress is expresses as local pressure multiplied by a constant. When load constant can be accurately determined, component SFA's accuracy approaches that of full SFA and is capable to directly account for phase shifts between load components.
Full SFA – stochastic fatigue analysis where fatigue stress is determined using FE analysis, typically a combination of global and local model, with directly applied complex hydrodynamic load cases. Full SFA is considered state of the art analysis, but it can be surpassed by time domain analyses.

REFERENCES

[1] NORSOK N-001, Integrity of offshore structures, Edition 8, September 2012.

[2] DNVGL-OS-C102, Offshore Standard, Structural design of offshore ship-shaped units, Edition July 2018.

[3] DNVGL-RP-C210, Recommended Practice, Probabilistic methods for planning inspection for fatigue cracks in offshore structures, Edition November 2015.

[4] Lotsberg Inge, Fatigue Design of Marine Structures, Cambridge University Press, 2016.

[5] DNVGL-RP-C203, Recommended Practice, Fatigue design of offshore steel structures, Edition April 2016.

[6] Dalane Jan Inge, Fatigue Reliability – Measured Response of the Heidrun TLP Tethers.

[7] DNVGL-RU-OU-0300, Fleet in service, Edition July 2018.

[8] PROFAST software V2.2.06.

SNAME
THE INTERNATIONAL COMMUNITY
FOR MARITIME AND OCEAN PROFESSIONALS

A COMPREHENSIVE FRAMEWORK TO DEVELOP RESILIENT OFFSHORE STRUCTURES

ALBERTO MORANDI
GUSTOMSC, US INC.

Proceedings of the 24th Offshore Symposium, February 20th 2019, Houston, Texas
Texas Section of the Society of Naval Architects and Marine Engineers

ABSTRACT

The marine and offshore industries face uncertainty in oil prices as well as concerns over climate change and energy transition. These are challenges but also opportunities as the industry moves from a culture seeking evolutionary concepts to a culture seeking efficiency through innovation. Safety and efficiency are key to supporting future progress and such goals are achieved through optimization of hull forms, load paths and material efficiency. On the structural discipline, advanced methodologies exist to cover the key concepts of shape optimization through concept models of hull form, definition of loads and load paths, modeling of randomness and uncertainty, customization of structural reliability through Structural Reliability Analysis (SRA), decision making metrics, robustness & resiliency and computational multi-criteria multi-objective optimization. This paper proposes to integrate such concepts into a conceptual framework to develop resilient offshore structures. The concepts described have wider application to several types of structures on land, marine and offshore-based.

Keywords: Reliability, Robustness, Resiliency, Structures, Offshore

INTRODUCTION

The marine and offshore industries continue to address our needs and ambitions [1], Figure 1:

- Growing population seeking increased standards of living
- Sustainable harvesting of natural resources at sea (fishing, sea mining, solar energy, oil & gas, wind, tidal and sea kelp harvesting for energy)
- Efficient, reliable and safe working environment at sea

The next 20 years present challenges and opportunities, both in terms of market developments and technical evolution. Based on current trends, Table 1 articulates scenarios for the future development of the marine and offshore industries.

In Scenario #1, activity in the oil and gas industry remains depressed and focused on lower break-even cost and enhanced production of existing fields. Political motivation to change towards renewable energy sources remains limited. Offshore wind installed capacity continues to expand progressively at about 20% on a yearly basis but from a relatively small percentage of the total energy matrix, having reached 14,300 MW installed in 2016 (89% in Europe, 11% in Asia) and 16,100 MW by June 2017. Projects go ahead in Europe without government subsidies as offshore wind becomes economically competitive. In the US, individual states take the lead in supporting offshore wind with some support from the federal government. Large-scale offshore fish farms continue to develop based on projects now active in Norway.

Figure 1 – Developing Natural Resources at Sea [1]

Scenario	Oil Price	Energy Transition	Opportunity
1	Low	Slow	Enhance production from existing fields Scrap, Re-evaluate, Refurbish, Reuse (CO_2 Injection)
2	High	Slow	Develop Complex and Demanding Areas Ultra-Deep Water, Harsh Environment, Remote Locations (Arctic)
3	Low	Fast	Oil & Gas Cos. accelerates transformation into Energy Cos. Renewables, Energy Storage and Transmission / Transport
4	High	Fast	Green and Advanced Technical Solutions Oil & Gas + Renewables

Table 1 – Where will you be in 2039?

Scenario #2 is a recovery in oil prices combined with continued effort to reduce project costs, leading to a return to complex and demanding areas for oil and gas production. Offshore wind and fish farming will continue to grow.

Scenarios #3 and #4 consider a higher political motivation to develop renewable energy. Oil and gas companies transform themselves into energy companies. In 2017, for example, the Danish Oil Natural Gas (DONG) company divested from hydrocarbons and focused entirely on renewables, rebranding itself as Ørsted. STATOIL is also positioning itself as a leader in renewable energy, rebranding itself as Equinor. Advances will continue to take place in fish farming and energy generation from sea kelp [2], for example.

On the structural discipline, maturing and emerging technologies will integrate key areas historically treated in an intuitive and somewhat fragmented manner:

- Structural Form
- Development of Load Paths and Material Allocation
- Constructability, Durability and In-Service Monitoring and Inspection Techniques
- Addressing uncertainties due to structural and non-structural factors through the three R's: Reliability, Robustness and Resiliency.

Key enabling technology developments will occur around shape optimization through concept models of hull form, definition of loads and load paths, modeling of randomness and uncertainty, customization of structural reliability through Structural Reliability Analysis (SRA), decision-making metrics, robustness and resiliency and multi-criteria multi-objective optimization.

The objective of this paper is to follow on recent keynote lectures at the Instituto Superior Técnico, Lisbon [3] and at Texas A&M University, College Station, Texas [4] as well as a presentation in the Robustness Seminar at the SNAME Offshore Symposium of 2018 [5] to articulate a comprehensive framework for optimal structural safety, reliability and efficiency.

RELIABILITY, ROBUSTNESS AND RESILIENCY

Figure 2 illustrates the key concepts. The reliability of a system undergoes small fluctuations under limited perturbations during normal operations. It can be affected by an extreme event to an extent that may be mitigated by a robust design, which allows first responders and further repair and recovery efforts to restore the system within a reasonable time frame and without cascading into a catastrophic failure.

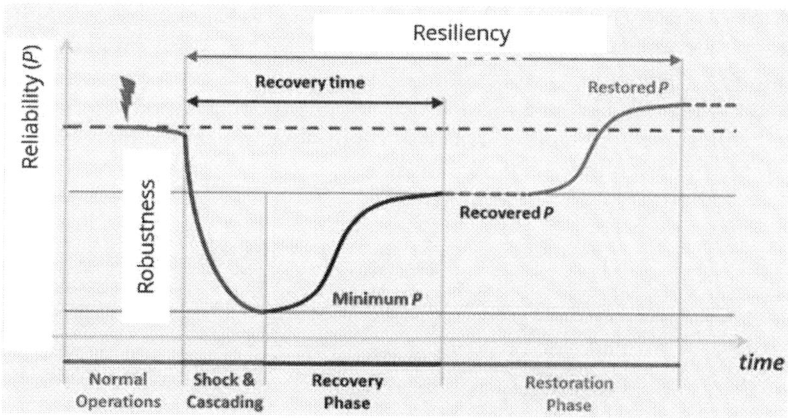

Figure 2 – The Three R's: Reliability, Robustness and Resiliency

Everything starts with reliability at the design stage. Figure 3 illustrates key concepts in reliability based design and decision making: the vertical axis is the benefit B minus total cost Z(p), composed of C(p) which is the fabrication and operation cost and D(p) which is the risk cost of damage or failure. These are functions of a design parameter p (such as plate thickness, number and size of stiffeners, etc.). D(p) is the probability of failure multiplied by the economic consequence: $D(p) = p_f(p) * C_q(p)$. Probability of failure quantifies uncertainty and variability in load and resistance variables.

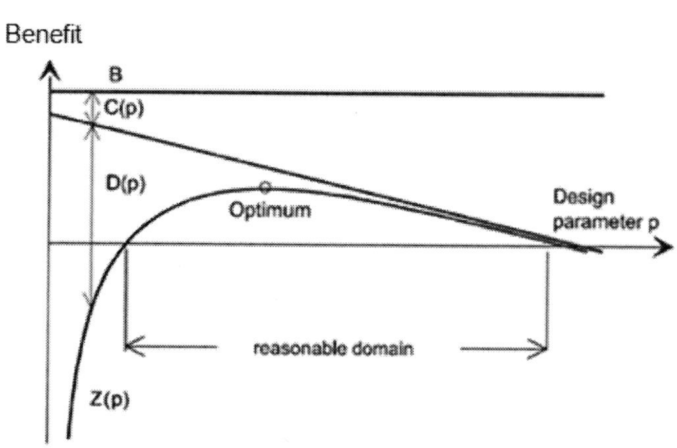

Figure 3 – Decision-making Metrics [6]

In order to deal with uncertainty and variability in structural behavior, in the last century engineers across different industries established breakthroughs in a new field: Structural Reliability Analysis (SRA). It is not surprising that aerospace engineers were pioneers in this field as they used lightweight materials without the well-defined endurance limit of mild steel and under relatively higher stress levels. In the UK the concept was elegantly described by Pugsley in 1942 [7]: '*In aeronautical work the general engineering outlook in regard to the specification of strength factors has been supplemented by an appreciation of the unusually variable nature of the loads coming upon the airplane structure. The two questions have generally been treated as separate ones, and it is only recently that efforts have been made to bring them together by relating both to accident rates. In doing so we have on the one hand to allow for the frequency of occurrence of loads of different magnitudes and on the other hand, to allow for the variation in strength among airplanes produced to a given design.*'

In the USA, Freudenthal in 1945 [8] demonstrated that through the application of probability theory the concept of safety can be rationalized. Pioneering work was also conducted in Europe on both sides of the 'Iron Curtain' [9], [10], [11]. By the beginning of the 1960's a comprehensive probabilistic theory was available first with simple concepts and then with an increasingly more sophisticated format. The First-Order Second-Moment method allowed multiple design factors to be generated from statistical data and the use of probabilistic theory to probe the low-probability edges of system behavior. A second-moment format for evaluation of structural reliability was proposed [12] which generates a "safety index" calculated from the mean (or bias) and variances (the variance is the second moment of a distribution) of the parameter distributions. The safety index is used to derive partial safety factors (PSF) applied to load and resistance variables in a method that became the foundation of probabilistic design theory [13]. The main principles are illustrated in Figure 4. This was a milestone; reliability analysis was now related to accepted civil engineering methods of design through the PSF format.

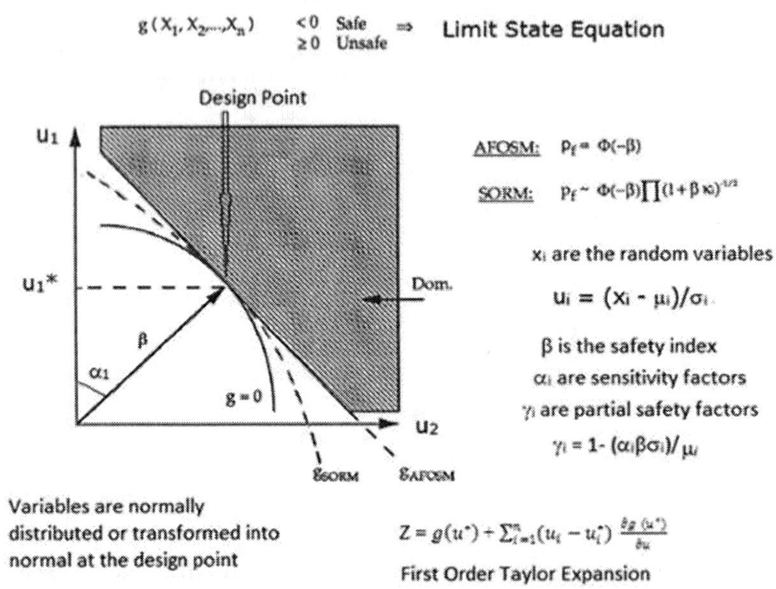

Figure 4 – Advanced First Order Second Moment Method (AFOSM) & Second Order Reliab. Method (SORM)

The PSF format provides the linkage to probabilistic approaches of structural design that can reconcile safety margins with observed rates of failure and enable risk-based design criteria. Another advantage is exploiting the ultimate load bearing capacity of a structural component considering its plastic behavior by adopting a limit state approach. The ISO series of offshore standards adopts a partial safety factor format for the following general limit states [14]:

- Serviceability Limit States (SLS, related to fitness for use);
- Fatigue Limit States (FLS, related to durability and structural integrity);
- Ultimate Limit State (ULS, related to the ultimate bearing capacity of the structure);
- Accidental limit state (ALS, related to abnormal or accidental events).

SRA theory has progressed covering multiple types of probability distributions, non-linear failure surfaces, correlation between basic variables and limit states, load combinations, etc. However, these more advanced methods have often found resistance from many engineers who prefer the proven deterministic allowable stress design format.

Modeling limit states often requires advanced analysis of failure processes; statistics and probability theory are not familiar to many practicing engineers and require large amounts of data to quantify low probability events. Even when such data is available, users need to carefully consider if the underlying phenomena is ergodic (statistical properties constant over samples) and stationary (constant over time). The results of probabilistic structural analysis are often too sensitive to assumptions for the underlying mean and variance of the key variables and the payoff is not always convincing in terms of structural efficiency. In addition, the goal of reconciling design safety margins with observed failure rates is complicated by non-structural factors that trigger structural failures that are not related to the structural safety margins considered in design.

There are cultural issues at play and in 1954 Freudenthal [15] realized a problem that still exists today: '*...the concept of safety is deeply rooted in engineering design, whereas the notion of finite (no matter how small) probability of failure is repulsive to a majority of engineers.*' In litigation-prone countries, prescriptive design provides a safer framework for engineering practice than the risk-based approach that underpins probabilistic based structural design.

The challenge of dealing with processes that may not be ergodic and / or stationary is complex and apparent by the unfortunate and unprecedented destruction caused by hurricane Harvey's rainfall, Figure 5.

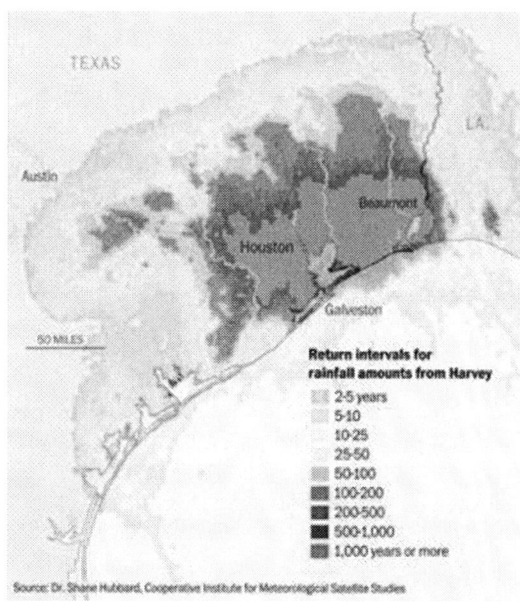

Figure 5 - Hurricane Harvey Rainfall Effects [16]

As heavy rain continued and water started to overflow on the emergency spillway of the Addicks reservoir, the US Army Corps of Engineers decided to perform controlled releases from both the Barker and Addicks dams. Sections of Buffalo Bayou and its tributaries did not have enough capacity to flow the released water downstream of the dams causing the disastrous flooding of neighborhoods that were supposed to be safely outside established flood plains (like the 100-year or higher) and causing billions of dollars of damage.

Key to this discussion is that changes in circumstances, internal structure or external environment may create an unsafe bias where safety margins that were previously consider sufficient are no longer sufficient. For example, some climate models suggest that climate change may influence hurricanes to be more intense globally and have higher rainfall rates [17], although the subject is very complex. Another example is that extensive and uncontrolled housing development may affect land drainage. Finally lack of investment in our waterways can be another factor in preventing more effective drainage in an extreme situation. It is noted that prior to Harvey in 2017, Houston suffered with the Memorial Day floods in 2015 and Tax Day floods in 2016. Today about 40% of the US population leave in coastal areas and many in less than 10ft above sea level. Some estimates place total insured property values in US coastal areas at 10 trillion dollars.

Offshore structures face similar challenges due to extreme storms as well as changes to existing structurers such as adding topsides weight and upgraded functionality, adding elements that attract metocean loading, foundation subsidence, etc. [5]. Integrated concepts such as floating production systems are composed of many subsystems generally designed according to the codes and practices of different disciplines. Perturbations within such individual subsystems and procedures can have a cumulative effect (Swiss Cheese Model, Figure 6) and ultimately trigger a structural failure not really related to the load cases considered in design.

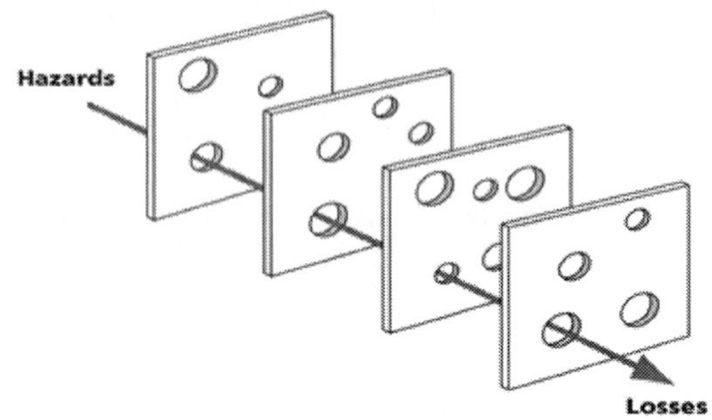

Figure 6 - Swiss-Cheese Model

The blowout of the Macondo well was the worst disaster in the history of the Gulf of Mexico. The US Chemical Safety Board investigation pointed to a structure failure – buckling of the riser pipe due to differential pressure (effective compression) rendering the Blowout Preventer (BOP) unable to shear the off-center buckled pipe and seal the well [18]. The reports by the US Chemical Safety Board and by the US National Academy of Science, Engineering and Medicine [18], [19] are an illustration that non-structural factors can ultimately trigger a structural failure not really related to the load cases considered in design.

STRUCTURE ROBUSTNESS AND RESILIENCE

In addition to wider industrial risk assessment and mitigation techniques, within the structural discipline robustness is a developing concept to mitigate the challenges identified in the previous section [5]. For example, Haver and Winterstein [20] formulate control of both the ULS and ALS in terms of avoiding a bad-behaving problem where a disproportional consequence follows an increase in loading, Figure 7.

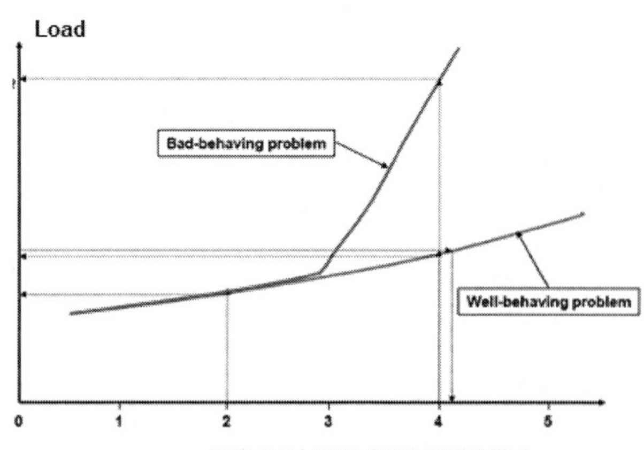

Figure 7 – Illustration of a Load Problem [20]

Proactive measures are key in developing robustness and resilience:

- Reduction of the probability that the action occurs or reduction of the action intensity (e.g. prevention through monitoring, inspection, disconnection, etc.);
- Reduction of the effect of the action on the structure (e.g. protection through ice management, vessel heading changes, floater draft adjustment and slacking of mooring lines, etc.);
- Mitigation of the consequences of failure (evacuation, shutting down wells, etc.)

Offshore structures in the Gulf of Mexico were impacted by hurricanes such as Andrew (USA, 1992), Roxanne (Mexico, 1995), Ivan (USA, 2004), Katrina (USA, 2005), Rita (USA, 2005) and Ike (USA, 2008). A study of floating production system performance under hurricane Lili (USA, 2002) [21] indicated that the design bases for these novel systems (including metocean criteria) were in some cases challenged and needed to be further studied. Requalification criteria emerged for vintage structures realizing that these structural systems often possess an inherent reserve strength that is greater than the strength of the individual components [22], [23], [24] [25]. Reserves from different sources provide safe side bias, a key one being the nonlinear structural interaction between components through plastic deformation and load redistribution, which in redundant systems allows mobilization of alternative load paths after an initial damage.

Design codes have generally defined reliability and safety with reference to each structural component. A key concept in developing robustness and resiliency is to understand the system behavior and accept some member damage as long as the total damage is limited and does not lead to progressive collapse until repair and remediation take place. The concept of system measures of strength gained acceptance within the offshore industry in the 1990s [26], [27] and has continued to evolve [28]. For example, systems considerations indicated that X-bracing was a more efficient structural form for fixed platforms with implications for robustness and in-service inspection [27]. The same considerations apply to jack-up legs and X-bracing forms the basis of GustoMSC jack-up leg designs, for example.

ISO 19902 [29] states that '*A structure shall incorporate robustness through consideration of the effects of all hazards and their probabilities of occurrence, to ensure that consequent damage is not disproportionate to the cause. Damage from an event with a reasonable likelihood of occurrence shall not lead to complete loss of integrity of the structure. In such cases, the structural integrity in the damaged state shall be sufficient to allow a process system close down and a safe evacuation*'. A very similar statement is included in ISO 19900 [14]. The ISO 19902 definition is elegant as it bounds the perturbation level to events with a reasonable likelihood of occurring.

ISO 19900 and 19902 state that: '*Robustness is achieved by either: a) designing the structure in such a way that any single load bearing component exposed to a hazard can become incapable of carrying its normal design actions without causing collapse of the structure or any significant part of it; or b) ensuring (by design or by protective measures) that no critical component exposed to hazard can be made ineffective; or c) a combination of a) and b), above*'.

ISO 19902 elaborates on the structural aspects of robustness: '*Abnormal design situations are used to provide robustness against events with a probability of exceedance of typically between 10^{-3} and 10^{-4} per annum by avoiding, for example, gross overloading*'. It provides guidance on robustness considering vessel collisions, dropped objects, fire and explosion and discusses the impact of robustness on in-service inspections so that damage either propagate slowly or have little effect on overall structural reliability.

FORM, LOAD PATHS AND CONSTRUCTABILITY

Figure 8 illustrates offshore structural forms, which can be bottom founded, anchored or dynamically positioned (DP).

Figure 8 – Offshore Design Concepts [1]

Mobile Offshore Units (MOUs) are used when drilling for hydrocarbons as well as for offshore construction (heavy lifting, pipelaying, wind farm installation, civil construction, etc.). Their design generally centers on the utilization of deck space, overall load carrying capacity as well as safe and efficient operations around the well centers or construction site. Self-elevating units like jack-ups provide ample deck space, load carrying capacity and a stable platform for water depths up to 175m (GustoMSC CJ80 design) in a wide range of environments. For mild environment and deep water, the ship-shaped vessel is often the preferred solution offering ample deck space, load carrying capacity and accurate positioning utilizing a dynamic position (DP) system. For harsh environment, column-stabilized units such as semisubmersibles offer a more refined naval architecture solution where buoyancy develops from volume positioned below the sea surface minimizing wave action for optimal performance. Self-propelled vessels and semisubmersibles can also move off location in case of extreme storms such as hurricanes, typhoons, etc.

Energy production units for oil and gas (or wind) are deployed long-term at the same site and can be: a) fixed to the sea bottom such as piled fixed platforms (monopiles, caissons, tripods, jackets, etc.), compliant towers and tension leg platforms (TLPs); or b) moored such as Floating Production, Storage and Offloading (FPSO) vessels, semisubmersibles, spars and floating wind farms. There are water depth limitations to the use of structures fixed to the sea bottom so floating systems are preferred for deeper water. Floating production systems are generally moored, although some early production systems on DP have been used.
The design of these more permanent units is centered on safe and efficient flow of hydrocarbons from the subsea reservoir towards the topsides processing facilities and then towards the onshore infrastructure. There is an increasing realization that standardization offers more cost efficient solutions, key to deep water:

- A standardized FPSO (and riser system) for more remote locations without a developed pipeline infrastructure and requiring hull storage capacity.
- A standardized low cost semi-submersible solution with flexible risers and subsea wells (wet trees), where no significant hull storage is necessary.
- A customized TLP or spar (and riser system) where no significant hull storage is necessary and dry trees at platform deck more efficiently access closely spaced wells.

Concept modeling and topology optimization processes are under development allowing for the global optimization of a vessel structure to meet structural performance and safety criteria, facilitate optimization for production, and simultaneously enhancing structural life-cycle, in-service performance, and generate total ownership cost reductions [30]. Figure 9 illustrates free form structural systems:

- Form-active structures, which balance loads altering their configuration. Example: mooring systems.
- Vector-active structures in which the redirection of forces occurs by splitting forces into vectors along compressive and tensile elements in triangular configurations. Example: jacket structures, jack-up legs.
- Surface-active structures, which redirect forces through surface tension and compression. Example: shell structures.
- Section-active structures, which redirect forces through sectional forces such as shear and torsion in addition to tension and compression. Example: beams and plates.

Figure 10 and Figure 11 demonstrate their intuitive application in early airplane design and in semisubmersible design [31].

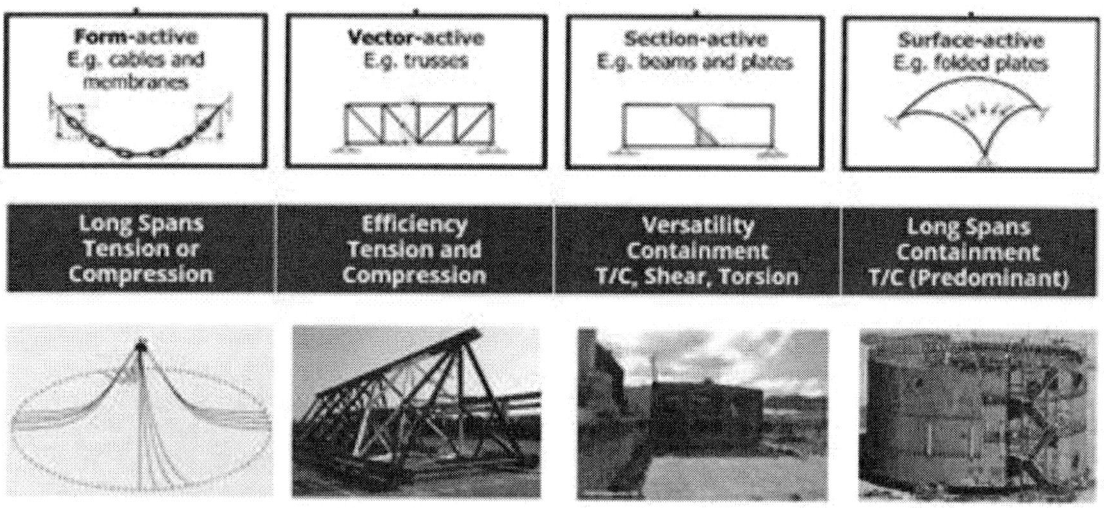

Figure 9 – Free From Structures

Figure 10 – Load Paths in Early Airplanes (Santos Dumont's 14-bis in the detail)

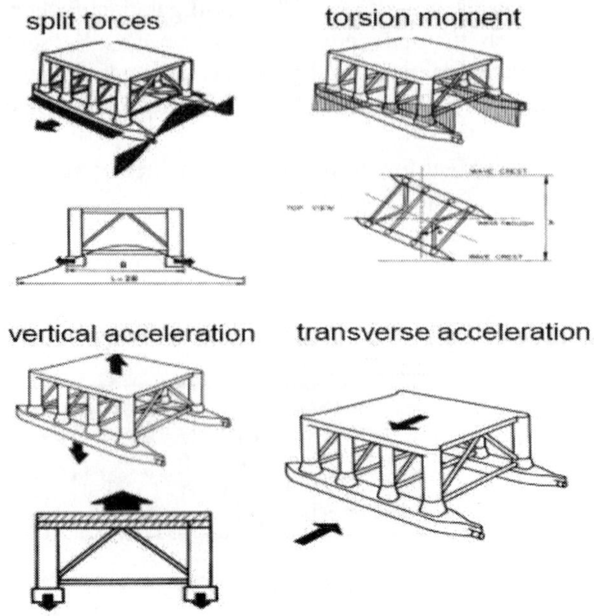

Figure 11 - Load Paths for Semisubmersibles

The essence of structural design (and robustness / resiliency) is to develop adequate load paths and allocating materials to transfer loads to the foundations and support the main design decisions:

- Topsides modules and equipment layout
- Marine systems layout, weight and CoG
- Structural layout, constructability and weight
- Mooring, anchors and risers layout (as applicable)
- Dynamic Positioning and propulsion (as applicable)
- Route and site specific considerations such as environmental impact, metocean conditions, clearances, seabed layout, soil conditions, etc.
- Global load matrix (combinations of functional and environmental loads to be considered in design)
- Global performance: air gap, stability and watertight integrity, offset, stability, hydrodynamics, motions and accelerations, mooring / anchor loads and riser loads
- Interface structural layout and in-service inspection plans
- Pre-service conditions (loadout, transportation, lifting, installation, etc.).

Maintaining structural integrity requires attention to material selection and corrosion protection as well as the design of structural details, both in terms of fatigue design, accessibility and in-service inspection. Typical areas for careful verification include topsides-deck integration, deck-hull connection, column-pontoon connection (Figure 12 and Figure 13), hull-riser interface and mooring foundations.

A good understanding of the structural layout and load paths will identify likely stress concentration areas and help develop smooth transitions of sufficient strength [31]. Careful Finite Element (FE) modeling is important in modern structural detailing; Figure 14 illustrates key sources of uncertainty based on generic solid modeling. At a global level, risk based inspection planning allow cost optimization while maintaining the required design safety levels [32].

Figure 12 – Column-Pontoon Connection [31]

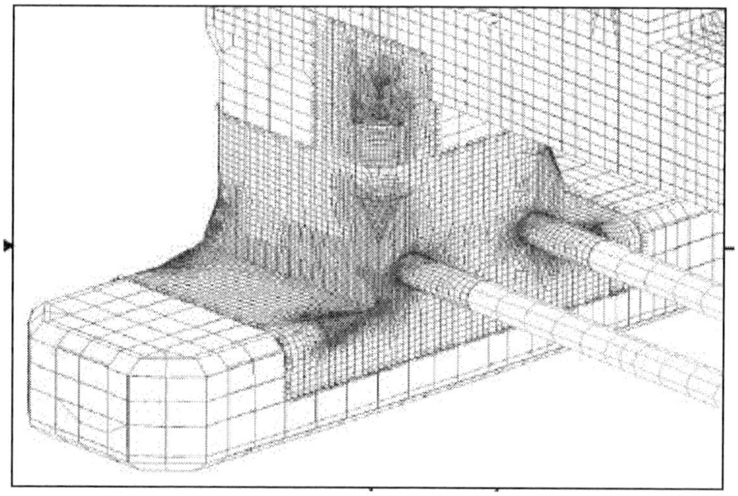

Figure 13 – FE Mesh for Column-Pontoon Connection

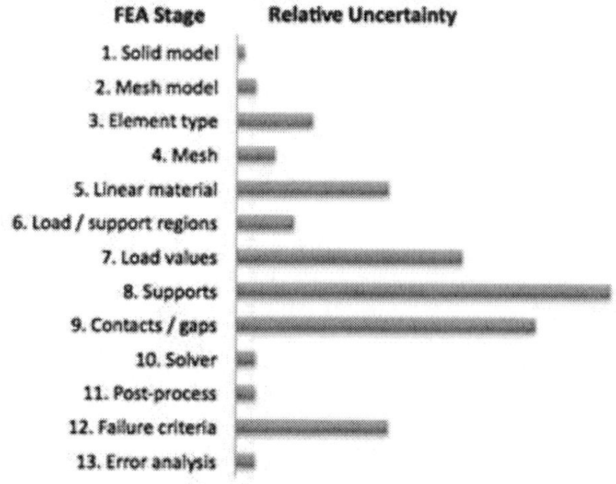

Figure 14 – Major Uncertainties in FE Analysis [33]

MULTI-CRITERIA MULTI-OBJECTIVE OPTIMIZATION

Powerful insights are obtained from computer models calibrated against model tests to establish a larger number of data points and a deeper insight into the physics than possible from limited and expensive model testing [34] [35] [36]. Structural real-time monitoring during operations provides important validation to design assumptions and computational models [37]. Validated computational models can be used to generate computational analogs and digital twins of real sytems from where response surfaces can be established defining the limit state equations used in SRA [26].

The next logical step is to obtain computer-aided reliability based optimal solutions. The development of the concepts of Figure 3 into more elaborated decision-making metrics can be an input to multi-criteria, multi-objective optimization based on non-dominated solutions [38] [39] [40] as illustrated in Figure 15. A *dominated design solution* is inferior in all design attributes to at least one other feasible solution. A *non-dominated design solution* is superior to all other solutions in at least one attribute.

Fast computing determines non-dominated solutions at initial levels, which are further refined through adaptive sampling. The Pareto Optimal Front is determined containing the final non-dominated solutions.

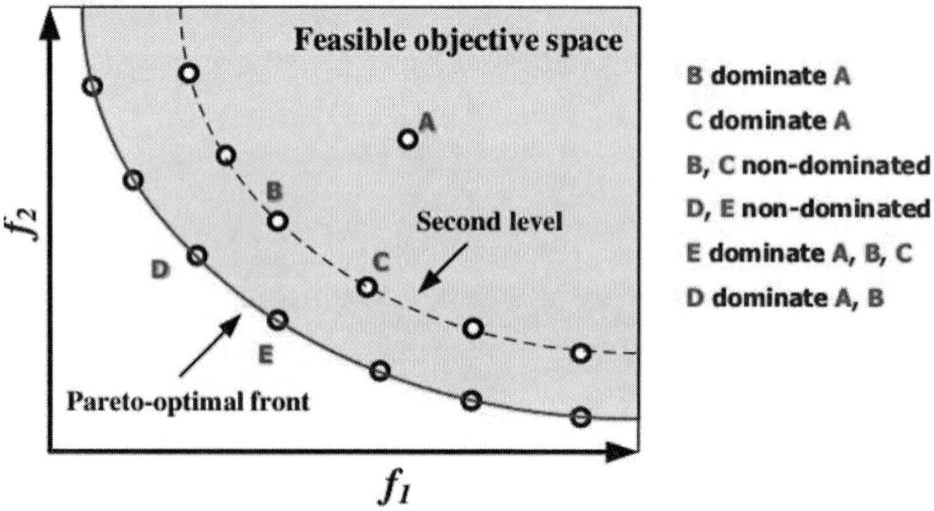

Figure 15 – Pareto Optimal Front [41]

CONCLUSION AND FUTURE DEVELOPMENT

Safety, reliability and efficiency are key to supporting progress as marine and offshore engineering play an important role in supporting higher standards of living. The author hopes to stimulate further investigation into achieving such goals through optimization of hull forms, load paths and material efficiency.

Modeling of variability and uncertainties at low probability levels remains a challenge but significant amounts of data will be available from increased digitalization and real-time performance monitoring, following a trend in other industries where Big Data is having a significant impact on performance. A firm handle on failure data and on modeling of randomness and variability will enable the vision of pioneers such as Pugsley and Freudenthal: safety criteria will be confidently related to observed rates of failure.

Computer-aided reliability-based design will allow integrated concept models of hull form and structure to be numerically simulated to find optimized solutions, Figure 16. Artificical Inteligence (AI) concepts are ideally suited to find such optimized solutions, releasing more time for designers to find and investigate innovative solutions.

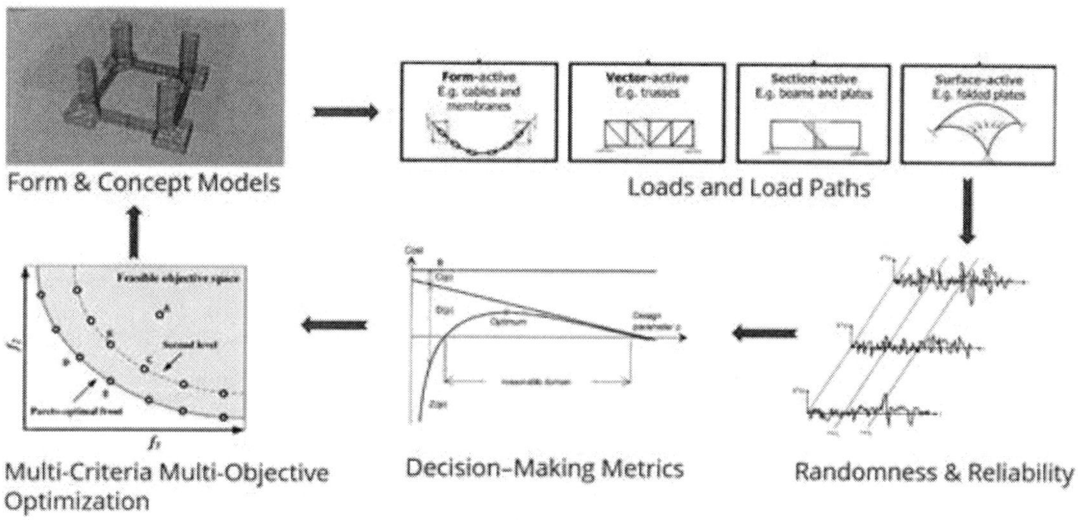

Figure 16 – Theory of Everything for Structural Design

According to DNV-GL, by 2025 the internet of things (IoT, the network of all physical objects that can be connected to the internet or a local web) is expected to encompass 0.5-1 trillion devices – with a potential economic impact of USD 2.7-6.2 trillion annually [42]. Examples of IoT-enabled applications are remote monitoring and control of homes, and personal health and fitness tracking [42].

Similar digitalization will occur in the marine and offshore industries [43] and support the concepts of reliability, robustness and resiliency discussed in this paper, Figure 17. Performance monitoring in real time will allow identification of early signs with opportunities for intervention and mitigation of failure induced by load and resistance issues or by non-structural issues.

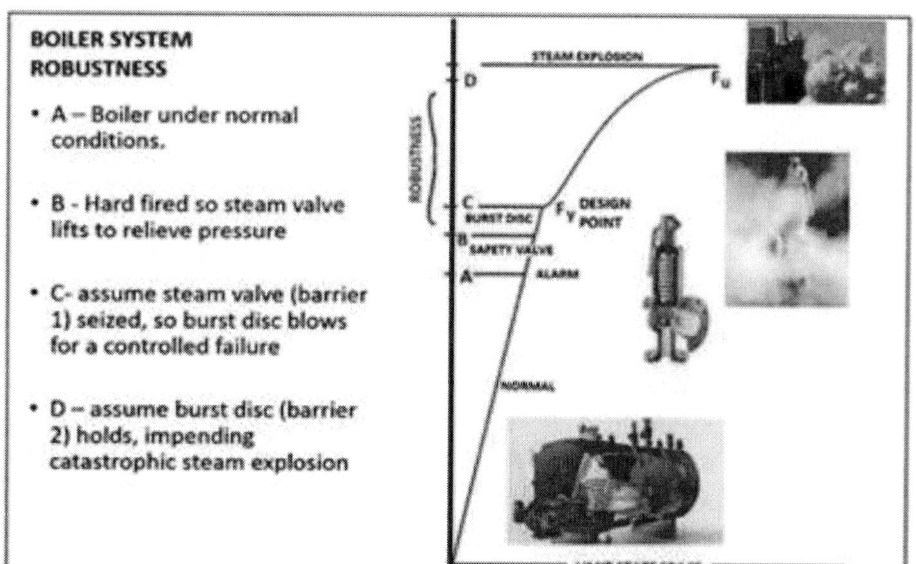

Figure 17 – Robustness Generalized [44]

REFERENCES

[1] GustoMSC, "Complete and Secure Solutions drive Safety and Efficiency," *GustoMSC Inside Magazine,* November 2017.

[2] ARPA-E, Advanced Research Projects Agency-Energy, "ARPA-E Mariner Program," 2017. [Online]. Available: https://arpa-e.energy.gov/?q=arpa-e-programs/mariner.

[3] A. Morandi, "Keynote Lecture - Challenges in the Strucutral Design of Jack-ups and semsibmersible Rigs," in *6th International Conference on Marine Structures*, Lisbon, 2017.

[4] A. Morandi, *Ocean Engineering Seminar - Challenges in the Structural Design of MOUs,* College Station, Texas, 2017.

[5] A. Morandi, "A Brief Review of Robustness Efforts in the Offshore Industry," in *SNAME Offshore Symposium*, Houston, 2018.

[6] N. Kuschel and R. Rackwitz, "Optimal design under time-variant reliability constraints," *Structural Safety,* vol. 22, no. 2, pp. 113-127, 2000.

[7] A. Pugsley, "A Philosophy of Aeroplane Strength Factors," British Aeronautical Research Committee - Report 1906, 1942.

[8] A. Freudenthal, "The Safety of Structures," *Transactions - American Society of Civil Engineers,* 1945.

[9] E. P. A. Torroja, "Safety Factor Calculation," in *4th Congress, Int. Ass. Bridge and Struct. Eng.*, Cambridge, 1952.

[10] J. Ferry-Borges, "Critical View of the Methods for the Probabilistic Calculation of Safety," in *4th Congress, Int. Ass. Bridge and Struct. Eng.,*, Cambridge, 1952.

[11] K. Szalai and P. Lenkei, "Hungarian Experience in Structural DEsign Coding - Historical Antecedents to EUROCODE-2," *Periodica Polytechnica Ser. Civil Engineering,* vol. 36, no. 3, pp. 339-353, 1992.

[12] C. Cornell, "Bounds on the Reliability of Structural Systems," *Journal of Structural Engineering,* vol. 93, no. 1, pp. 171-200, 1967.

[13] A. Hasofer and N. Lind, "Exact and Invariant Second-Moment," *Journal of the Engineering Mechanics Division, ASCE,* 1974.

[14] ISO, "ISO 19900:2013 - Petroleum and natural gas industries -- General requirements for offshore structures," International Standards Organization, 2013.

[15] A. Freudenthal, "Safety and the Probability of Structural Failure," *Transactions of the American Society of Civil Engineers ,* 1954.

[16] J. Samenow, "Harvey is a 1,000-year Flood Event Unprecedented in Scale," *Washington Post, 31st August 2017,* 2017.

[17] NOOA, "Global Warming and Hurricanes - An Overview of Current Research Results," 26 October 2017. [Online]. Available: https://www.gfdl.noaa.gov/global-warming-and-hurricanes/.

[18] CSB, "Drilling Rig Explosion and Fire at the Macondo Well," U.S. Chemical Safety and Hazard Investogation Board, 2010.

[19] NAS, "Interim Report on Causes of the Deepwater Horizon Oil Rig Blowout and Ways to Prevent Such Events," The National Academies of Science, Engineering and Medicine, Washington, D.C., 2010.

[20] S. Haver and S. Winterstein, "Environmental Contour Lines: A Method for Estimating Long Term," *Transactions, Society of Naval Architects and Marine Engineers ,* vol. 116, pp. 116-127, 2009.

[21] A. Morandi, "Assessment of Performance of Deepwater Floating Production Facilities," Bureau of Safety and Environmental Enforcement (BSEE) - Dept. of the Interior, 2004.

[22] K. Digre, W. Krieger, D. Wisch and C. Petrauskas, "API RP 2A Draft Section 17 - Assessment of Existing Platforms," in *BOSS - Behaviour of Offshore Structures*, 1994.

[23] API, "API Recommended Practice 2SIM - Structural Integrity Management of Fixed Offshore Structures," American Petroleum Institute, Washington, D.C., 2014.

[24] R. G. Bea, "Human and Organizational Factors in Reliability," *Risk Analysis,* vol. 22, no. 1, pp. 19-35,

2002.

[25] R. Bea, E. Jin, O. Valle and R. Ramos, "Evaluation of the Reliability of Platform Pile Foundations in the Bay of Campeche," in *OMAE - Offshore Mechanics and Arctic Engineering*, 1998.

[26] P. Frieze, A. Morandi, M. Birkinshaw, D. Smith and A. Dixon, "Fixed and Jack-up Structures: Basis for Reliability Assessment," *Marine Structures,* 1996.

[27] J. Gebara, H. Westlake, S. DeFranco and P. O'Connor, "Influence of Framing Configuration on the Robustness of Offshore Structures," in *Offshore Technology Conference - OTC*, Houston, 1998.

[28] P. Frieze, A. Morandi, R. Snell, T. Moan and M. M., "Uncertainty Assessment in ISO 19900," in *OSRC 2016 - The 3rd Offshore Structural Reliability Conference*, Stavanger, 2016.

[29] ISO, "ISO 19902:2007 - Petroleum and natural gas industries -- Fixed steel offshore structures," International Standards Organization, 2007.

[30] R. G. Keane, T. McNatt and J. E. Beach, "Reducing Total Ownership Cost: Designing Robust Ship Structures," *Naval Engineers Journal,* vol. 17, no. 4, pp. 41-57, 2017.

[31] B. Boon, "Revisit of a 1970s semi-submersible pipe layer," [Online]. Available: https://www.gustomsc.com/sites/default/files/Downloads/Article/Revisit%20of%20a%201970s%20semi-submersible%20pipe%20layer.pdf.

[32] Y. Gorbatov and C. Guedes Soares, "Cost and reliability based strategies for fatigue maintenance planning of floating structures," *Reliability Engineering and systems Safety,* vol. 73, pp. 293 - 301, 2001.

[33] J. Akin, "Finite Element Analysis Concepts via Solidworks," 2009. [Online]. Available: https://www.clear.rice.edu/mech403/HelpFiles/FEAC_final.pdf.

[34] A. Morandi, P. Das and D. Faulkner, "Finite Element Analysis and Reliability Based Design of Externally Pressurized Ring Stiffened Cylinders," *Transactions of the Royal Institution of Naval Architects,* 1996.

[35] A. Morandi, P. Das and D. Faulkner, "Reliability-Based Design of Externally Pressurized Vessels," *J. Offshore Mech. Arct. Eng.,* vol. 120(3), pp. 149-153, 1998.

[36] A. Morandi, P. Das and D. Faulkner, "Ring Frame Design in Orthogonally Stiffened Cylindrical Structures," in *Offshore Technology Conference - OTC-7801-MS*, Houston, 1995.

[37] A. Morandi, D. Karunakaran, A. Dixon and M. Baerheim, "Comparison of Full-Scale Measurements and Time-Domain Irregular Sea Analysis for a Large Deepwater Jack-Up," in *Offshore Technology Conference*, Houston, 1998.

[38] A. Morandi, V. Zanic, P. Das and D. Faulkner, "An Investigation on the Application of Alternative Materials to Externally Pressurized Vessels," in *The Institute of Materials - Structural Materials in Marine Envrironments*, London, 1994.

[39] P. Das, C. Tolikas, A. Morandi, V. Zanic and N. Warren, "Multiple Criteria Synthesis Technique applied to the Reliability Based Structural Design of the Hull Components of a Fast SWATH Ship," in *2nd Intl. Conf. Fast Sea Transportation*, Yokohama, 1993.

[40] C. Tolikas, A. Morandi, P. Das and D. Faulkner, "Reliability Based Multiple Criteria Optimization of a Fast SWATH Ship," in *Computational Stochastic Mechanics*, Athens, 1995.

[41] D. E. Goldberg, Genetic algorithms for search, optimization, and machine learning, Reading, MA: Addison-Wesley, 1989.

[42] DNV-GL, "INNOVATION DRIVERS - DIGITALIZATION," 2017. [Online]. Available: https://to2025.dnvgl.com/innovation-drivers/digitalization/.

[43] ABS - American Bureau of Shipping, "Condition-Based Survey," [Online]. Available: https://ww2.eagle.org/en/innovation-and-technology/digital/condition-based-survey.html.

[44] B. Cheater, "A DEFINITION AND A FRAMEWORK FOR THE ASSESSMENT OF ROBUSTNESS IN FLOATING PRODUCTION UNITS," in *SNAME Offshore Symposium*, Houston, 2018.

INVESTIGATING IMPACT OF REVISED DNVGL-OTG-13 GUIDELINE ON AIR GAP PREDICTIONS FOR SEMISUBMERSIBLES

GAURAV SINGHAL
FLOATING SYSTEMS TEAM LEAD, WOOD

JUN CHENG
SENIOR NAVAL ARCHITECT, WOOD

JOÃO PESSOA
SENIOR NAVAL ARCHITECT, WOOD

ANDERS MARTIN MOE
PROJECT DEVELOPMENT MANAGER, WOOD

Proceedings of the 24th Offshore Symposium, February 20th 2019, Houston, Texas
Texas Section of the Society of Naval Architects and Marine Engineers

ABSTRACT

In December 2015, a deepwater drilling rig operating in North Sea was struck by a steep wave, leading to death of one crewmember, several injuries and extensive damage to the living quarters. Following the incident and subsequent investigation, DNVGL issued a new guideline in September 2016 (DNVGL-OTG-13 "Prediction of air gap for column stabilized units"; revised March 2017) regarding the air gap calculations and requirements. Revised guidelines recommend including mean and low frequency contributions to the air gap predictions in addition to the linear wave-frequency motions and wave asymmetry factor considered in the simplified approach (DNVGL-RP-C205). This paper investigates the impact of revised guidelines for semisubmersible design by comparing air gap predictions from simplified approach, revised OTG13 guidelines, and model test data.

Keywords: SNAME, Offshore, Symposium, Air Gap, DNVGL-OTG-13, DNVGL-RP-C205

INTRODUCTION

Minimum air gap is a critical design parameter for semisubmersible designs and governed primarily by the extreme wave condition at the location of interest. Several class rules specify minimum air gap requirements for production semisubmersible design (e.g. ABS requires minimum 1.5m air gap for 100-year design wave in Gulf of Mexico). Estimating air gap is quite complicated and typically involves knowledge of free surface elevation as well as dynamic vessel motions. There are several published studies that provide guidance on air gap calculations (Manuel and Winterstein 1999, Sweetman 2001 & 2004, Kvaleid et al. 2014, etc.). Some of these studies outline approaches that are not easy to use for practical applications, but some other studies provide simplified methods that enable sufficiently accurate air gap computations in an efficient manner. It is the simplified methods that the current industry codes recommend for air gap calculations (particularly, DNVGL-RP-C205) and these have been the industry norm for past several years.

Recently, DNVGL published revised guidelines (DNVGL-OTG-13; revised 2017) for assessing air gap for column stabilized units in the wake of a drilling rig accident in the North Sea. This paper investigates the impact of revised guidelines on design air gap by comparing with model test measurements and the simplified approach from DNVGL-RP-C205 (2014). The overall objective is to conclude whether the revised guidelines have an impact on column height selection for semisubmersible design. The paper is structured as follows - air gap calculation methodology is first summarized, then the numerical test cases considered for this paper are presented along with the details of relative motion assessment results and conclusions.

AIR GAP CALCULATION METHODOLOGY

DNVGL-RP-C205 (hereafter RP-C205) provides a simplified numerical approach for computing air gap for semisubmersible platforms. In this approach, the free surface elevation and the motions of the platform are calculated using linear diffraction-radiation theory. The surface elevation is modified by a coefficient (α) to account for the asymmetry of crests and troughs. The air gap using simplified approach is defined by

$$a = \alpha\eta^{(1)} - z \qquad (1)$$

where $\eta^{(1)}$ is the first-order surface elevation (including incident, radiation and diffraction effects) and z is the vertical motion of a point on the vessel. Typically, asymmetry factor of 1.2 is used for standard semisubmersible concepts as recommended in the RP.

In the recently released DNVGL-OTG-13 (hereafter OTG13), an updated methodology for air gap calculation was provided. In OTG13, air gap is defined as the distance between the bottom of the deck (or other relevant point) and the relative wave elevation ("upwell"). The OTG recommends that "upwell" should include low-frequency motions due to wind and waves and a mean contribution from vessel inclination (apart from linear wave-frequency motions and non-linear free surface elevation). The mean contribution should reflect the actual loading conditions of the vessel and should additionally include a 1° inclination in the most critical direction to account for uncertainties in ballasting. The upwell according to the new guidelines is thus given by wave-frequency, low-frequency and mean contributions. Subsequently, the total air gap can be calculated by subtracting the total upwell from static freeboard:

$$a\,(x, y, t) = a_0\,(x, y) - \chi\,(x, y, t) \qquad (2)$$

where a_0 is the static freeboard and χ is the total upwell. Total upwell is the distance between the free surface and the floater mean water line. The contributions to total upwell are:

- Wave frequency upwell (χ_{WF}): OTG13 recommends using 90% percentile level following Gumbel maxima to compute the wave frequency upwell.
- Low frequency upwell (χ_{LF}): OTG13 recommends using the most probable maximum value using contributions from low frequency wind and wave motions.
- Mean upwell due to mean floater inclination (χ_{mean})

Assuming uncorrelated low- and wave-frequency upwell, the total upwell is given as:

$$\chi = \chi_{mean} + \sqrt{\chi_{WF}^2 + \chi_{LF}^2} \qquad (3)$$

where $\chi_{mean} = z_{mean}$ (the mean vertical displacement), $\chi_{WF} = \alpha\eta^{(1)} - z_{WF}$, and $\chi_{LF} = z_{LF}$. The wave frequency and low frequency vertical motions (i.e. z_{WF} and z_{LF}, respectively) include contributions from heave, roll and pitch motions.

DESCRIPTION OF NUMERICAL TEST CASES

The present study performs calculation of relative wave elevation (or upwell) for semisubmersible floater concept (hereafter, semi). Two cases, considering a production and a drilling semi, are selected for this study (Figure 1).

The production semi is a generic four-column configuration with ring pontoons. The vessel is modeled with panel model mesh for the submerged surface as shown in Figure 1. Free-surface elevation is calculated using linear diffraction and radiation theory using panel method. To account for viscous effects, a Morison model of columns and pontoons is also included (represented by vertical lines for the columns; pontoon Morison members not shown). The Morison elements in columns and pontoons are scaled down so that only drag loads are included (no buoyancy or added mass effects).

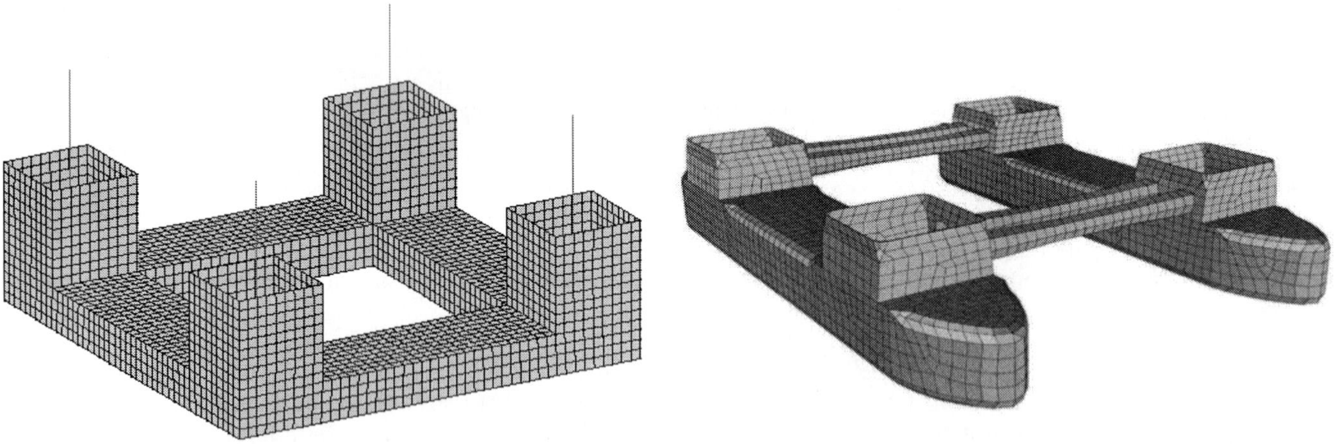

Figure 1 Test Cases: Production Semi (left) and Drilling Semi (right)

The drilling semi used in this study (Figure 1) is described in detail in Pessoa et al. (2018). In their paper, Pessoa et al. (2018) considered operational and survival draft conditions for the drilling semi. For purposes of this paper, only the survival draft condition results are considered as these are more relevant for extreme sea states and were also found to be governing due to proximity of pontoons to the free surface at survival draft (Pessoa et al. 2018).

RELATIVE WAVE ELEVATION ASSESSMENT

Relative wave elevation (or upwell) is assessed for selected test cases using methodologies described in RP-C205 and OTG13. Further, model test data are used to compare the numerical results for both production and drilling semis considered in this study.

For the production semi, the relative wave elevation data were measured at several wave probes as shown in Figure 2. Numerical predictions for relative wave elevation are estimated at the grid locations indicated in Figure 2. Numerical and model test results are obtained for the design storm condition for a range of wave headings namely head sea, quartering sea, beam sea, and following sea. For the drilling semi, model tests and numerical results were obtained at three wave probe locations as shown in Figure 3. Several wave height and wave period combinations, comprising of regular and irregular waves, were numerically analyzed as well as measured in wave basin at these locations (for details, refer to Pessoa et al. 2018).

As discussed in "Air Gap Analysis Methodology" section, OTG13 recommends inclusion of various terms and factors to the simplified approach from RP-C205, i.e. consideration of mean and low frequency contributions, updated wave asymmetry factor, and calculating wave frequency contribution using 90% percentile level in the Gumbel distribution. This paper uses a systematic approach to highlight the impact from each of these terms. For the production semi case, the wave asymmetry factor is set as 1.2 (same as simplified approach) while the wave frequency contributions are calculated using extreme value from the Rayleigh distribution as well as the Gumbel distribution (90% level as recommended by OTG13). For the drilling semi case, impact from higher wave asymmetry factor (1.2 and 1.3) is also investigated in addition to the wave frequency contributions following Rayleigh and Gumbel distributions. In addition, mean and low frequency contributions are compared to the wave frequency contributions.

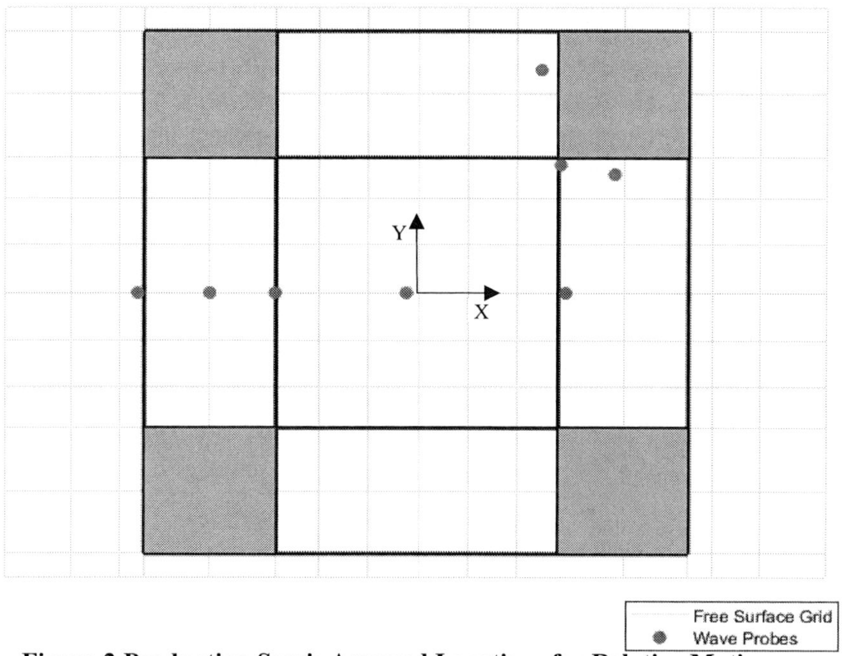

Figure 2 Production Semi: Assessed Locations for Relative Motion

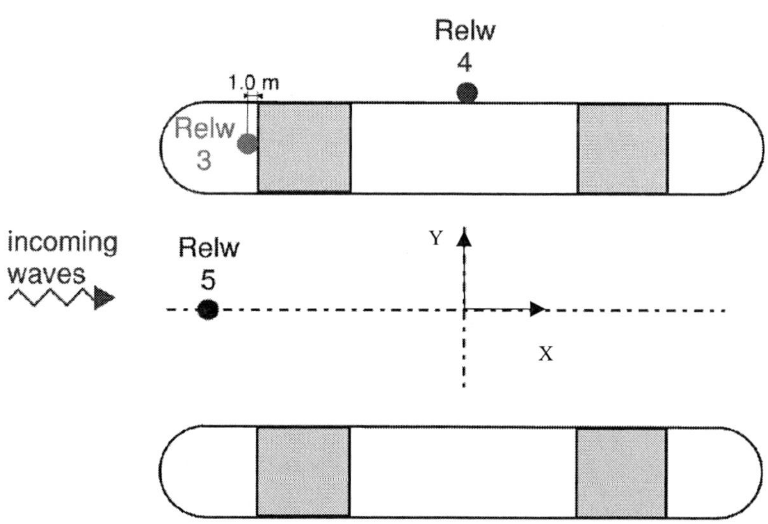

Figure 3 Drilling Semi: Assessed Locations for Relative Motion (Taken from Pessoa et al. 2018)

DISCUSSION OF RESULTS

Production Semi - First, the production semi results are presented for total upwell normalized with respect to the design wave height. Figure 4 shows normalized total upwell for wave measurements along the X-axis of the semi for various wave headings. Extreme values are presented based on a 3-hour storm duration following Rayleigh distribution using RP-C205 as well as OTG13. In general, the numerical results follow similar trend compared to measurements using model tests. It is found that numerical predictions from OTG13 are slightly more conservative than those obtained using the simplified approach from RP-C205. The minor differences between two approaches are mainly from the mean and low frequency components (as the wave frequency contribution is same for both approaches as mentioned above). The comparison of analytical results with model test data does not appear to favor either numerical approach (i.e. OTG13 or RP-C205) as results are similar for measurement locations along the X-axis. For 270° heading, the results from RP-C205 and OTG13 approaches overlap as the analysis points lie along the neutral axis (perpendicular to the wave heading) thereby resulting in no contributions from mean and low frequency components.

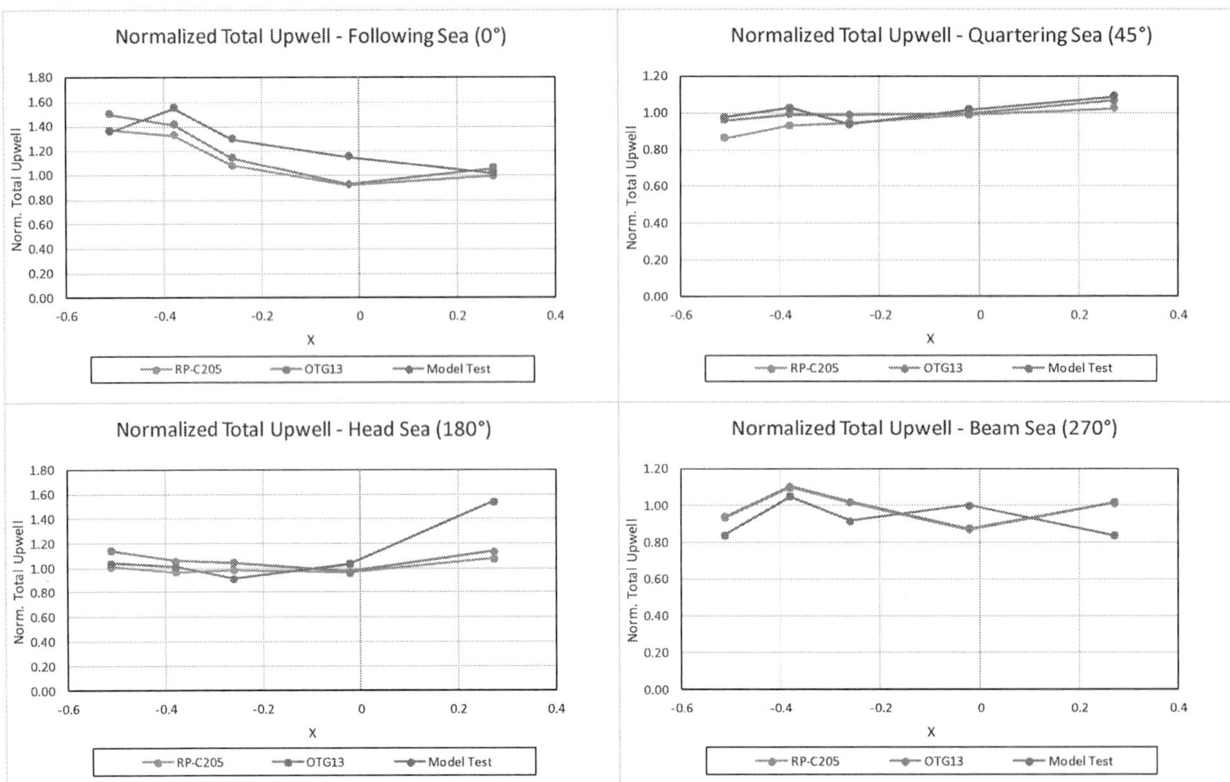

Figure 4 Comparison of Relative Wave Elevations between Numerical and Measured Data along X-Axis for Production Semi

Predicted and measured data are presented in another fashion in Figure 5, which shows mean, low frequency and wave frequency upwell components by combining data from all measurement locations for the analyzed wave headings. For this comparison, wave frequency contributions using OTG13 approach are obtained using the recommended 90% percentile level in the Gumbel distribution (orange points); while those using RP-C205 are based on Rayleigh distribution using extreme value in 3-hour storm duration (blue points). There is significant data scatter for mean and low frequency components, while the wave frequency upwell seems to be clustered closer to the line with slope 1. Also, the wave frequency contribution is significantly larger than the mean and low frequency contributions. It is noted that wave frequency upwell predictions using OTG13 approach are found to be larger than those predicted using traditional approach and model test measurements. This is primarily due to usage of 90% percentile level in the Gumbel distribution for OTG13 predictions. As the column height for production semi is selected based on maximum upwell (i.e. minimum air gap), using OTG13 approach would result in a conservative column height during initial semi sizing and design, thereby impacting the overall hull weight and cost.

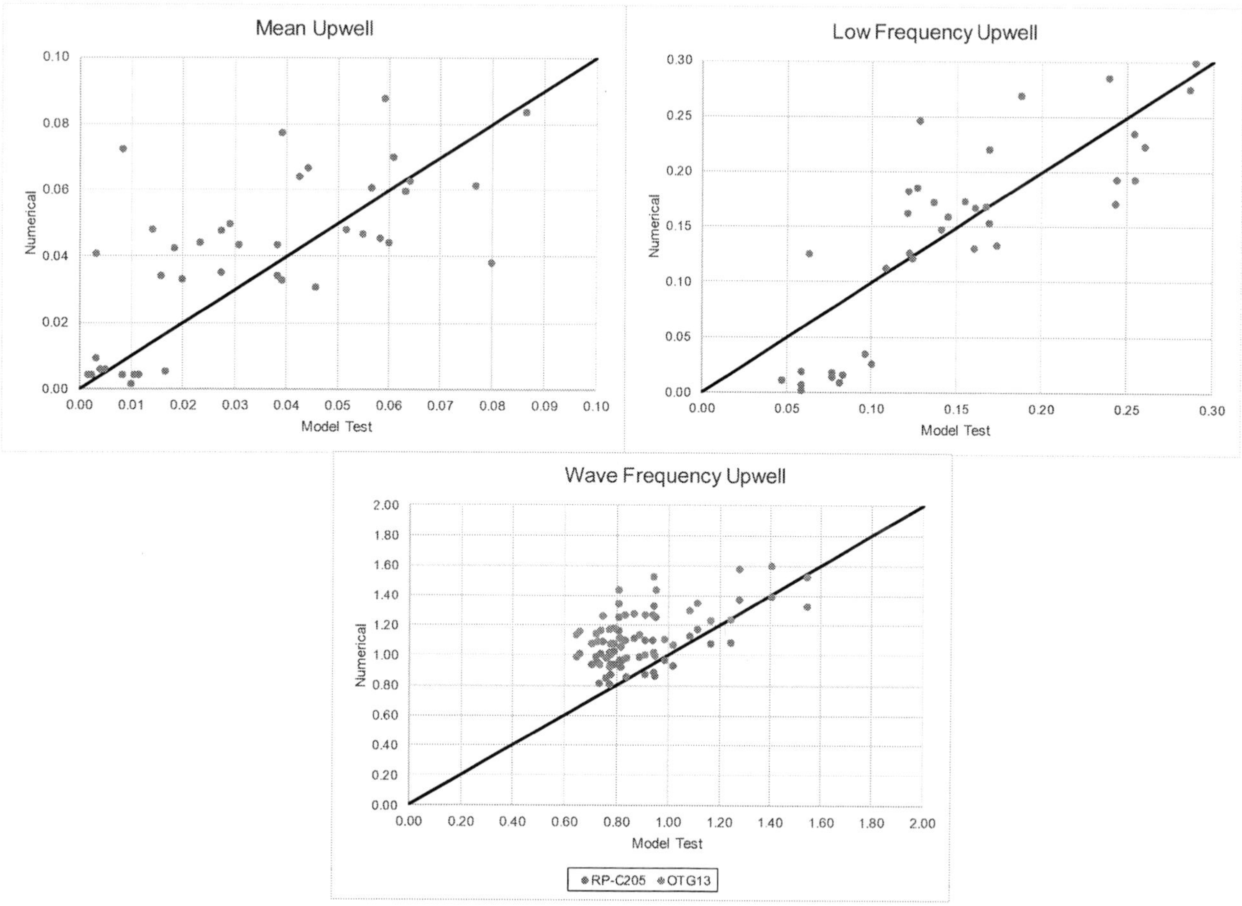

Figure 5 Mean, Low Frequency, and Wave Frequency Upwell Components for Production Semi

Drilling Semi – Comparison of normalized total upwell (with respect to largest wave height) for a drilling semi is shown in Table 1 for 0° wave heading. Analytical calculations using OTG13 approach are performed using different considerations (i.e. Rayleigh distribution with α=1.2 similar to that prescribed in RP-C205 as well as using 90% Gumbel maximum with α=1.2 and α=1.3) to understand the impact of various factors. Using an approach similar to RP-C205 (Rayleigh distribution with α=1.2) but accounting for mean and low frequency contributions, numerical predictions from OTG13 are found to be slightly conservative than those obtained using RP-C205 and model test results. Upon applying 90% Gumbel maximum without changing the asymmetry factor value, the predictions using OTG13 approach become larger as expected. With higher asymmetry factor (α=1.3) at gauges 3 and 5, the predictions using OTG13 get even larger compared to model test results.

Although limited model test data are available for the selected drilling semi, our findings indicate that the total upwell analytical calculations are influenced more (i.e. tending towards higher conservatism) by the 90% Gumbel maximum and higher asymmetry factor rather than the contributions from mean and low frequency components.

Table 1 Normalized Total Upwell Comparisons

Wave Gauge 3					
Hs	Model Test	RP-C205	OTG13		
		Rayleigh; α=1.2	Rayleigh; α=1.2	Gumbel; α=1.2	Gumbel; α=1.3
2.0	0.83	0.70	0.95	1.03	1.09
3.5	1.27	1.13	1.33	1.47	1.57
4.5	1.59	1.61	1.79	2.00	2.14
7.0	1.65	2.03	2.20	2.46	2.64
Wave Gauge 4					
Hs	Model Test	RP-C205	OTG13		
		Rayleigh; α=1.2	Rayleigh; α=1.2	Gumbel; α=1.2	Gumbel; α=1.3[1]
2.0	0.29	0.29	0.29	0.33	NA
3.5	0.51	0.45	0.45	0.51	NA
4.5	0.67	0.73	0.73	0.82	NA
7.0	1.03	1.06	1.06	1.20	NA
Wave Gauge 5					
Hs	Model Test	RP-C205	OTG13		
		Rayleigh; α=1.2	Rayleigh; α=1.2	Gumbel; α=1.2	Gumbel; α=1.3
2.0	0.66	0.42	0.79	0.82	0.85
3.5	0.81	0.59	0.90	0.96	1.00
4.5	1.19	1.10	1.33	1.46	1.55
7.0	2.32	1.38	1.59	1.76	1.88

Note1: Higher asymmetry factor not applicable for Wave Gauge 4

CONCLUSIONS

The paper investigates the impact of revised guidelines by DNVGL (OTG13) on air gap predictions for semis. The paper uses a comparative approach utilizing numerical predictions as well as model test measurements for conventional semis (production and drilling) to conclude whether there were shortcomings in the simplified approach commonly adopted by the offshore industry for semi designs.

Our limited investigation shows that, in general, the revised OTG13 guideline is conservative in comparison to simplified approach from RP-C205 as well as model test measurements. While the OTG13 approach incorporates better physics (via inclusion of mean and low frequency effects) and therefore arguably more accurate, comparison of numerical predictions with model test data (for selected test cases) indicates that use of 90% Gumbel maxima and updated asymmetry factor influences the total upwell calculations more than the mean and low frequency contributions. As such, incorporating mean and low frequency terms may make analytical computations more cumbersome with little benefit. Our assessment concludes that simplified approach is still sufficiently robust when compared to model test data and therefore still applicable for deciding appropriate column height for initial semi design. In later design stages, the selected column height should be verified using model tests to confirm sufficient air gap as well as to recommend structural enhancements for top-of-column equipment to sustain impact loads from wave run-up.

ACKNOWLEDGEMENTS

The authors acknowledge the support of Wood management in publication of this paper. Authors are also grateful to the two reviewers for their constructive comments that led to considerable improvement of this paper.

REFERENCES

1. DNVGL-RP-C205 Environmental Conditions and Environmental Loads, DNV-GL, Aug. 2017.
2. DNVGL-OTG-13 Prediction of air gap for column stabilised units, March 2017.
3. Manuel, L. and Winterstein, S.R., "Air gap response of floating structures under random waves: analytical predictions based on linear and nonlinear diffraction". Oral presentation at the 18th International Conference on Offshore Mechanics and Arctic Engineering, St. John's, Newfoundland, Canada, July 11-16, Paper No: 99-6041, 1999.
4. Stansberg, C., "Nonlinear wave amplification around column-based platforms in steep waves. Proc., ASME 33rd International Conference on Ocean, Offshore and Artic Engineering, San Francisco, California, June 8-13, OMAE2014-24569.
5. Sweetman, S., "Air Gap Analysis of Floating Structures Subject to Random Seas: Prediction of Extremes Using Diffraction Analysis Versus Model Test Results". PhD thesis, Stanford University, Stanford, California, August 2001.
6. Sweetman, S., "Practical Airgap Prediction for Offshore Structures", Journal of Offshore Mechanics and Artic Engineering, 126 (2): 147-155, 2004.
7. Pessoa, J., C.T. Stansberg, N. Fonseca, and M. Laranjinha, "Experimental and numerical study of the Free Surface Elevation Over the Pontoons of a Semisubmersible Platform in Waves", Proceedings of the ASME 2018 37th Conference on Ocean, Offshore and Arctic Engineering, Madrid, Spain, June 17-22, 2018, OMAE2018-78009.